International Commodity Market Models

INTERNATIONAL STUDIES IN ECONOMIC MODELLING

Series Editor

Homa Motamen-Scobi

Executive Director
European Economics and Financial Centre
London

Economic Modelling in the OECD Countries
H. Motamen-Scobi
Modelling the Labour Market
M. Beenstock
Input-Output Analysis
M. Ciaschini
Models of Disequilibrium and Shortage in Centrally Planned Economies
C. Davis and W. Charemza
International Commodity Market Models
O. Güvenen, W. C. Labys and J. B. Lesourd
Recent Modelling Approaches in Applied Energy Economics
O. Bjerkholt, Ø. Olsen and J. Vislie
Economic Modelling at the Bank of England
S. G. B. Henry and K. D. Patterson

International Commodity Market Models

ADVANCES IN METHODOLOGY AND APPLICATIONS

Edited by

Orhan Güvenen

President of the State Institute of Statistics of Turkey
Professor of Econometrics
Bilkent University

Walter Labys

Professor of Resource Economics
West Virginia University

and

Jean-Baptiste Lesourd

Research Professor in Economics
University of Law, Economics and Science
Marseilles

SPRINGER-SCIENCE+BUSINESS MEDIA, B.V.

First edition 1991

Originally published by Chapman & Hall in 1991
Softcover reprint of the hardcover 1st edition 1991

Typeset in 10/12 Sabon by
KEYTEC, Bridport, Dorset

ISBN 978-94-010-5367-9

British Library Cataloguing in Publication Data
International commodity market models: advances in methodology and application. (International studies in economic modelling)
1. Commodity markets
I. Güvenen, Orhan II. Labys, Walter C. III. Lesourd, Jean-Baptiste
IV. Series
332.644

ISBN 978-94-010-5367-9 ISBN 978-94-011-3084-4 (eBook)
DOI 10.1007/978-94-011-3084-4

Library of Congress Cataloging-in-Publication Data
International commodity market models: advances in methodology and application/edited by Orhan Güvenen, Walter Labys, and Jean-Baptiste Lesourd. —— 1st ed.
p. cm. —— (International studies in economic modelling)
Includes bibliographical references and index.
ISBN 0–412–35690–2
1. Commodity exchanges—Econometric models—Congresses.
2. Commodity futures—Econometric models—Congresses. I. Güvenen, Orhan, 1939–. II. Labys, Walter C., 1937–. III. Lesourd, Jean-Baptiste. IV. Series.
HG6046.I454 1991
332.64'4'015195—dc20 90-41430
CIP

Contents

Contributors

F. Gerard Adams	Economics Research Unit, University of Pennsylvania, USA
Ahmad Afrasiabi	Allegheny College, USA
Patrick Artus	Caisse des Dépôts et Consignations, France
Sveinbjörn Blöndal	Country Studies II Division, OECD, France
Timothy J. Considine	Pennsylvania State University, USA
Jean Pierre Daloz	University of Nice, France
Vytis Didziulis	Economics Research Unit, University of Pennsylvania, USA
Bruce L. Dixon	University of Arkansas, USA
Martine Durand	General Economics Division, OECD, France
Orhan Güvenen	State Institute of Statistics and Bilkent University, Turkey
Andrew J. Hughes Hallett	University of Strathclyde, UK
Walter C. Labys	Department of Mineral and Energy Resource Economics, West Virginia University, USA
Jean-Baptiste Lesourd	Centre National de la Recherche Scientifique, and Université de Droit, d'Economie et des Sciences d'Aix-Marseille, France
Montague J. Lord	Inter-American Development Bank, USA

Mark Newton Lowry Christensen Associates, Madison, USA

Charles D. Kolstad University of Illinois, USA

Eugene A. Kroch Economics Research Unit, University of Pennsylvania, USA

Lars Mathiesen Norwegian School of Economics and Business Administration, Norway

T. Gordon MacAulay Department of Agricultural Economics and Business Management, University of New England, USA

Donald O. Mitchell International Commodity Markets Division, The World Bank, USA

Massoud Moallem Rockford College, USA

Franz C. Palm Department of Economics, University of Limburg, The Netherlands

Jacques Percebois Faculté de Droit et des Sciences Economiques, Montpellier, France

Gonzague Pillet Université de Genéve and Université de Fribourg, Switzerland

Gordon C. Rausser California Agricultural Experiment Station, Giannini Foundation of Agricultural Economics, USA

Jean-Michel Ruiz Université de Droit, d'Economie et des Sciences d'Aix-Marseille, France

Takashi Takayama Department of Economics, University of Western Australia, Australia

M. Elton Thigpen International Commodity Markets Division, The World Bank, USA

Noel D. Uri Resource Technology Division, US Department of Agriculture, USA

Ben Vogelvang Department of Economics and Econometrics, Free University, The Netherlands

Nicholas Walraven California Agricultural Experiment Station, Giannini Foundation of Agricultural Economics, USA

Foreword

RONALD C. DUNCAN

During the 1980s, substantial advances were made in the global modelling of commodity markets in several areas, advances which were reflected in many of the papers delivered to the Applied Econometrics Association meeting held at the World Bank in Washington, DC, in October 1988. The several areas where I see advances being made, some of which the International Commodity Markets Division of the World Bank has taken part in, are the following: (a) in the theoretical specification of commodity price behaviour; (b) in the increased emphasis on modelling imperfect markets; (c) in the incorporation of the interrelationships between macroeconomic and commodity market variables; (d) in the specification of supply response, particularly in respect of perennial crops; and (e) in the realization of complementarity between time series analysis and econometrically estimated structural models.

Improvements in the specification of the commodity price formation process have probably been the most important of the above advances. Until the early 1980s, prices were modelled as a simple linear function of stocks. Gilbert has played an important role in introducing the rational expectations hypothesis into the specification of commodity prices. Recent work by Gilbert, Trivedi, and Deaton and Laroque offers the possibility of non-linear specification of the relationship between prices and stocks within an expectational framework and of thereby capturing the phenomenon of sharp run-ups in commodity prices.

Gilbert has also played an important role in clarifying the interrelationships between macroeconomic variables and primary commodity prices. The need to include these relationships in commodity models became more important with the increased variability in exchange rates and interest rates since the early 1970s. The relationship between the large build-up in the external debt of developing country producers of primary commodities and primary commodity markets and the relationships between commodity prices and structural adjustment programmes of primary commodity producers are current areas of interest where the links between commodity markets and macroeconomic variables are poorly understood and where, no doubt, research effort will be concentrated in the next few years. The interest in the use of commodity price movements as an early indicator of general price movements has also lent impetus to study of the interrelationships with macroeconomic variables.

The assumption of perfect competition of primary commodities has, until recently, been the standard for commodity models. It has been an interesting development that, in the past 10–15 years, several metals markets have become more competitive—aluminium, copper, nickel and tin, in particular. These are markets that have become much more competitive, either by reason of an increased number of firms, (whether national entities or private firms), which have entered the market (aluminium, copper and nickel) or through a breakdown of an international agreement (tin). However, the nickel and tin markets as well as, for example, the grains, coffee, cocoa, iron ore, rubber and petroleum markets cannot be thought of as perfectly competitive. Modelling of some of these markets as imperfectly competitive should lead to more realistic specifications. Progress in this area has been initiated, for example, by the use of a game-theoretic price formation of the iron ore industry by Priovolos.

The modelling of supply response, particularly of the long-run response, is, I believe, at present the least well-developed area in commodity modelling. This is particularly true of energy and metals commodities, unlike another group of commodities, perennial crops, where use of the vintage capital approach (given data on tree cohorts and age–yield profiles), integrated with short-run capacity-utilization variables, gives a robust supply specification. I commend here the work done by Akiyama and Trivedi.

Finally, I make reference to the recently discovered complementarity between time series analysis and econometric modelling of structurally specified markets. Previously, there was a strong sense of rivalry between these techniques. In large part because of the work of Hendry, time series analysis of economic variables (such as cointegration) is now seen as a necessary step before going on to specification and econometric estimation of a system of equations. This is not, of course, a development of importance only to the modelling of primary commodities; but it is, I believe, of such importance that it is worth giving it a special mention here.

As the reader will see, the recent advances in commodity modelling which I have briefly noted are reflected strongly in the papers presented to the Conference and reproduced here. I highly recommend this volume of papers to anyone interested in research into and application of primary commodity models.

Editors' Note

This book is composed of revised versions of some of the most important methodology-oriented papers presented at the 25th International Conference of the Applied Econometrics Association on International Commodity Market Modelling, which took place at the World Bank, at Washington, DC, on 24–26 October 1988. This Conference was organized by the Commodity Chapter of the Applied Econometrics Association.

The editors wish to thank all members of the Scientific Committee and the Organization Committee who have been extremely helpful in the preparation of this Conference.

The editors also wish to thank the following organizations, firms and personalities who have been exceptionally generous in their support:

– The World Bank (Ronald C. Duncan, Chief, International Commodity Markets Division);
– The Chicago Board of Trade (Patrick Catania);
– The Marché ă Terme International de France – M.A.T.I.F. (Bernard Bricheux);
– The Société Générale;
– The Honorable Robert Byrd (U.S. Senate);
– The West Virginia University (Neil Bucklew, President);
– The Université de Droit, d'Economie et des Sciences d'Aix-Marseille (Lucien Capella, President, and Jacques Metzger, Vice-President);
– The Centre National de la Recherche Scientifique – C.N.R.S. (Maurice Claverie, Directeur du Programme Interdisciplinaire de Recherche sur l'Energie et les Matières Premières – P.I.R.S.E.M.).

25TH INTERNATIONAL CONFERENCE OF THE APPLIED ECONOMETRICS ASSOCIATION

The World Bank, Washington, DC, 24–26 October 1988

Scientific committee

F. Gerard Adams (University of Pennsylvania)
Antonio Aznar Graza (University of Saragoza)
Joel Clark (Massachusetts Institute of Technology)
Marcel Dagenais (Université de Montréal)
Ronald C. Duncan (World Bank)
Christopher Gilbert (University of Oxford)
Orhan Güvenen (Université de Paris-Dauphine)
Andrew Hughes Hallett (University of Newcastle)
Walter C. Labys (West Virginia University)
Jean-Baptiste Lesourd (Centre National de la Recherche Scientifique, and Université de Droit, d'Economie et des Sciences d'Aix-Marseille)
Jean Paelinck (Erasmus University)
Louis Phlips (Université Catholique de Louvain)
André Sapir (Université Libre de Bruxelles)
Henri Serbat (Chambre de Commerce et d'Industrie de Paris)
Horst Siebert (University of Constance)
Jean-Marie Viaene (Erasmus University)
Jean Waelbroeck (Université Libre de Bruxelles)

Organization committee

F. Gerard Adams (University of Pennsylvania)
Brigitte Bocum (West Virginia University)
Philippe Chalmin (Conservatoire National des Arts et Métiers)
Ronald C. Duncan (World Bank)
Angela Durham (West Virginia University)
Orhan Güvenen (Université de Paris-Dauphine)
Gabriel Hatchikian (Université de Paris-Nanterre)
Jane R. Labys (West Virginia University)
Walter C. Labys (West Virginia University)
Jean-Baptiste Lesourd (Centre National de la Recherche Scientifique, and Université de Droit, d'Economie et des Sciences d'Aix-Marseille)
Montague J. Lord (Interamerican Development Bank)
Daniéle Meulders (Université Libre de Bruxelles)
François Rigaud (Université de Droit, d'Economie et des Sciences d'Aix-Marseille)

Jean-Michel Ruiz (Université de Droit, d'Economie et des Sciences d'Aix-Marseille)
Henri Serbat (Chambre de Commerce et d'Industrie de Paris)

Introduction

LAWRENCE R. KLEIN,

NOBEL LAUREATE

Economists have long been interested in, and even fascinated by, commodity market indicators, especially price quotations. The relevant markets have several features that make them attractive for quantitative study and analysis:

1. the goods are well-defined primary products;
2. the scope of the markets is generally global;
3. there is usually a high degree of competition in the market processes;
4. there are associated futures markets that reflect *expectations* as well as actual transactions.

There are great advantages to the research economist in dealing with primary product markets. The units of measurement are meaningful and easy to handle. Bushels of wheat, tons of steel, pounds of copper, ounces of gold, or barrels of oil are much more suited to statistical analysis than billions of dollars of expenditures on some basket of items, whether in current or constant prices. There is, essentially, no index number problem for quantitative units and prices in a single market. This, of course, is an overstatement because there are grades of quality and type among primary commodities, but diversity of units of measurement is a problem of second-order importance. Not only are units easy to handle, but the data are rich for research purposes. In the economy at large, price indices have almost always risen for more than four decades. There is a troublesome problem in de-trending such statistical series, but commodity prices go *up and down*. Gold has been as high as $800 per ounce and as low as $35 per ounce. The latter figure was a controlled valuation, and even after the freeing up of gold prices it has not been unusual for gold to fall in price to $300 per ounce or less. Oil prices, grain prices, industrial commodity prices have all been up and down in recent memory.

The fact that prices rise and fall in primary commodity markets means that variability is high, and this contributes to interesting results in statistical estimation. When variance is low, standard errors of estimated coefficients in statistical relationships may be high, or the relationships themselves may be uninteresting. For statistical purposes, variability of observed data, if it is not purely random, is to be desired. It is the high degree of competition in free markets with relative homogeneity of product that contributes to the variability that is observed. At the extreme, competition should be 'perfect' in the economist's sense of the term, but in practice it is imperfect, yet strong enough to produce interesting and challenging databases for economic analysis.

Many of the commodity markets, but not all, have futures markets, i.e. contracts for delivery to buy or sell at prices that fluctuate continuously in the daily market sessions. These futures prices carry a great deal of information that is not otherwise available. They show what people think about the future course of prices of individual commodities, with financial commitment attached to these personal views. The published futures prices show what *average* market opinion thinks about price movements. In an economic analysis that uses the important concept of expectations, we usually lack directly observed statistical series, but here we have concrete cases where expectations are measured. These data may be used for demand and supply equation estimation, but they are also available for testing theories about the way expectations are formed. These are extremely valuable and potentially informative bodies of data for the economic testing of expectations theory and possibly for improving forecasts of parts of the economy.

THE EMPIRICAL BASE

Data on commodity markets have been collected over long periods of time. Adam Smith, in the *Wealth of Nations*, reported on grain prices in England from the thirteenth century through the eighteenth. His were basically annual data. Now we have high frequency reports on daily, hourly or virtually continuous trading activities, all around the world. It is not difficult to construct long-time series of both prices and quantities quoted in primary commodity markets. Statistical techniques that are data intensive can draw upon fairly large samples and thereby use more satisfactory inference techniques.

There is a high degree of speculation in these markets—not simply competition that drives price up and down, but wide speculative swings that occasionally push prices far outside their customary limits. Statistics and scientific probing not only like variability in order to make sharp inferences, but they also like extremes of behaviour, swings that we might call pathological or *outliers*. It is interesting to study these highly unusual cases for insight into peoples' economic behaviour. When traders attempt to corner or control specific markets, we see such extreme movements.

For the most part, modern prices are quoted in US dollar units. Not so for Adam Smith. In his era, he naturally considered grain prices in pounds (£), shillings (s) and pence (d). The facts that markets are global, that telecommunications are virtually instantaneous and that primary goods tend to be homogeneous mean that quotations can be studied from different perspectives for any particular country's interests. It is possible to study the sterling, yen or deutschemark prices of commodities as well as the US dollar price. Fluctuations of exchange rates among the currencies of interest may yield different statistical patterns for the different currency denominations. If exchange rates were fixed, it would not matter what currency or denomination was used for the analysis. But the combined effects of currency movements and primary price movements can be useful for each country's assessment of its own situation. It has recently been suggested that commodity price movements carry important advance signals of price and currency movements. For implementation of stabilization policy in individual countries, it is important to have timely analyses of the measurements of primary commodity prices denominated in the country's own currency units. Further, to coordinate appropriate policy movements among countries, it is important that they are all consulting a common database that has global characteristics so that every policy maker can, in principle, be informed about the statistical series that partners are using to guide their policies.

CROSS-DISCIPLINARY ASPECTS OF PRIMARY COMMODITY ANALYSIS

Economic and statistical analyses are obviously of great importance in studying commodity markets. A deep knowledge of market-clearing processes, the institutional structures of the industries related to each commodity market—whether on the supply or demand side—and the statistical methods of data handling for inference purposes are all needed in order to make good sense of the wealth of information on commodity market data. In addition, a technological understanding of the economic processes underlying each market is necessary. The agronomy of crop production, the techniques of crop distribution from harvest to end-use, the contributions of meteorology, the engineering of metallurgy, the engineering of processing factories, the combating of oil spills, the control of pollution and many other technological aspects of the different markets are essential for a good understanding of the forces at work in each case.

Also legal and political factors play roles in the markets and require some specialized knowledge of their effects. Almost every market is different; and so a specialized technological background is required, but that adds much substance to the research. By fitting together appropriate cross-disciplinary bodies of information in commodity market studies, a high degree of interest and analytical challenge can be attained.

For all these reasons I am enthusiastic about the subject of this volume and want to recommend its scholarly pursuit. This conference volume did not tell us everything that we want to know about particular commodity markets or ways of studying them, but it did take us a sizeable step along the way to informative knowledge in an important area of the economy. It accomplished in fine style just what is stated in the title of the conference and in this resulting volume.

PART ONE

Advances in Modelling Methodology

1

New horizons in international commodity market modelling

WALTER C. LABYS, JEAN-BAPTISTE LESOURD, NOEL D. URI AND ORHAN GÜVENEN

1.1 INTRODUCTION

Most international commodity flows take place within a market framework. We are concerned here with new horizons in efforts to model these markets. Since previous reviews in this area such as those by Adams and Behrman (1978) or Labys (1975), many new modelling efforts have been made or are under way in the hope of improving our understanding of these markets. This includes not only how these markets affect economic development in commodity exporting nations but also how they influence industrial processing and manufacturing activities in commodity consuming nations. Some of these modelling efforts have also dealt with how international commodity policies might improve the functioning of such markets. International commodity markets are complex organisms, often with a large number of submarkets and of actors directly or indirectly involved in production and consumption. The modelling of such organisms is not easy and to a certain extent few modelling breakthroughs have been made. At the same time certain modelling problems have been solved and progress can be witnessed, since this modelling area was launched more than two decades ago.

This chapter provides an overview of the modelling developments and offers prescriptions for the future. It attempts to go beyond the previous modelling reviews which have already appeared in works by, for example, Behrman (1978), Adams (1978), the World Bank (1981), Labys and Pollak (1984), Labys and Wood (1985), Labys *et al.* (1988), Takayama and Labys (1985) and Labys *et al.* (1989). This goal is accomplished by concentrating on substantive issues which will shape new modelling efforts in the future.

The chapter consists of the following parts: Section 1.2 on recent changes in modelling needs, Section 1.3 on modelling developments including the modelling process, modelling market structure and modelling market interactions and Section 1.4 on the future.

1.2 RECENT CHANGES IN MODELLING NEEDS

It has long been recognized that commodity modelling does not constitute a unique professional area but consists of an amalgamation of modelling attempts stemming from agricultural economics, energy economics, mineral economics, marine economics and less directly commodity futures and financial economics. While these models do feature a common core of modelling methodologies, the methodologies themselves are varied. They include econometric methods, mathematical programming, input–output analysis and systems simulation theory and methods. Most of our modelling needs have grown out of attempts in each of these commodity-related disciplines to analyse related commodity issues, to determine the impacts that particular policies might have on commodity markets and industries, and to provide short-, medium- and long-term commodity market and price forecasts. In order to discover new horizons in international commodity market modelling, we begin by examining important needs in agricultural modelling, energy modelling, mineral modelling, futures and financial modelling and economy-interaction modelling.

1.2.1 Agricultural modelling

Agricultural commodity models are many and varied, but they all embody certain similar characteristics. First, the accumulation of evidence suggests that the aggregate demand for food is price inelastic. Hence a small increase in the quantity available will lead to a larger proportionate decline in prices and vice versa. For a review of the literature available in this area, the interested reader is referred to Tweeten (1967), Adams and Behrman (1976), Stern *et al.* (1976), Seeley (1985), Fox (1986), Huang (1985), Roe *et al.* (1986) and Sullivan *et al.* (1989). Most of the studies cited in these references use single equation econometric techniques to estimate the price elasticity of demand for food. Second, particularly in developed countries, the aggregate income elasticity for food is small. Limits on human capacity for food mean that, as incomes rise, individuals in general do not increase the number of meals in a day. The study of food consumption behaviour depends on the estimation of the income elasticity of the demand for food. Examples of surveys of such studies include those by US Department of Agriculture (1965), Ahalt (1977), Goreau (1978), Huang and Haidacher

(1983), Marks and Yetley (1987), Smallwood *et al.* (1987), Haley and Dixit (1988) and Sullivan *et al.* (1989). Typically, these studies involve specifying and estimating single equation (and, in a few instances, simultaneous equation) econometric models of the demand for food. The major implications of these studies (for developed countries) are that the rate of growth of the aggregate demand for food is primarily delimited by the rate of growth in the total population.

In addition to these demand characteristics, the related agricultural markets are distinguished by competitiveness (in the neoclassical microeconomic sense), technological change and the fixity of assets. With regard to competitiveness, agricultural production is characterized by a relatively large number of firms, a homogeneous product (at least within fairly broad categories) and few barriers to entry (Guither, 1963; Quance and Lu, 1978; Harrington and Manchester, 1985; Webb *et al.*, 1985, Teigen, 1988). Rapid technological change is indicated by the extent of mechanization and electrification of farms, extensive genetic changes (such as hybrid wheat and corn and artificial insemination) and the extent of chemical utilization associated with fertilizers and pesticides (including herbicides, insecticides and fungicides) (Kislev and Peterson, 1982; Weaver, 1983; Ball, 1985; Reilly, 1988). Asset fixity means that land and durable capital goods (such as buildings, equipment and machinery) are so specialized that their value in other (non-agricultural) uses is low (except for land located adjacent to expanding urban areas) (Rosenberg, 1976; Solnick, 1975; Johnson and Quance, 1982; Chambers and Vasavada, 1983).

These characteristics of aggregate food systems—inelastic demand for food, slow growth in total demand, competitive market structure, significant technological change, and the tendency of resources to become fixed within the agricultural sector—serve as the basis for constructing models of agricultural commodity markets. (Note that Hathaway (1963) has a much more extensive discussion—replete with examples—of these five characteristics.) The movements in prices of agricultural commodities can thus be said to be inexorably intertwined with these five characteristics. For example, a transitory increase in export demand will increase wholesale or border prices. This will cause (Granger-cause, 1969) an increase in farm product prices, resulting in a rise in the factor inputs used in agricultural production. When export demand falls back to its initial level, however, the inelastic demand for food results in lower prices. Since the factors of production drawn into agriculture as a consequence of the initial increase in demand become more or less fixed once committed, they remain in agriculture and their value changes in response to changes in the prices of agricultural commodities.

One of the primary implications of the characteristics of aggregate food systems is instability. Large fluctuations in prices and quantities are frequently the consequence of these economic characteristics (Waugh, 1966;

Rausser and Hochman, 1979; Chavas, 1983; Conway *et al.* 1987; Uri and Jones, 1988). When these characteristics are coupled with climatic and weather uncertainties, the potential for instability becomes even more significant. The inherent instability of agriculture is offered as the justification for active intervention by governments in the agricultural and food systems (Johnson *et al.*, 1985; Langley *et al.*, 1987; Gardner, 1987; Uri, 1989).

Each of these factors is important in agricultural sectoral modelling in the aggregate. However, when examining market behaviour at the individual commodity level, the factors must be modified and additional characteristics must be taken into account. The additional characteristics relate to agronomic or biological peculiarities of a given commodity and include commodity growth habits, climatic conditions, perishability and nutritional value. They vary widely across the major agricultural commodity groups: meat (cattle, hogs and sheep), grains (corn, wheat, grain, sorghum etc.), dairy products (milk, butter and cheese), fruits and vegetables (apples, peaches, tomatoes etc.), vegetable protein sources (soybean, peanuts etc.) and poultry (eggs, broilers and turkeys). The reader should consult Bailey (1987), Bickerton (1987), Crom (1988) and Webster and Williams (1988) for more elaborate discussions of this issue.

An important differentiating characteristic of individual agricultural commodities is the nature of their inventories. For most commodities, constant market valuation implies that the value of the commodity does not fluctuate with time. In the case of living (or biological) inventories, such as young beef cattle, eggs in an incubator and seedlings in a plant nursery, the situation is different. These inventories do change in value with time even with constant final output market valuations. Hence, for constant market valuations, some agricultural commodities at various stages in the life cycle are 'inventory-in-process'. These inventories-in-process undergo physical changes that augment the value of the inventory (Antle, 1984; Feinerman and Siegel, 1988; Ryan, 1985; Shoemaker, 1986).

The 'growing' nature of the inventory of many agricultural commodities is associated with the life cycle of those commodities whose reproductive traits directly influence production and supply and hence impact on market prices. For example, for beef cattle about one year is required for breeding to weaning of a female offspring. One additional year is needed before the heifer can be bred and a third year is required before the heifer will wean an offspring intended for slaughter. Finally, an additional 15–16 months are required to fatten offspring for delivery to the market. Thus an identifiable lag exists between the time when the requisite signal to increase production is perceived, for example an increase in price, and the time when the herd is expanded to produce more animals for slaughter (Trede *et al.* 1977; Crom, 1988). The stage at which maturity occurs (not only for beef cattle but for most other agricultural commodities as well) is largely a biological one that can be influenced only minimally by economic factors (Stillman, 1987).

The dynamic lags between a specific market variation and the response to that variation are thus influenced by biological and physical constraints as well as economic factors associated with uncertainty and partial adjustment. The interaction of these elements frequently results in both price and quantity cycles for many agricultural commodities (Rausser and Hochman, 1979; Mueller and Jansen, 1988; Vasavada and Ball, 1988). The length of the biological process does not uniquely dictate these cycles but it is a major determinant. Consequently, agricultural commodity cycles are not of fixed length. Rather, their length is affected (and altered) by market (economic) considerations and technological change as well as by biological factors.

From the perspective of modelling individual agricultural commodity markets, the five characteristics denoted earlier are modified. Consumers vary their consumption of meat with changes in price more than they vary their consumption of grains (Huang, 1985; Marks and Yetley, 1987). That is, the price elasticity of demand is (absolutely) greater for the former group of commodities than it is for the latter. Substitutability in demand among various agricultural commodities also impacts on the elasticity of demand. For example, the price elasticity of demand for many fruits and vegetables is (absolutely) relatively large because of the availability of substitutes (Vansickle and Alvarado, 1983; Sun, 1987). A similar situation holds for red and white meats (Crom, 1988).

Concerning the income elasticity characteristic, substantial evidence indicates that there is a shift in consumption from grain (cereal) diets to diets consisting of more meat, dairy products and fruit and vegetables as incomes rise (Regier and Goolsby, 1980; Pinstrup-Anderson and Caicedo, 1978; Marks and Yetley, 1987).

On the supply side, asset fixity between agricultural commodities does not appear to be a major constraint. Land, equipment and machinery, and labour all move from the production of one crop to the production of another and from one owner to another even though they remain within the agricultural system. These sorts of reallocations from one commodity to another in response to changes in relative commodity prices make the supply relationship for individual commodities seem elastic (Chambers and Vasavada, 1983; Vasavada and Ball, 1988; Nielsen *et. al*, 1989).

Finally, commodity interactions or substitution at the final consumer demand and producer supply levels increase the responsiveness of demand and supply to fluctuations in prices and complicate the commodity modelling process even more. Moreover, interactions between plant and animal commodity groups affect the dynamic behaviour of particular agricultural commodities (Harlow, 1960; Maki, 1962; Stillman, 1985). Thus, for example, feed grains such as corn and grain sorghum are inputs into the production of meat. An increase in the price of feed grains results in more beef cattle being fed on range land as opposed to feedlots. This lowers the demand for feed grains leading to a reduction in feed grain prices and ultimately in a reduction in feed grain acreage planted (see Crom (1988) for an extensive discussion of this case).

1.2.2 Energy modelling

Energy models represent an important application of commodity market modelling techniques. The evolution of the international oil market during the past 20 years, particularly since the first 'oil shock' of 1973–4, has triggered important modelling needs concerning that particular market. This extensive proliferation of oil models can be justified because oil and oil products have become the major energy source since the Second World War (whereas coal and solid fuels were the dominant energy source of the nineteenth century and of the first half of the twentieth century). The impression that oil reserves were declining, of course, has somewhat eroded since that oil shock as well as the shock of 1979–80. The comparatively high prices prevailing for oil and refined oil products have caused substitutions toward other energy sources such as natural gas, coal and nuclear-generated electricity, and toward other economic resources or production factors such as labour or capital, through energy-saving actions or investments. Nevertheless, oil remains dominant and crude oil prices constitute leading indicators of other energy prices (the prices of refined oil products and natural gas prices are closely linked to oil prices, and the same is true to a lesser extent of coal and even electricity prices). Below we plan to discuss energy modelling needs initially in terms of such oil market developments and later in terms of energy-economy models, where energy market behaviour is interlinked with macroeconomic adjustments.

Starting with sectoral models, the highly erratic nature of oil market behaviour has spawned the consideration of time series modelling techniques (Pepper, 1985). Typical of new developments in time series oil modelling has been the 'gross product worth' model of the Rotterdam spot market developed by British Petroleum (Drollas and Barthes, 1985). In this model, finite-difference time lag equations are replaced by continuous differential equations. It is well known (Gondolfo, 1981) that first-order, second-order and higher-order finite-difference equations describing one-period, two-period and higher lag dependences are equivalent to, respectively, first-order, second-order and higher-order differential equations. While first-order equations will in general generate smooth time solutions such as exponential functions, second- and higher-order equations are more interesting because they will be able to depict oscillatory and even more 'chaotic' types of behaviour more typically observed in time series.

The formulation of models embodying these properties can be seen in well-known time series statistical approaches such as linear and non-linear trend fitting, autoregressive integrated moving-average (ARIMA) methods, transfer functions and vector autoregression (Granger and Newbold, 1986). These methods have been used mostly for the analysis of oil prices on the Rotterdam spot market and other short-term oil price forecasts. They may, of course, be implemented by more complex models in which the time behaviour of several variables now explains the behaviour of the dependent variable, using multivariate ARIMA techniques. Short- and long-run behaviour of this variable can be further modelled using cointegration techniques (Engle *et al.*, 1989).

One additional feature of time series models is that they can be specified in continuous time; another and even more important feature is that they can be specified as being non-linear. This helps to explain very important price variations as the result of a comparatively small shock, instead of the dampened harmonic behaviour that would be the solution of a well-behaved linear differential equation system. This permits researchers to introduce non-linearities in time series models of energy markets, because most standard models have difficulties in explaining the 'oil shocks' of 1973–4 and 1979–80, as well as the reverse 'oil shock' of early 1986. Of course, in the end, no mathematical system can predict such shocks and the best hope is to construct models that provide rapid adjustment to such shocks, once they occur. Such models, however, are difficult to apply, because of difficulties encountered in estimating the parameters of non-linear equations and systems. Because of the recent development of oil futures markets and oil option markets, the need has arisen to extend the application of these time series methods to the analysis of oil futures market behaviour (Bopp and Sitzer, 1988) and oil option market behaviour. Of some interest has been the Black and Scholes method (Albouy, 1987).

Another need which has arisen to deal with these dramatic market changes has been to explain so-called 'energy gaps'. In such models, the behaviour of oil markets is explained on the basis of empirical supply and demand equations that are estimated separately for different political and economic environments (Gately, 1979). However, there is usually no underlying theoretical framework which ensures the overall consistency of such models, which makes them difficult to apply in forecasting applications. In fact, these models typically have not advanced to the point where they can be represented by a formal mathematical or econometric framework.

To meet the needs of explaining energy demands and supplies, standard commodity models (SCMs) have been applied because they usually employ a set of refined theoretical assumptions which underlie their equilibrium properties, while their dynamics usually rest on some lagged variable dependences that are more or less complex depending on the selected model details (Labys, 1973; Labys and Pollak, 1984). As discussed in several in-depth articles by Gately (1984) and Baldwin and Prosser (1988) among others, long-term oil models based on the SCM methodology can be classified into two broad categories: (a) recursive simulation models, in which economic agents' behaviour results from their 'memory' of past and present events and economic variables, and (b) intertemporal optimization models, in which oil producers maximize their discounted future rents according to Hotelling's well-known theory of exhaustible resources (Hotelling, 1931).

As shown in recent review articles (EMF, 1982; Gately, 1984; Hogan and Leiby, 1985; Labys and Wood, 1985), there has been a need to model oil supply correctly. Both recursive models and time optimization models have difficulties in accounting for the price collapse of 1986, although recursive models are able to describe both the 1973–4 and 1979–80 'oil shocks'. They behave better than intertemporal optimization models, however, because the latter have difficulties in explaining both the 1979–80 'oil shock' and the 1986 'price collapse'.

Among some eight prominent models reviewed by the EMF (1982) and by Gately (1986), only two were able to explain the significant price collapse from 1986 to 1990 when they were used for projections according to an optimistic scenario. Furthermore, minimum prices reached during the price collapse were largely overestimated by these two successful projections, with the IEES–OMS model predicting a minimum price of oil of $16 in 1981 and the model of Gately predicting a minimum price of $27 (1981) against an actual minimum of $7 to $10 in 1986.

This failure to describe the dramatic 1986 oil price decrease satisfactorily can, in our opinion, be ascribed to two main needs arising in oil models. The first need is to provide a better than crude description of the oil market structure. To correct this deficiency, several authors such as, for example, Griffin (1985) and Choe (1984) have developed more realistic econometric representations of market behaviour, based on the earlier works of Salant *et al.* (1981), Aperjis (1981) and Hnyilicza and Pindyck (1976). In these works, more complex market representations such as Nash–Cournot models or Stackelberg duopoly or oligopoly models have been proposed, thus allowing, especially in the Stackelberg approach, for a 'price war' through price cuts aimed at increasing market shares. Thus the introduction of complex market structures is one of the new directions toward which oil models are now oriented.

Another need that arises with oil models is that, as noted recently by Gately and Rappoport (1988), the potential for energy savings through energy conservation policies or energy conservation management at the microeconomic level has not been adequately captured. The short-term impact of energy conservation actions between 1974 and 1979 has probably been overestimated, thus contributing to the failure of models to represent the second 'oil shock' of 1979–80. Conversely, the longer-term impact of energy conservation actions between 1974 and 1985 has probably been underestimated, thus contributing to difficulties in representing the reversed 'oil shock' of 1986 in most models.

A more realistic description of energy savings and energy conservation behaviour, therefore, is one of the important needs of oil modelling in particular, and of energy models in general. It is somewhat difficult, of course, to describe energy conservation management correctly at the micro-economic level. This is because, as noted by Consonni and Lesourd (1986) and by Jacques *et al.* (1988), energy management often rests on variables such as training and corporate management that are not easy to represent in quantitative terms. However, some new directions in which energy conservation might be incorporated into modelling can be sought in engineering or process descriptions on the demand side, such as the MEDEE demand model developed by the EEC and IEPE at Grenoble (Château and Lapillone, 1980). The 'latent equipment' energy demand models developed by Carlevaro and Spierer (1983, 1984), which have been applied to energy demand models for Switzerland, are also interesting in this context; more recently, Carlevaro *et al.* (1988) developed a general equilibrium model which is oriented toward a long-term assessment of energy conservation policy in Switzerland.

Other engineering-related models have also been developed, mainly on the energy supply side, using linear and non-linear programming methods. Such models concentrate on the transformation of fuels and other energy primary inputs into refined fuel products and other oil products. They may be used for demand analysis, explaining demand through analysis process descriptions. These models use techniques and representations that are of interest for the engineering operation of oil refineries and other petroleum processing units, and so they may also be labelled technico-economic models. One of the first important models developed along these lines was that of Adams and Griffin (1972), while the model of Kennedy (1974) made use of non-linear programming formulations. The attempt of programming models to model oil supply can be found in the works of Eckbo *et al.* (1979), in the MORE model developed by Shell International (Minguet, 1985) and in the MAREN model developed by Total-C.F.P. (Compagnie Française des Pétroles-French Petroleum Company) (Bourdaie, 1988); the two last models take into account energy sources other than oil.

New directions in which programming models might develop in the future are essentially the same as those generally described for the other forms of commodity models. The directions mentioned include advanced programming methodologies such as non-linear programming, mixed integer programming and linear complementarity programming. For a recent explanation of linear complementarity programming, see Labys *et al.* (1989).

The need also exists to provide more detailed and wider based models of natural gas markets. These models have received less attention than oil models. This is because, as noted by Lesourd *et al.* (1988) and Percebois

(1989), the development of gas markets at the international level is a relatively recent phenomenon and because the international gas market is quite small relative to the oil market. Most trade flows and price levels also have been regulated, rather than adjusting in concordance with competitive market conditions. Recent models of gas markets deal mainly with the North American market (US Federal Power Commission, 1972; McAvoy and Pindyck, 1972, 1975; Rose *et al*., 1985). However, more recent modelling efforts have focused on methodological developments which reflect important new research areas in this field. These areas include the modelling of gas contracts by Hoel *et al*. (1986), Boucher and Smeers *et al*. (1988), Golombek *et al*. (1987) and Lesourd *et al*. (1988).

The requirement to model the relative price competitiveness of international coal markets has not been met. At the national or domestic level, several regional coal models have been developed; they concern mainly the North American markets. Among these models, one can mention the econometric and non-linear programming models developed by Kolstad (1989) and by Labys and co-workers (1979, 1980, 1982, 1989) and the international coal model of the US Energy Administration (1984) can be mentioned. The development of a broad-based international coal market model to meet all the required needs, following Kolstad (1989), would be an important research area in itself, where coal market structure, trade, transportation costs, policy variables and environmental concerns would be the primary focus.

1.2.3 Mineral modelling

Mineral modelling possesses many of the same modelling needs as agricultural and energy modelling. For example, risk and uncertainty issues arise not only because of the uncertainty surrounding the nature of geological deposits but also because of the uncertainty associated with the changes in exogenous variables, such as the impacts of inflationary and recessionary conditions on mineral demands. The modelling of price expectations is important as is the modelling of markets in disequilibrium, particularly since mineral markets have a complex array of stock–flow interactions. Readers interested in a major review of mineral modelling can consult, for example, Labys *et al* (1985) or Labys *et al*. (1989).

A first major modelling need in this area is to explain better the intertemporal linkages or the dynamic adjustment processes that exist between the price signals or expectations and the responses that occur in exploration, mine development, process capacity expansion etc. These long-run relationships were explored to some extent in a study conducted by Harris and Burrows of Charles River Associates (1978). While the study concentrated on reserve formation and mine development in the copper industry, very little progress has subsequently followed with attempts to explain capacity adjustments in mining and processing across a variety of mineral industries. One major problem appears to be that mineral investors do not exhibit the same producer price response, for example, as that displayed by planters of perennial tree crops. Mineral investors appear instead to focus on market share objectives or perhaps some more complex undiscovered profitability goals. One other study of dynamic adjustments has employed linear complementarity programming in an intertemporal context to model mineral investment behaviour. Hashimoto and Sihsobhon (1981) in their model of the world iron and steel industry (WISE) incorporated market expectations based on forward information and market dynamics.

A second need requiring attention is that of modelling the imperfect market structures which typically exist in mineral markets. While pure monopoly is rare and pure competition is found only occasionally, most mineral markets possess intermediate market structures whose behaviour rests somewhere between these two extremes. One problem is that such structures often approach that of bilateral oligopoly; here economic theory does not provide much help in explaining where price and equilibrium quantities should merge. A first need, to try to quantify the nature of mineral market structure, has been attempted by Hannan (1988). Where does a particular mineral market exist on a behavioural scale ranging from monopoly and competition (supply) or monopsony and competition (demand)? Some progress in this area has been made by energy modellers such as Salant *et al.* (1981) and Kolstad (1982). Less research has been conducted by mineral modellers. For example, Pindyck (1978) has explored copper and aluminium market response under conditions of resource cartelization, and Soyster and Sherali (1981) have explored the complexities of modelling imperfect copper market structures. But in general no extensive modelling of imperfect mineral market structures has taken place.

Regarding the explanation and prediction of minerals demand, this modelling activity has been hampered because of the complex nature of mineral substitution patterns. Non-fuel minerals are frequently substituted one for another. In addition, non-mineral and mineral substances can be substituted for one another in production and consumption. Substitution is therefore an important aspect of mineral modelling. If producers are cost minimizers, they will substitute one input for another when the relative price change. The ease with which that can be accomplished, however, depends on the feasible production possibilities. In some production processes, input substitution possibilities are very limited, whereas in others inputs can be varied with considerable ease. In the former case it will generally take large relative price changes to induce material substitution, whereas small changes in relative prices may lead to sizeable shifts in use when such shifts are less costly.

When substitution is costly, the response to changed factor prices may be asymmetric. That is, when relative prices resume their initial level, the process may not be exactly reversed. A good example of this is the switch from zinc to plastics that occurred in many decorative automotive components in the mid-1970s following the very high zinc prices of 1973–74. Although there was a major improvement in the price competitiveness of zinc in subsequent years, these markets did not recover because, having made the materials switch, the automotive industry and its suppliers had replaced outmoded facilities with new capital equipment with a useful life of many years.

Modelling methods that are based on statistical inference from past data tend to be quite good at capturing and predicting material substitution decisions that are marginal and that lie within the range of historical experience. They tend to be relatively weak, however, at handling substitution decisions that occur as a result of larger changes in price and availability of materials and where responses are asymmetric for technological, institutional or consumer preference reasons. Forecasting methods based on more explicit engineering assumptions tend to be better at capturing the latter forms of substitution. However, engineering-oriented methods tend to be weaker at capturing the magnitude of substitution possibilities that derive from relative price changes, because price is typically difficult to integrate into the engineering analysis. There have been a number of examples in recent years that testify to this phenomenon, not the least being the systematic underestimation of the response of energy demand to higher prices.

Another important mineral modelling need is to recognize that the mineral commodities used and traded within industrial requirements normally involve different stages of process or production (Kovisars, 1976). These can include mining, ore treatment (milling and concentration), reduction (smelting), purification (refining) and consumption by fabricators. Recycled material is an important process input for many mineral flows and may enter the supply flow at several stages. While Manne and Markowitz (1963) describe the early development of these models as an application of process analysis, they also report on the integration of this approach with spatial equilibrium analysis. Such models have been formulated in terms of production technologies that account for the important inputs (labour, energy and materials requirements) at each stage of the production sequence. This has required constructing mathematical equations that represent the various technological or engineering production possibilities. Usually the overall production flow has been disaggregated into elementary process routes, and input–output parameters are derived from each stage of each route on the basis of engineering data. For each process, programming techniques can be used to select from among the various production possibilities those that optimize a pre-set goal.

A process model can also lend itself to demand analysis by permitting a linkage to be made between final product demands and derived material demands. National economic activity can be used to explain final product demand. The process model is thus capable of explaining the product transformation process, i.e. how primary products derived from materials can be transformed into secondary and then tertiary products (or greater) until final product demand is met. Examples of process models which have attempted to meet these needs include those of Dammert and Palaniappan (1985), Ray and Szekely (1973), Tsao and Day (1971), Clark and Church (1981) and Kovisars (1975, 1976).

Mineral modelling has also had to meet the need of modelling how industrial economies might react to a sudden imposed reduction in strategic mineral suppliers. Non-fuel mineral ores are found in many parts of the world. Some deposits are controlled by national or multinational corporations, whereas others are controlled by producer country governments. Privately and publicly owned reserves may be exploited according to widely differing rules. In addition, the stage of economic development as well as the political philosophy of producer country governments may vary greatly from country to country. Some countries may be politically unstable, and others may be hostile to Western industrialized economies. Because Western industrialized countries are dependent on imports of many non-fuel minerals to support industrial production, modellers of mineral demand must consider the demand consequences of both short and prolonged supply disruptions.

Supply cutbacks may have many causes, including mining strikes, political problems in producer countries and deliberate reduction in offerings by members of producer cartels. The ease with which users can adjust to supply disruptions depends on the availability of substitutes and on the time required for adjustment to take place. It also depends on whether the cutbacks manifest themselves in the form of higher prices that will still clear the market or in the form of producer allocation or rationing, for example allowing each customer a specified percentage of the previous year's consumption.

Different policies have been proposed to deal with the issue of mineral supply security, including (1) the build-up of strategic and economic stockpiles; (2) government support, through tax or other incentives, for domestic production of minerals available in a country but not being mined; (3) encouragement of conservation; and (4) changes in the materials used in the fabrication of consumer products. Decisions on policies that would promote the security of mineral supplies through these mechanisms are improved by an understanding of mineral demand relationships over time, particularly the responsiveness of demand to changes in price, as well as the time required for the market to respond to such changes and the degree of reversibility of that response. Among previous efforts to model mineral supply restrictions, readers should consult the multicommodity study conducted by Charles River Associates (1975).

Mineral modelling is also closely tied to inter-industry behaviour patterns, i.e. how minerals serve as inputs to a number of industries that generate fabricated or manufactured outputs. This need can be met by embodying input–output techniques within a modelling framework.

Though input–output techniques and their applications have become more complex over the years, this approach basically endeavours to account comprehensively for the gross output of any given economy by dividing it into a relatively large number of product sectors. The output of each such sector is fully accounted for by being distributed as an input to all other such sectors in the form of 'intermediate demand' as well as 'final demand'. Final demand, in the aggregate, is the same as gross national (or regional) product, and it may be broken down into the familiar national accounting categories of consumer purchases, government purchases, investment and exports. For mineral industry studies, however, it must also be broken down into the mineral industry detail for which it is desired that there be an input–output account.

What makes input–output useful (though not sufficient by itself) for mineral modelling purposes is the fact that the matrices can be treated as a series of producing-sector-requirements equations. These requirements are in each case dependent, according to parameters stated in the matrix, on exogenous final demand variables as well as on other intermediate-industry variables further up the production chain. In the case of mineral demand modelling, some of the producing sectors are specifically mineral production and/or processing sectors. The effect of assuming different values for the parameters can also be investigated. However, owing to the requirement for complete consistency between mineral inputs, outputs and final demand, only the final demand 'bill of goods' may in fact be treated as a fully independent set of variables. Examples of input–output applications to meet mineral modelling needs can be seen in the work of Bingham and Lee (1976), Krueger (1976), Leontief *et al.* (1983) and Ridker and Watson (1980).

A final mineral modelling need has been to move away from conventional commodity modelling techniques to include the engineering characteristics of mineral production and consumption activities. Such models depend on the acquisition of parameters from engineering data and can lend themselves to modelling in a systems context (US National Academy of Sciences, 1982). In its simplest form, a system simulation model is nothing more than a set of hypotheses of interrelationships among variables, typically dynamically specified, in which the time behaviour of the interrelationships is revealed, based on model simulation. Thus a system simulation model can be distinguished most clearly from the other causal modelling approaches as one not requiring hypothesis testing in the statistical sense or the attribution of rational economic decision making to the actors being modelled.

Typically, a system simulation model will interconnect submodels that capture generally accepted behaviour on a small scale in an attempt to simulate overall performance on a large scale. As practised in some forms, the model will often articulate global constraints, generally accepted to exist but usually quantified only approximately because of limitations of knowledge that have important implications for the time behaviour of the model variables. For example, constraints that trigger turning points, or changes from growth to decline or vice versa, or changes from growth to steady state are favourite targets of analysis. One of the greatest strengths of the system simulation methodology for mineral modelling is its flexibility. One is not constrained by data or economic rationality in constructing the model. Because of this flexibility, however, the human resource requirements for system simulation modelling are unformly stringent. An example of a minerals system application can be reviewed in the work of Clark and Church (1981).

1.2.4 Financial interactions modelling

The modelling of certain commodity markets and commodity variables requires shifting the market environment to include what might be considered externalities to commodity supply and demand transactions. That is, needs have arisen particularly in modelling commodity price behaviour to consider influences occurring simultaneously in commodity futures markets and in related financial markets. These needs stem to a large extent from a series of events which began in 1972. These include the sudden jumps in petroleum prices from 1973 as well as changes in the world monetary system (caused by a transition from pegged to floating exchange rates for the major currencies). During this period the world also experienced several sharp increases in inflation which led many investors to consider commodity futures contracts as well as commodities themselves as a hedge against inflation.

To postulate a causal model of commodity price behaviour, price analysts thus began to consider the impact of financial and macroeconomic factors on important commodity variables. This impact was seen as (a) acting through commodity demand such as changes in industrial activity, own prices and prices of substitutes, monetary or liquidity conditions and exchange rates; (b) acting through supply such as factors of production and their relative efficiency, own lagged prices, financial and capital conditions reflected in interest rates and exogenous supply conditions such as weather and exchange rates; (c) acting through international economic conditions such as inflation, recession and changes in relative purchasing power; and (d) acting through related commodity futures market activity such as speculation and hedging. In providing short-run explanations of this type in particular, the monetary, exchange rate and futures market conditions appear to be most important.

Among attempts which have been made to meet this special need in explaining commodity price behaviour, Labys and Granger (1970) earlier considered the underlying price determinants according to whether they originated in the physical market or in an accompanying futures exchange. Employing cross-spectral and stepwise regression techniques, they found that futures as well as financial market conditions exerted a significant influence on commodity price fluctuations, particularly in the short run. With the floating of the major currencies and the rising inflation in the early 1970s, Labys and Thomas (1975) again sought to decipher the price impact of futures, financial and inflationary conditions. Like Labys and Granger, they dealt with the behaviour of individual commodities and concentrated on the importance of financial factors interacting with price formation on futures exchanges. But the financial variables considered were expanded to include exchange rates, money stocks, treasury bills and the Euro–dollar and Euro–sterling rates. The impact of these variables on commodity price behaviour was found to be important, but correlation tests were employed instead of a formal price model.

More recently price analysts have examined this problem at the macro-economic level, and thus have studied commodity prices only in the form of aggregate price indices. Cooper and Lawrence (1975) were among the first to discover the importance of the influence of international economic conditions and industrial activity on aggregate prices, particularly as these influences affected commodity demand. Bosworth and Lawrence (1982) also confirmed the significance of positive relationships between commodity prices and international economic activity. Supply-side variables such as production and inventories were also shown to influence aggregate price behaviour.

As this research expanded, needs have shifted to formulating more complete models of price formation based on the theory of excess demand which views price determination as a consequence of market equilibrium adjustments between demand and supply. Variations on this approach have included disequilibrium theories which emphasize the presence either of inventories (Hwa, 1979) or of dynamic expectations, learning and lagged adjustments (Labys, 1981). Most important in such model constructions was the reduction of possible structural specifications to a single reduced-form price equation. This equation has since provided a basis for related econometric tests. Enoch and Panic (1981) were among the first to apply such a model to aggregate commodity price behaviour. Interest rates were theorized to affect commodity demand because of their influence on holding inventories and on the relative returns from alternative forms of asset holding. However, econometric tests of their price equation did not prove interest rates to be significant.

Grilli and Yang (1981) employed a similar reduced-form approach but included a greater variety of monetary influences. As a long-run influence, increases in international liquidity were theorized to deter commodity-producing countries from 'distress sales' at times of low market demand and to stimulate other market participants to hold commodities as assets, which adds to the industrial demand for them. As a short-run influence, increases in interest rates were hypothesized to increase the opportunity costs of inventory holding, and variations in exchange rates were thought to create shifts between liquid and commodity assets. However, interest rates proved to be an important determinant only for aggregate non-food and metal prices, while the variability in exchange rates was stronger in explaining aggregate food and metal prices.

Chu and Morrison (1984) examined longer-run price behaviour, also by employing a reduced-form approach. They found that variations in commodity prices were influenced by variations in industrial production, dollar exchange rates *vis-à-vis* other currencies, and supply shocks. However, only the level of interest rates influenced commodity prices significantly, and this occurred during the early 1980s and not during the mid-1970s. Their approach was later adopted by the US Federal Reserve Bank of New York (1985) who tested its applicability to four of the major aggregate commodity price indices, i.e. the International Monetary Fund (IMF), Economist, United Nations Centre for Trade and Development (UNCTAD) and World Bank indices. Econometric regression tests confirmed the importance of interest rates in explaining commodity prices. In addition, they showed that supply-side factors become important when market conditions return to long-run 'equilibrium'.

Since then Gilbert (1986), in concentrating on the explanation of fluctuations in the World Bank commodity price indices, developed a more rigorous analysis of the impact of exchange rates on these indices, particularly as this linkage is influenced by the indebtedness of developing countries. Aggregate non-food agricultural and metal price indices were found to respond rapidly to exchange rate adjustments. Here the econometric method employed simulated the impact of interest rate shocks, instead of testing the statistical significance of coefficients in regression equations. Aggregate food price indices were found to lag behind these shocks, with a lag of almost 2 years.

Few of these studies, however, have provided econometric tests of the importance of commodities and commodity futures contracts as a means of hedging to diversify international portfolio asset holding. Labys and Thomas (1975) did examine commodities as assets and considered the returns from holding commodities compared with other financial investments as part of the overall commodity price determination process. Labys *et al.* (1989) have updated this work and show the importance of meeting this need to provide a broader explanation of commodity price behaviour. While some progress has been made in general in explaining the impact of the financial futures variables, the next step must be to model these relationships on an individual commodity level rather than an aggregate level.

1.2.5 Economy-interaction modelling

The need to improve the modelling of the linkages or interactions between commodity markets and the surrounding macroeconomy was mentioned above. Linkage in this case involves the coupling together of two or more models in a systematic fashion. For example, an input–output model of the energy sector can be linked with an econometric model of the macroeconomy. Analysing feedback effects, such as the impact of energy prices on economic growth, is important in this case. The response that has occurred to meet this need can be witnessed in the different forms that have emerged.

Most well known among these has been the linking of input–output models of the energy sector with macroeconomic models. For example, the Hudson and Jorgenson (1974) model consists of a macroeconometric growth model of the US economy integrated with an inter-industry energy model. The growth model consists of submodels of the household and producing sectors with the government and foreign sectors taken to be exogenous, and it determines the levels and distribution of output valued in constant and current dollars. It has been coupled with an energy-based input–output model to forecast long-term developments in energy markets within the framework of a consistent forecast of macroeconomic and inter-industry activity. The model has also been used to analyse the impact on energy demands of alternative tax policies, including a uniform British thermal unit tax, a uniform energy sales tax and a sales tax on petroleum products.

Among other energy-economy models, Groncki and Marcuse (1980) report on the BESOM model which combines a programming model of energy supply with a long-run macroeconomic growth and input–output model. One example of the use of the model has been to analyse the economic impacts of fuel scarcities. Constraints can be placed on the availability of fuels and resources, and the required fuel substitutions are determined. Coefficients in the input–output model are revised to reflect the new fuel mix, and the input–output model is again solved with the revised mix. Several iterations are required between the two models in order to obtain a solution in which the energy demands and fuel mix are consistent in the two models. Impacts on the macroeconomic growth and various industrial sectors in the economy are then evaluated.

While these models analyse energy-economy interactions in a domestic context, the need also exists to model energy-economy interactions at the international level. One development in this direction has been the models of Chichilnisky (1981) which are based on trade and balance of payments theory. Their model consists of a two-sector balance of payments model which analyses the relations between oil prices and output, employment and prices of goods in industrial economies. The industrial or developed country region in a competitive market economy that produces two goods (consumption and industrial goods) with three inputs (capital, labour and oil). It trades industrial goods for oil with a developing oil exporting region which acts as a monopolist. The general equilibrium solution of the model determines endogenously the principal variables of the industrial region: output and prices of industrial goods as well as employment of their factors and their prices. Simulations of the model are used to determine the impact of different levels of oil prices in the two regions.

Another need met by energy-economy modelling has been that of the large-scale international models which examine trade between major regions of the world and attempt to include energy production and consumption within that model. Scenarios are generated with computer simulation models which explore the impact on international trade and growth of changes in energy conditions. The latter include rising oil prices as well as increased oil depletion and energy conservation. Some of these models include energy as only a minor sector of the whole model, while others consider energy as the driving sector of the model. Examples of these include the Hughes and Mesarovic (1978) world integrated model (WIM), the Herrera and Scolnik (1976) version of the Bariloche model, the Linneman (1976) MOIRA model, and the International Institute for Applied Systems Analysis sets of models described by Basile (1979).

1.3 MODELLING DEVELOPMENTS

There is no easy way of organizing the many recent commodity modelling developments into a unified framework. The categorization that has been adopted here is thus arbitrary; it has been divided according to the modelling process, the modelling market structure and modelling market interactions.

1.3.1 The modelling process

The response to commodity modelling needs is discussed in terms of recent advances in economic theory which have prompted a number of innovative approaches to the modelling process. The discussion of these developments is based on a recent study by Johnson (1986).[1]

[1]Portions of the following presentation on risk and uncertainty, intertemporal decision making, aggregation, rational expectations and regulation have been drawn from Johnson (1986).

(a) *Risk and uncertainty*

Many commodity policy decisions are made when outcomes are uncertain. This uncertainty evolves from the market structure of the industry, the international instability of commodity markets, the environment, changes in technology and a host of other factors. Uncertainty has only recently been recognized as a feature in policy analysis and a complexity for commodity policy modelling. Important advances that deserve attention in policy analysis and modelling recognize choice under uncertainty. The correspondence between decision making at individual and market levels and the relatively strong assumptions on the utility functions required have been analysed by Newbery and Stiglitz (1981).

Risk and uncertainty have been considered by commodity modellers by introducing stochastic processes. Adams (1981) has defined four commodity modelling areas where stochastic processes can come into play. The first of these concerns the assumption of profit maximization that underlies the purely competitive and non-competitive models that describe both the selling and the purchasing sides of international commodity markets. Instead of risk neutrality, utility maximization may involve expected profits and higher moments of the distribution of profits. With the assumption of quadratic utility functions, for example, the variance can be considered as well as the mean. Some empirical evidence exists for risk analysis in the production of primary agricultural commodities (Behrman, 1968; Hazell, 1983).

Secondly, stochastic processes can describe the impact of market fluctuations on inventories held by oligopolists who are cooperating in settling prices. Thirdly, stochastic processes also have been related to the modelling of futures price behaviour. This behaviour has often been linked to the speculative motive for holding inventories when the expected price increase is greater than the interest rate plus storage costs (or for going short in the opposite case). A related activity is the practice of hedging. The modelling of commodity markets has recently attempted to incorporate these phenomena, including the possibility of destabilizing speculation due to 'bandwagon' effects.

Another possibility for incorporating stochastic processes involves the incorporation of concentration and market power in commodity models (Hannan and Labys, 1988b). Assume that the population of sellers in an industry is fixed, that each firm starts with an equal market share, and that the normal distribution of growth rates facing each firm is the same, independent of firm size and past growth history. These assumptions satisfy Gibrat's law of proportional growth. The resulting log-normal distribution often fits firm size data reasonably well in the industrialized economies. Thus stochastic factors may supplement the explanations offered later concerning the causes of concentration and market power.

(b) *Intertemporal linkages and adjustment processes*

Many commodity resource allocation and production problems have dynamic underpinnings. For example, the dynamic aspects of reserve depletion and exploration policy have long been recognized in natural resource economics. In addition, the actual physical aspects of expected profitability and capacity formation reflect neoclassical adjustment processes. In the case of investment in commodity-producing capacity, the adjustment problem is related to that of filling the gap between actual and desired levels of the fixed factor. In other words, given a difference between the desired and the actual stock of the fixed factor, what factors determine the investment flow to fill that gap? As Adams (1978) and other have emphasized, if a firm has no market power (as is assumed in the neoclassical theory), the costs of investment goods which it faces are constant and the optimal rate of investment to fill any shortfall between desired and actual capital is first infinite and then zero. Lucas (1967), however, has provided a rationale for a steady flow of investment in such a situation by positing internal increasing costs of adjustment as the rate of investment rises owing to problems of integrating the new fixed factors into the production process. Most recent empirical efforts to estimate investment functions posit an adjustment to the discrepancy between desired and actual fixed factors, perhaps with some representation of uncertainty and of capital market imperfections (by such characteristics as liquidity and credit availability).

For many of the commodity markets of interest, unfortunately, direct data on capital stocks or investment are not available. Therefore the process of adjusting to the discrepancy between desired and actual fixed factors often is substituted back into the short-run supply function to obtain a long-run function with the investment decisions collapsed therein (Adams and Behrman, 1978; Behrman, 1968). More recently, attempts have been made in fact to model long-run capacity adjustment processes. Examples of modelling capacity formation can be witnessed in research on mining capacity formation by Harris and Burrows (1978) and on perennial tree crop supply by Akiyama and Trivedi (1987). Intertemporal decision analysis has also been widespread in the modelling of oil depletion and reserve formation by Farzin (1986).

(c) *Price expectations*

The existence of stochastic processes and adjustment processes means that expected prices replace the actual prices in many of the relations discussed above. Opening a mine, for example, often requires a long gestation period with many important decisions being made long before the relevant mineral sales price is known. In addition, speculative inventory holdings often depend basically on the difference between the current price and some expected future price. French (1984) has thus emphasized the importance of including expected future prices in commodity policy modelling. One possibility is to use future prices from an appropriate futures market. Because expectations often influence spot and futures market behaviour simultaneously, this alternative has seldom been explored in modelling international commodity markets, particularly when only annual observations are available. Labys and Granger (1970) earlier undertook some initial explorations using data from future markets with monthly observations. More recently some progress has been made by Gilbert and Palaskas (1988).

The dominant representation of expected prices in studies to date, however, has been largely one of a polynomial function of past and current prices. Most common have been simple lags or the geometric distribution, but sometimes higher-order polynomials have also been used. These expectation equivalents are subject to the general criticisms of distributed lags, as summarized by Griliches (1967). They also do not incorporate extra information often available in the real world, e.g. climatic conditions in major agricultural nations.

Most recent advances in the formation of price expectations in commodity models have addressed the rational expectations notion of Muth (1961). From the rationality theory, expectations of agents are modelled as if determined by the framework or the true parameters of the underlying commodity model. Since rational expectations are essentially forward looking, models must be based on conditioning variables that are ultimately projectable. If the conditioning variables cannot be projected, the associated complex models can be simplified in terms of characterizing the behaviour of economic agents. Finally, rationality and regulatory or policy regimes are not inseparable at the conceptual level. Implications of rationality for future model specifications are important. Research which has proved valuable involves the solving of dynamic rational expectations models.

(d) *Aggregation and linkage*

Unfortunately, commodity models destined for aggregate policy analysis have a limited number of microeconomic variables which can be easily linked with macroeconomic variables (see, for example, Johnson, 1981). The experience with large-scale econometric models has emphasized the importance of understanding the microfoundations of macroeconomic models. One development in this area is the hybridization or linkage of different modelling algorithms, an approach which can serve to improve the modelling of market interactions. Hogan and Weyant (1980), for example, offer a 'combined' energy model approach for reaching equilibrium solutions with a combined set of models.

(e) *Non-linearities and functional forms*

The theory underlying a commodity model will suggest the functional form of the structural equations. The typical relationships are estimated either in linear form or in a form that is linear in the parameters, for example log-linear or some other transformation. While these forms generally conform to a theoretically formulated model, Adams (1978) cites instances where the functional form only broadly approximates the theoretical specification. This reflects the difficulty and cost of non-linear estimates when short sample periods are involved. While simplified functional forms, even linear approximations, may be suitable for many simulations covering a 'normal' period, the difficulty is that the abnormal or extreme periods are precisely the points when the non-linearities of the model should become effective. More recent attention to such non-linearities can be found in the work of Rausser *et al.* (1983).

(f) *Controlled markets*

Concerns about the environment, internalization of externalities, commodity cartels and embargoes, natural resources and other factors emphasize the importance of market control. Recently, the modelling of market control has been advanced to improve the specification of models that characterize producer and consumer behaviour under different regulatory structures and to evaluate the performance of regulated markets (Newbery and Stiglitz, 1981). Because many regulations are designed to counter impacts of uncertainty and to condition expectations, these developments in modelling regulation have far-reaching implications. They are important to the modelling of international markets for agricultural or energy commodities, where outcomes depend on highly regulated output, input and even exchange rate markets.

While the competitive market model can be adapted to include the influence of market regulation, it still remains a competitive model. This predicament can lead to serious consequences when modelling energy markets whose structure not only involves regulatory policy but also tends to be non-competitive (Labys, 1980). That is, their structure may vary from complete contol in the form of monopoly to lesser degrees of non-competitive behaviour such as that of duopoly or oligopoly. The principal transformation that must come about in describing market behaviour in controlled or non-competitive markets is to consider price determination from the point of view of the actions of individual market participants rather than of the workings of the market as a whole. Developments in the modelling of monopolistic markets can be seen in the work of Blitzer *et al.* Modelling more complex imperfect market behaviour, for example, can be seen in the exploration of the Stackelberg model by Gilbert (1978) and the Nash–Cournot model by Salant *et al.* (1981) and by Kolstad (1982).

(g) *Substitution*

Among the different types of substitution, 'material for material' substitution is the most well known, implying the replacement of one material for another in the production of a particular good. Secondly, 'other factors for material' substitution results from the applications of new technologies which economize by replacing material inputs into production processes by other inputs such as labour, capital and energy. Thirdly, 'technological substitution' arises when an advance in technology reduces the amount of material needed to produce a particular product. It can also occur when those advances lead to the creation of new final and intermediate products, some of which can be substituted for the raw material itself. Two other forms of substitution exist which are of a less direct nature. 'Quality for material' substitution results when there is a desire to change the character of a product. It usually leads to a cost reduction but not necessarily. More radical is 'inter-product' substitution which results in a change in the mix of goods needed to achieve a particular function or purpose. Here a product including its material is replaced by an entirely different means of achieving a desired function.

Recent attempts to improve the modelling of substitution have followed diverse rather than uniform directions. Most of these recognize that while price differentials are important factors that influence substitution in the short run, other determinants, such as technology, play a greater role in the long run. Previous advances by Berndt and Wood (1975) using the translog approach have been extended to explaining interfuel substitution in regulated markets by Fuss (1980) and Wood and Spierer (1984) and to mineral substitution by Hazilla and Kopp (1984) and Moroney and Trapani (1981). Nordhaus *et al.* (1987) have explained the long-run demands for copper incorporating competitive metal activities within an intertemporal programming framework. Finally, Hannan and Labys (1988a) have expanded the traditional logistic substitution model to include relative prices and other factors influencing mineral and materials substitution.

(h) *Time series analysis and forecasting*

A final commodity modelling development involves the application of time series methods to explain and to forecast basic commodity market variables. Traditionally time series methods have been univariate, but more recently time series methods of a multivariate models have the capability of dealing with several commodity variables at once. Univariate models have been applied to commodity variables to obtain a representation of the generating mechanism of a commodity market variable such as prices. Typically the behaviour of a weakly stationary time series can be formulated as embodying autoregressive (AR), moving-average (MA) or ARMA processes. These models have been used as an alternative to structural models for forecasting individual commodity variables such as prices. Although the former have been proved to generate forecasts superior to those of structural models in some cases, they often fail to take account of all possible explanatory information, i.e. the influences of exogenous influences on the endogenous variable of interest.

A vector of purely non-deterministic stationary economic time series can most generally be represented by a multivariate ARMA process. In such models any type of causal relationship among commodity variables is possible. However, because of the complexity of identification and estimation procedures (e.g. non-linearity in coefficients), simplifying assumptions have to be imposed to obtain models which are more plausible for practical purposes.

The vector autoregressive model (VAR) is one of these simplified versions and assumes that the polynomials in the matrix of polynomials of the error process are of degree zero. Again, a feedback relationship among the variables is often possible. VAR models are particularly useful for commodity forecasting purposes, but the estimated coefficients often have no economic interpretation attached to them. Sometimes the commodity variables employed in these models are cointegrated. As mentioned above, cointegration modeling developed by Granger (1986) and Engle and Granger (1987) should be applied instead.

The general multivariate ARMA model can be expressed in the form of a dynamic simultaneous equation model. This is possible when commodity time series can be distinguished as input compared with output series using prior information, i.e. that available from economic theory. The final form of such a model, which expresses the output series as a function of present and past values of input series as well as an ARMA process of the error term, is also referred to as a simultaneous transfer function model. The one-way causality relationship found beween the variables in this model can be tested and the lag structure of the variables and the error term can be specified using time series techniques. The transfer function model can be employed for both prediction and policy analysis (see, for example Chapter 8 later).

Finally, the classical linear regression model can be set up in the frequency domain by applying finite Fourier transforms to the variables. This generates sets of observations which are indexed in the frequency domain rather than in the time domain. The spectral regression is thus the regression of the transformed dependent variables on the transformed independent variables. This regression method facilitates the application of 'band spectrum regression' in which regression is carried out in the frequency domain with certain wavelengths omitted. Band spectrum regression is particularly attractive in commodity applications because of its effectiveness in dealing with seasonality, errors in variables and the problem of serial correlation in the error terms.

It is well known that structural change occurs frequently in commodity markets with the consequence of causing variations in the parameters of commodity models. In addition, such variations may also be justified for econometric reasons. First, coefficient variations may occur as a result of imposing an incorrect functional form. As stated by Rausser *et al.* (1983), 'the approximation of highly nonlinear "true" relationships by simpler functional forms along with observations outside the narrow samples range, provides perhaps the strongest motivation for a varying parameter structure'. Second, coefficient variation may be due to the omission of the relevant variables. Duffy (1969) has shown that changes in coefficients related to omitted variables will produce variation in the coefficients of included variables in the regression, unless the omitted variables are uncorrelated with both the dependent and independent variables.

Third, the use of proxy variables for unobservable explanatory variables may cause parameter variation. This is because proxy variables will only partially capture changes in the behaviour of the true variables; the relationship between them may change over time. Finally, the use of aggregate data can be another source of coefficient variation. The aggregate data are measured by weighting the relative importance of the various sets of micro units. Because the relative importance of these micro units can be expected to vary over time but the parameters in the estimated aggregate equation will remain constant, it will cause coefficient variation.

The time-varying coefficient model is an alternative approach that allows one to deal with the problem of parametric variation or the instabilities in commodity relationships by assuming that a unit change in one of the independent variables, all else equal, will not have a constant expected effect on the dependent variable at all points in time. Commodity modellers have thus explored the use of stochastic coefficient estimation based on a non-stationary or a time-varying random process to overcome some of the above-mentioned instability problems. An example of the stochastic coefficient estimation technique applied to agricultural prices can be seen in the work of Conway *et al.* (1987).

(i) *Trade and spatial equilibrium*

Explaining commodity trade flows has become an issue of increasing importance. Commodity policy analysis is concerned with trade impacts of different trade measures on exporting as well as importing countries. Given the complexity of trade arrangements negotiated, for example, between the USA, the EEC and Japan, it is not surprising that model developments should lean in this direction. Most recently, Monke and Taylor (1984) have developed a model which classifies countries into groups as commodity producers or consumers according to their trade price responsiveness. The model features a cross-section time series data set for major cotton producing and consuming countries, where quantitative controls on international trade interact with this response or adjustment. Particular attention is paid to how non-price-responsive countries react to the impacts of government trade policies.

Takayama and Labys (1985) reviewed applications of spatial and intertemporal price allocation (STPA) models to regional and trade issues. Increases in these modelling applications can be seen in numerous studies conducted by a number of policy-making groups including national governments, international organizations and private enterprises. The reasons underlying these efforts have been the increased instability in these markets, the various oil crises and the growing international debt burden. An important modelling development in this area deals with the market configuration where multiple final product markets exhibit asymmetry in demand behaviour; this requires improving demand equations and demand parameter estimates to reflect this pattern. One early attempt in this direction was the nine commodity and two-region world food trade and stabilization model developed by Hashimoto (1977).

Nguyen (1977) also used this approach to explain world trade and reserve policy evaluation by expanding the number of world regions to ten. A robustness test was conducted for 1974, and then a series of trade-related policy evaluations were pursued for 1976. Following these key developments, Whitacre (1979) studied Japanese agricultural trade policies by employing a nine-commodity and five-world-region model (USA, Canada, Japan, Australia and the rest of the world). Similar spatial and intertemporal studies were also pursued by Shei and Thompson (1977) and Puri *et al.* (1977) for wheat and pork respectively.

Another spatial modelling development has been the design of linkage mechanisms which link spatial equilibria over time not only in a continuous sense but also in a discontinuous sense of productive capacity suddenly coming on-line. This can be seen as important for the solution of mineral or energy models in two ways which explain (a) stock adjustment in the form of physical inventories and (b) stock adjustment in the form of reserve or productive capacity adjustments. Because of the long time lags necessary for mineral project development, mineral modellers have been more concerned with using STPA models for investment analysis. While Uri (1976) accomplished this task within the quadratic programming framework, Kendrick (1967) independently chose to modify linear programming in the direction of mixed integer programming (MIP). Like recursive programming, it represents an application of linear programming; only the integer characteristic is introduced to accommodate combinations of 0–1 variables which reflect the non-existence or existence of a mining or production facility.

Applications and advances of the mixed integer programming approach are fully described in the work of Kendrick and Stoutjesdijk (1978) who explain its use in analysing industrial investment programmes in regional economic development. This work has stemmed from applications by the World Bank to study fertilizer sector planning (Choksi *et al.* 1980). Other applications of this approach can be seen in the work of Dammert (1980) who employed it to study copper investment, allocation and demands in Latin America. Dammert also showed how reserve levels and reserve limits could be coupled with depletion in the mining and processing of different ore grades. A similar approach has been taken by Brown *et al.* (1983) in their modelling of worldwide investment analysis in the aluminium industry.

Further research on demand asymmetry has resulted in the development of linear complementarity programming, as explained by Labys *et al.* (1989). The above-mentioned work by Hashimoto and Sihsobhon (1981) dealt with product demand in a model (WISE) of the world iron and steel industry. This model achieved two results. First, it replaced the short-term stock adjustment mechanism of STPA models with the investment adjustment mechanism proposed by Uri (1977). Second, it modified the concept of expectations based on backward information to include that of expectations based on forward information. The latter permitted a more complex description of investment decision making by analysing market situations in which these decisions are based on rational expectations compared with those based on actual industry plans. More recently Yang and Labys (1985) have employed linear complementarity programming to include natural gas in their Appalachian regional energy trade model.

Regarding future developments, very little work has taken place to model inventory behaviour using the STPA framework. This does not necessarily imply that inventory or stock adjustments are not important in these industries. On the contrary, they have played an extremely important role, for example, in the recent history of the petroleum (1981–2) market and the tin market (1982). Rather, the inventory adjustment issue in these industries that requires a large sum of capital investment for their expansion is a short-term operational issue. In comparison with the short-term issue, the most important single issue in intermediate or long-term modelling is that of stock adjustment in the form of changes in capacity required over, say, 30 years in the future.

(j) *Model evaluation*

The theoretical issues involved in model evaluation have been much discussed and debated. The most often cited literature includes Gass (1980) and Greenberger *et al.* (1976), both of whom distinguish two fundamental aspects of model evaluation: validation and verification. Validation refers to the correspondence of the model to the underlying processes being modelled. This form of validation will include three elements: (a) the structural features of the model; (b) the inclusion of relevant variables such as policy instruments and concepts of importance for the issues to be analysed; and (c) the predictive capability of the model. Structural evaluation is, of course, the essence of scientific analysis. It is based upon (a) the conceptual specification of the model, (b) the specification and application of the measurement process by which the model data are generated or obtained, (c) the specification and analysis of the scientific hypotheses derived from theory underlying the model and to be tested via analysis of the model data and (d) the selection of the final model best supported by the scientific laws, principles, maintained hypotheses and tested hypotheses which emerged from the research process. Validation of the model structure normally involves measurements and hypothesis testing, but also includes analysis and/or counter-analysis involving the variables and concepts integral to the policy issues for which the model was intended. Many of the developments that have taken place in the above methods of model evaluation have stemmed from the MIT Energy Model Assessment Program (Labys and Wood, 1985). Attempts to deal only with parametric and non-parametric validation techniques in energy models are discussed by Labys (1977). Their extension to evaluating commodity models in general appear in Labys (1990) and Labys and Pollak (1984).

1.3.2 Market structure

The model of pure competition is often a good approximation for many commodity markets, including in particular many of the agricultural markets of the important internationally traded commodities. Price competition models are comparatively simple, and they lend themselves relatively easy to quantitative modelling. However, they rest on underlying assumptions which in many cases are oversimplified that is: (a) there are many sellers and buyers, none of whom can influence market prices significantly, (b) the commodity is a homogeneous product which precludes market segmentation on the basis of quality differences, or on other bases such as geographical location, and (c) prices are perfectly flexible, so that the market is able to react instantaneously to any variation in some of the variables influencing it.

Modelling departures from the competitive hypothesis is one of the most important new directions in which commodity modelling is now engaged. Indeed, the other extreme behaviour regarding the number of agents acting on the market, which is the monopoly or monopsony situation, is quite easy to model, because a simple description of such a situation has been well known since Cournot (1838). However, the description of markets in terms of either perfect competition or, at the other extreme, monopoly or monopsony is not, generally speaking, a good approximation. In the case of the processing of most commodities, and for most of the actual selling and buying directly on international commodity markets, the presumption that the entities involved have an imperceptible influence on market prices is often not valid. Instead these markets are characterized by varying degrees of market power, i.e. the capacity to influence market prices. Modelling this phenomenon requires consideration of market structure, the number of firms in the industry and the degree of cooperation between them. The most frequently encountered modelling complication results from the fact that the structure of the markets in which mineral commodities are bought and sold is rarely perfectly competitive (Labys, 1980). These markets can be characterized by few sellers. A single firm can be dominant in one market (e.g. a single producer dominates the molybdenum market) and a single producer country can dominate another (e.g. Zaire dominates the cobalt market).

Conversely, just as mineral or energy markets are seldom perfectly competitive, they are also rarely controlled by a single monopolist or monopsonist. Most mineral markets are intermediate in structure. Intermediate market structures are the most difficult to model because neither simple monopoly nor perfect competition theories (market structures whose behaviour is best understood) are applicable. The result is that each mineral market must be studied individually to determine how prices are set and to understand the institutions that govern the way in which the market operates. A recent breakthrough has occurred with an attempt by Hannan (1988) to quantify the major determinants of different mineral market structures based on a factor analytic approach.

This work is important because, whenever a market structure rests somewhere between the extremes of monopoly competition, the number of equations and the number of variables that are required to describe the market are usually larger than in these limiting cases. In the most complex situations, the market has to be described in terms of numerous variables and equations. Here, factor analysis is indeed a good method to detect pertinent variable aggregates, leading the way to the selection of a satisfactory model in complex market situations. This complexity of market structure description in intermediate cases is illustrated by the fact that even a case that would at first sight appear simple to model in the case of duopoly (two sellers) is, in reality, already quite complex.

Whereas a competitive market model requires, in principle at least, only one aggregate demand equation and one aggregate supply equation together with and an equation for prices and a model-closing inventory identity, a duopoly market model even in the simplest theoretical description would be much more complex to describe. There is not even one single reference model in this case. It is well known that the Cournot duopoly approach in which each of the sellers considers the market share of the other seller as given is unable to describe many empirical duopoly situations, in which one of the sellers is trying to increase his market share, for instance, by engaging in a 'price war'. In such a situation, the market price is usually unstable in nature and is the result of a complex iteration process rather than the solution of a single equation, as appears from the analyses carried out by Von Stackelberg (1952). The application of Stackelberg models to actual commodity markets is indeed an important new direction in commodity modelling. Ulph and Folie (1980) have given more general descriptions of such applications. Development of Stackelberg models is also important because many market structures, in which one dominant producer or group of producers is faced with a smaller group of producers or a smaller producer, are encountered in practice. In this context, game-theoretic developments along the lines of the Nash–Cournot player cooperative game approach (Nash, 1953) often prove a useful tool, as shown by Pindyck (1978), among others.

Other types of intermediate market situations, as well as other types of departures from ideal competitive behaviour, are also worth studying and represent important new directions in commodity modelling. Mineral markets are characterized by various degrees of vertical integration which represent a significant departure from competitive behaviour. For example, some copper or bauxite mining firms are integrated forward into smelting, refining and fabrication, whereas others sell concentrates to custom smelters. The degree of vertical integration affects pricing and costs and, therefore, the quantity of a mineral demanded. Vertical integration makes it difficult to model mining without modelling metal manufacturing (smelting and refining), since the demand for the ore is essentially set at the refinery level. Market structure can also be non-competitive in a spatial context, for example, where one country or one region represents the dominant supplying source (Takayama and Labys, 1985).

Finally, another new direction concerns markets in which some kind of price rigidity is observed, such as markets characterized by bilateral contracts between a seller and a buyer, or markets characterized by a disequilibrium due to price rigidity. While the first case has been evaluated in some energy markets, the second case has been less extensively studied. Other various econometric methods which have been developed to take care of disequilibrium market behaviour (Gourieroux *et al.*, 1980a, b; Maddala, 1983; Quandt, 1985) could be applied fruitfully to a larger extent in commodity modelling, thus opening new directions and approaches.

1.3.3 Market interactions

Modelling the interaction within or between individual markets involves three steps. The first step requires identifying through economic theory and empirical techniques the presence or absence of interactions. Without an understanding of whether the components of commodity markets interact (e.g. are metals prices in international markets and agricultural commodities prices in domestic markets and energy prices in regional markets all interrelated?) and if so how (e.g. does one component have an identifiable undirectional influence on the others?), there is no guarantee that model linkages are correctly defined. In addition, the analytical properties of a combined model linking commodity market models are, in general, not deducible from the properties of the component commodity market models. In fact, the component models might have quite distinctive mathematical properties (Clark, 1975; Taylor, 1975; Zangwill and Garcia, 1981; Battern and Boyce, 1986; Hanson and Robinson, 1989). Consequently, this step in modelling market interactions involves formulating a theory and specifying the structure of a combined model. The third and final step in modelling commodity markets interactions requires solving the combined commodity markets model.

Given the theoretical constructs that suggest that there are commodity market interactions (either temporally across different commodities or spatially across the same commodity produced or sold in disparate locations), several empirical approaches are available to investigate the nature and extent of the suspected interactions. For example, causality tests have been successfully applied to identify both the temporal and spatial dimensions of commodity market interactions (Uri and Rifkin, 1985; Uri, 1989). More recent efforts in this area have relied on cointegration tests to investigate whether commodity market variables are interrelated (Granger, 1986; MacDonald and Taylor, 1988).

The second step in modelling commodity market interactions focuses on the combination of commodity market models where optimizing behaviour on the part of all participants in the respective commodity markets is presumed. This is consistent with neoclassical microeconomic theory, where optimizing behaviour on the part of both consumers and producers is assumed. Following Moroney (1986), there are several useful criteria for delineating a combined commodity market model. First, the structure of the combined model should follow the natural organization of the commodity data for each of the commodity markets separately. That is, the variables included in the combined model should be consistent with the available data on the quantity demanded of the various commodities, the quantity supplied, the stocks, the prices etc. Second, individual commodity market models should be separable. This ensures that specific market nuances, such as institutional considerations, changing market structure and/or significant stochastic innovations, can be incorporated into the integrated structure. In addition, the separability of the individual component models guarantees modularity of the combined modelling structure. That is, once inputs and outputs for all commodity markets are determined, any internal changes in one or more of the commodity market models will leave the remaining component models unperturbed.

Additionally, as better specifications of individual commodity market models become available, old models and specifications can be replaced with the new improved versions without imposing significant redesign and respecification costs on the combined modelling structure. A third criterion useful in constructing a combined model is that such a model should be computationally efficient. That is, the combined model should be configured in such a fashion that it can be solved in a straightforward way with minimal cost. This not only facilitates simulation and forecasting exercises but allows for a wider range of policy initiatives to be explored (see Johnson *et al.* (1985) and Murata (1984) for elaboration of this point).

The third and final step in modelling market interactions with a combined model is the determination of an optimal solution. With the advent of more powerful computers in conjunction with more sophisticated solution algorithms and computer software, solving complex combined commodity market models has become much more tractable. Thus, for example, the work of Mathiesen (1985a, b) has resulted in a significant improvement in the speed and reliability of solving highly non-linear combined commodity market models (see Boyd and Uri (1989) for an application of this approach). This means that for commodity market models, enhanced complexity through the integration of a series of commodity-specific components accurately portraying reality is now practically feasible, whereas in an earlier period such an objective was only a dream.

1.4 THE FUTURE

Commodity modelling is being enhanced and extended by several new developments. These include (a) relying on more and better data, (b) integrating commodity models with other types of models, (c) incorporating policy considerations into the commodity modelling process and (d) examining the interrelationships between commodities and resource allocation and the quality of the environment. Each of these is very briefly discussed in what follows.

With regard to the first of these developments, the current state of computer technology has greatly increased our ability to access and process large amounts of data. This means that commodity modelling efforts that previously were too demanding computationally can now be undertaken. Hence, for example, specifications that previously were intractable, such as highly non-linear specifications, can now be investigated for forecasting and for policy analysis.

The second development involves the increased integration of commodity models with other (non-commodity) models. For example, we have reviewed above the various efforts to integrate commodity-specific models with macroeconomic models to identify and measure the nature and extent of the connection between specific commodity markets and the macroeconomy.

Other important modelling efforts not mentioned above include those of Hanson and Robinson (1989), Orcutt *et al.* (1986) and Shoven and Whalley (1984). The latter efforts extend modelling beyond econometric methods to encompass parameters estimated by other methods. For example, they employ engineering parameters or parameters which are the result of some Delphi H approach. Moreover, the approaches typically do not adequately reflect adjustment lags on the part of consumers and producers, nor do they attempt to capture the impact of technological innovations.

A third area of model development involves the endogenizing of policy considerations as an integral part of the modelling process. Historically, one finds commodity models are specified and estimated primarily on the basis of economic considerations. Thus, for example, the quantity of corn exported might be specified to be a function of the domestic price of corn, the price of a substitute, the stock of corn and the exchange rate. This is the specification commonly used by agricultual economists (Chambers and Just, 1979; Dunmore and Longmire, 1984; Houck, 1986). This type of model specification, however, is now being adapted to include government policies, such as market regulations or other modes of intervention, e.g. in the case of agriculture in the USA, via the Export Enhancement Program (US Department of Agriculture, 1988). Commodity modellers are thus explicitly incorporating policy variables such that more realistic characterization of commodity markets can be reached. This is an arduous task, since it is often difficult to quantify government policies and because policy initiatives are frequently erratic.

Commodity production and utilization also has a direct effect on resource use and on the quality of the environment. The use of land in the production of a commodity, for example, necessarily leads to the deposition of toxic chemicals or to erosion, which in turn impacts the future uses to which the land can be put. Because land as a factor of production has value, economic considerations dictate that the price of a commodity fully reflect the cost of disturbing the initial quality of the land in addition to the cost of the other factors of production (Pierce *et al.*, 1983; Williams *et al.* 1984; Sutton, 1988). Commodity modelling then, to reflect reality adequately and accurately, must (and is just beginning to) take such interrelationships into account in a quantitative way (Osteen and Kuchler, 1986; Setia *et al.*, 1988). Furthermore, since there are externalities associated with the production of commodities that impact the environment (and the quality of life), commonly modelling activities are beginning to incorporate the consideration of externalities explicitly into their analytical framework (Crosson and Brubaker, 1982; Huang and Uri, 1989).

A fifth area of development concerns the need to untangle more carefully the nature of market structures and the role played therein by different forms of market interventions such as regulated prices, subsidies, taxes and trade controls. (The following comments are based on Labys and Wood

(1985).) The potential of multidisciplinary analysis in this area is the least, unless the disciplines of quantitative political analysis and industrial organization advance in this direction. Recent attention to modelling market structure has been reviewed in terms of the seminal work of Pindyck (1978), Salant *et al.* (1981), Kolstad (1982) and others. We have thus seen our perception of the embodied structure of the petroleum market advance from simple monopoly to rather complex forms of oligopoly. A lesser amount of work has taken place regarding the role of market interventions, with the exception of studies by Fuss (1980) and Wood and Spierer (1984). Commodity models are thus likely to improve in the direction of offering a more realistic picture of these two related aspects of market structure.

Sixth, we expect increasing attention to be applied to the integration of behavioural and technical/engineering information and concepts in commodity models. There is already a suggestion on how this integration will take place. Brock and Nesbitt (1977) have provided a conceptual analysis of energy model integration and have presented several energy models which emphasize the fundamental linkages between supply, conversion and use in a macroeconomic context. Recently Hogan and Weyant (1980) have continued this early work, emphasizing that large-scale commodity models involve integration of component models describing separate economic and technical aspects of the energy system. The relations between the component models may be characterized as a network of process models, and their joint solution as a system equilibrium solution. The network formalization is quite powerful in identifying, modelling and analysing information flows, especially in clarifying measurement issues (both conceptual and point in time/space/system) and in developing solution algorithms and computational facilities.

An important operational implication is that the network formalism greatly facilitates understanding of how component models (processes) may be efficiently integrated. The high costs in both resources and time of large-scale model development mean that any advances which 'spread' modelling costs and reduce implementation time will greatly increase the potential utility of resulting integrated models. Further efficiency gains will also occur to the extent that common software useful for describing and solving such network models comes into wide use. Certainly anyone interested in large-scale commodity models for policy research and analysis will want to become familiar with this approach.

Another development relating to the integration of behavioural and technical/engineering models is due to Lau and Berndt (1983) and shows considerable promise in providing a more natural description of the behavioural–process interactions in commodity transformation systems. Lau considers the conceptual problem of incorporating characteristics of a factor input in measuring that input, e.g. technical information about capital (age, efficiency, fuel requirements etc.), education of labour or characteristics of materials. He finds that the conditions on the underlying production function consistent with such an input 'quality' adjustment are not very restrictive—most importantly that quality adjusting variables enter multiplicatively as the Cobb–Douglas and the translog.

The relevance of the quality adjustment approach in integrating economic and engineering models is that it provides a means of directly incorporating technical characteristics into models of commodity market behaviour. This is important in its own right since often we would like to know, for example, how changes in the energy efficiency of a capital good will affect costs of production. Even more interesting is that the incorporation of such technical information directly into commodity models will provide a natural linkage for the process network formulation considered by Hogan and Weyant. These techniques, then, will contribute to introducing more realistic descriptions of institutions, technologies and markets into formal commodity models.

REFERENCES AND FURTHER READING

Adams, F. G. (1978) *Stabilizing World Commodity Markets: Analysis, Practice and Policy*, Lexington Books, Lexington, MA.

Adams, F. G. (1981) Commodity exports and NIEO proposals for buffer stocks and compensatory finance: implications for Latin America. *Quarterly Review of Economics and Business*, **21**, 48–82.

Adams, F. G. and Behrman, J. (1976) *Economic Models of World Agricultural Commodity Markets*, Ballinger, Cambridge, MA.

Adams, F. G. and Behrman, J. (1978) *Econometric Modeling of World Commodity Policy*, Ballinger, Cambridge, MA.

Adams, F. G. and Griffen, J. M. (1972) An econometric linear programming model of the U.S. petroleum industry. *Journal of American Statistical Association*, **7**, 542–51.

Ahalt, J. (1977) Trends in food consumption, prices and expenditures, in *Food and Agricultural Policy* (ed. D. Paarlberg), American Enterprise Institute, Washington, DC.

Akiyama, T. and Trivedi, P. K. (1985) Specification of the price equation in a prototype model of a perennial crop, Working Paper, World Bank, Washington, DC.

Akiyama, T. and Trivedi, P. K. (1987) Vintage production approach to perennial crop supply: an application to tea in major producing countries. *Journal of Econometrics*, **36**, 133–61.

Albouy, M. (1987) Nouveaux instruments financiers et gestion du couple rentabilité–risque sur le marché du pétrole, in *Pétrole: Marchés et Stratégies* (eds R. Ayoub and J. Percebois), Economica, Paris, 56–76.

Antle, J. (1984) The structure of U.S. agricultural technology. *American Journal of Agricultural Economics*, **66**, 414–21.

Aperjis, D. G. (1981) *Oil Markets in the 1980's: Oil Policy and Development*, Ballinger, Cambridge, MA.

Bailey, K. (1987) *A Structural Econometric Model of the Canadian Wheat Sector*, TB

1733, Economic Research Service, US Department of Agriculture, Washington, DC.

Baldwin, N. and Prosser, R. (1988) World oil market simulation. *Energy Economics*, **10**, 185–98.

Ball, V. E. (1985) Output, Input, and Productivity Measurement in U.S. Agriculture. *American Journal of Agricultural Economics*, **67**, 475–86.

Basile, P. S. (1979) The IIASA Set of Energy Models, Working Paper, International Institute for Applied Systems Analysis, Laxenberg.

Battern, D. and Boyce, D. (1986) Spatial interaction, transportation, and interregional commodity flow models, in *Handbook of Regional and Urban Economics* (ed. P. Nijkamp), North-Holland, Amsterdam.

Behrman, J. R. (1978) International Commodity Market Structures and the theory underlying International Commodity Market Models, in *Econometric Modeling of World Commodity Policy* (eds. F. G. Adams and J. R. Behrman) Heath Lexington Books, Lexington, 9–46.

Berndt, E. R. (1983) Quality adjustment, hedonics, and modern empirical demand analysis, in *Price Level Measurement* (eds W. E. Diewert and C. Montmarquette), Proceedings of a Conference sponsored by Statistics Canada, Minister of Supply and Services, Ottawa, pp. 817–63.

Berndt, E. R. and Wood, D. O. (1975) Technology, prices and the derived demand for energy. *Review of Economics and Statistics*, **57**, 259–68.

Bickerton, T. (1987) *USSR Oilseed Production, Processing and Trade*, FAER 232, Economic Research Service, US Department of Agriculture, Washington, DC.

Bingham, T. H. and Lee, B. S. (1976) *Estimates of the Interrelationship Between Consumer Expenditures and Natural Resource Consumption*, Proceedings of the XIV Symposium of the Council for the Application of Computers and Mathematics in the Minerals Industry, Pennsylvania State University, University Park, PA.

Black, F. and Scholes, M. (1972) The valuation of option contracts and a test of market efficiency. *Journal of Finance*, **27**, 399–418.

Bopp, A. E. and Sitzer, S. (1988) On the efficiency of futures markets: another view. *Energy Economics*, **10**, 199–206.

Bosworth, B. and Lawrence, R. T. (1982) *Commodity Prices and the New Inflation*, The Brookings Institution, Washington, DC.

Boucher, J. and Smeers, Y. (1988) Economic forces in the European gas market. *Energy Economics*, **9**, 2–16.

Boyd, R. and Uri, N. (1989) Assessing the impact of an oil import fee. *Energy: The International Journal*, **14**, 29–44.

Bourdaie, J. M. (1988) *MAREN, Modelling Energy Markets in the Western World*. Communication, 25th International Conference of the Applied Econometrics Association, Washington, DC.

Brock, H. and Nesbitt D. (1977) Large scale energy planning models: a methodological analysis, Report to the Natural Science Foundation, Stanford Research Institute, Menlo Park, CA.

Brown, M., Dammert, A., Meeraus, A. and Stoutjesdijk, A. (1983) Worldwide investment analysis: the case of aluminum, World Bank Staff Working Paper 603, World Bank, Washington, DC.

Burbridge, J. A. and Harrison, A. (1982) Testing the effects of oil-price rises using vector autoregressions. *International Economic Review*, **25**, 459–84.

Campbell, J. Y. and Schiller, R. J. (1988) Interpreting cointegrated models. *Journal of Economic Dynamics and Control*, **12**, 505–22.

Carlevaro, F. and Spierer, C. (1983) Dynamic energy demand models with latest equipment. *European Economic Review*, **23**, 161–94.

Carlevaro, F. and Spierer, C. (1984) Nouvelles perspectives d'évolution de la demande d'énergie en Suisse jusqu'à l'an 2000. *Sweizirische Zeitschrift für*

Volkswirtschaft und Statistik/Revue Suisse d'Economie Politique et de Statistique, 483–520.

Carlevaro, F., Müller, T. and Antille, G. (1988) Repercussions Économiques d'un, Abandon du Nucléaire en Suisse, Analyse à l'Aide d'un Modèle d'Equilibre genéral, Département Fédéral de l'Energie, Bern, Switzerland.

Cerchi, M. and Havenner, A. (1988) Cointegration and stock prices. *Journal of Economic Dynamics and Control*, **12**, 333–46.

Chambers, R. and Just, R. (1979) A critique of exchange rate treatment in agricultural trade models. *American Journal of Agricultural Economics*, **61**, 249–57.

Chambers, R. and Vasavada, U. (1983) Testing asset fixity for U.S. agriculture. *American Journal of Agricultural Economics*, **65**, 761–9.

Charles River Associates (1975) *Economic Issues Underlying Supply Access Agreements: A General Analysis and Prospects in Ten Mineral Markets*, Charles River Associates, Cambridge, MA.

Château, B. and Lapillone, B. (1980) Long term energy demand simulation, in *Energy Models for the European Community*, IPC, London, pp. 120–8.

Chavas, J. P. (1983) Structural change in the demand for meat. *American Journal of Agricultural Economics*, **65**, 148–53.

Chichilnisky, G. (1981) Oil prices, industrial prices and outputs: a general equilibrium macro analysis, Working Paper, United Nations Institute for Training and Research, New York.

Choe, B. J. (1984) A model of world energy markets and OPEC pricing, Staff Working Paper 633, World Bank, Washington, DC.

Choksi, A. M., Meeraus, A. and Stoutjesdijk, A. J. (1980) *The Planning of Investment Programs in the Fertilizer Industry*, Johns Hopkins University Press for the World Bank, Baltimore, MD.

Chu, K. Y. and Morrison, T. K. (1984) The 1981–82 recession and non oil primary commodity prices. *IMF Staff Papers*, **31**, 93–140.

Clark, J. P. and Church, A. (1981) Process analysis modeling of the stainless steel industry. Paper presented at the Workshop on Nonfuel Minerals Demand Modeling held at Airlie House, Warrenton, VA, June 1–2 (unpublished). Board of Nonfuel Mineral and Energy Resources, Natural Academy of Sciences, Washington, DC.

Clark, P. (1975) Intersectoral consistence and macroeconomic planning, in *Economy-Wide Models and Development Planning* (eds C. Blitzer, P. Clark and L. Taylor), Oxford University Press, Oxford.

Considine, T. J. and Mount, T. D. (1984) The use of linear logit models for dynamic input demand systems. *Review of Economics and Statistics*, **66**, 434–43.

Consonni, A. and Lesourd, J. B. (1986) Industrial accounting and control systems: a survey. *Energy Conversion and Management*, **26**, 357–61.

Conway, R., Hallahan, C., Stillman, R. and Prentice, P. (1987) Forecasting livestock prices: fixed and stochastic coefficient estimation, Technical Bulletin No. 1725, Natural Resource Economics Division, US Department of Agriculture, Washington, DC.

Cooper, R. M. and Lawrence, R. T. (1975) The 1972–75 commodity boom. *Brookings Papers on Economic Activity*, **3**, 671–723.

Cournot, A. (1838) Research into the Mathematical Principles of the theory of wealth. (Translated N. T. Bacon), McMillan, New York (1897).

Crom, R. (1988) *Economics of the U.S. Meat Industry*, AIB 545, Economic Research Service, US Department of Agriculture, Washington, DC.

Crosson, P. and Brubaker, S. (1982) *Resources and Environmental Effects of U.S. Agriculture*, Resources for the Future, Washington, DC.

Dammert, A. (1980) Planning investments in the copper sector in Latin America, in *Commodity Markets and Latin American Development: A Modeling Approach* (eds W. C. Labys, M. Nadiri and J. Nunez del Arco), National Bureau of Economic Research, New York.

Dammert, A. and Palaniappan, S. (1985) *Modeling Investments in the Copper Sector*, University of Texas Press, Austin, TX.

Davidson, A. L. and Kim, S. (1988) OPEC stability; an empirical assessment. *Energy Economics*, **10**, 174–84.

Drollas, L. P. and Barthes, A. (1985) Gross product worth at the Rotterdam spot market, in *Energie: Modélisation et Econométrie*, (eds J. Fericelli and J. B. Lesourd), Economica, Paris, pp. 584–605.

Duffy, W. J. (1969) Parameter variation in a quarterly model of the post-war U.S. economy, Ph.D. Thesis, University of Pittsburgh.

Dunmore, J. and Longmire, J. (1984) *Sources of Recent Changes in U.S. Agricultural Exports*, AIB 121, Economic Research Service, US Department of Agriculture, Washington, DC.

Eckbo, P. L., Jacoby, H. D. and Smith, J. L. (1979) Oil supply forecasting: a disaggregated process approach. *Bell Journal of Economics*, **9**, 218–38.

Energy Modeling Forum (1977) Energy and the economy, EMF Report 1, Stanford University, Stanford, CA.

Energy Modeling Forum (1982) World oil, Working Paper, Stanford University, Stanford, CA.

Engle, R. F. and Granger, C. W. J. (1987) Co-integration and error correction: representation, estimation and testing. *Econometrica*, **55**, 251–76.

Engle, R. F. and Granger, C. W. J. and Hallman, J. J. (1989) Merging short and long term forecasts. *Journal of Econometrics*, **40**, 45–62.

Enoch, C. A. and Panic, M. (1981) Commodity prices in the 1970's. *Bank of England Quarterly Bulletin*, **1**, 42–53.

Farmer, R. (1986) Futures markets and petroleum supply, Working Paper EIA-0486, US Department of Energy, Washington, DC.

Farzin, Y. H. (1986) *Competition in the Market for an Exhaustible Resource*, JAI Press, Greenwich, CN.

Feinerman, E. and Siegel, P. (1988) A dynamic farm-level planning model for beef feedlot production and marketing. *American Journal of Agricultural Economics*, **39**, 413–25.

Fericelli, J. and Lesourd, J. B. (eds) (1985) *Energie: Modélisation et Econométrie*, Economica, Paris.

Fox, K. (1986) Agricultural economists as world leaders in applied econometrics. *American Journal of Agricultural Economics*, **68**, 381–6.

French, M. W. (1984) Should policy-making account for market expectations: an estimate of this impact on commodity policy. *Journal of Policy Modeling*, **10**, 243–58.

Fuss, M. A. (1980) The derived demand for energy in the presence of supply constraints, in *Energy Policy Modeling* (eds W. T. Ziemba, S. L. Schwartz and E. Koenigsberg), Martinis Nijhoff, Boston, MA, pp. 66–85.

Gardiner, W. and Dixit, P. (1987) *Price Elasticity of Export Demand*, FAER 228, Economic Research Service, US Department of Agriculture, Washington, DC.

Gardner, B. (1987) Causes of U.S. farm commodity programs. *Journal of Political Economy*, **95**, 290–310.

Gass, S. (1980) *Validity and Assessment Issues of Energy Modeling*, NBS-569, Natural Bureau of Standards, US Department of Commerce, Washington, DC.

Gately, D. (1979) The prospects for OPEC, five years after 1973–1974. *European Economic Review*, **2**, 360–79.

Gately, D. (1984) A ten year retrospective: OPEC and the world oil market. *Journal of Economic Literature*, **22**, 1100–14.

Gately, D. (1986) The prospects of oil prices, revisited. *Annual Review of Energy*, **11**, 513–38.

Gately, D. and Rappoport, P. (1988) The adjustment of U.S. oil demand to the price increase of the 1970's. *Energy Journal*, **9**, 93–108.

Ghosh, S., Gilbert, C. L. and Hughes Hallett, A. J. (1981) Optimal control and choice of functional form: an application to a model of the world copper industry, Discussion Paper 8109/E, Erasmus University, Rotterdam.

Ghosh, S., Gilbert, C. L. and Hughes Hallett, A. J. (1987) *Stabilizing Speculative Commodity Markets*, Clarendon Press, Oxford.

Gilbert, R. (1978) Dominant firm pricing policy in a market for an exhaustible resource. *Bell Journal of Economics*, **9**, 185–95.

Gilbert, C. L. (1985) A rational expectations price model for a continuously produced primary commodity, Paper presented at the World Congress of the Econometric Society, Cambridge, MA.

Gilbert, C. L. (1986) Exchange rates, LDC debt and commodity prices: a model of the World Bank Commodity Price Index, Discussion Paper, International Economics Division, World Bank, Washington, DC.

Gilbert, C. L. and Palaskas, T. (1988) Expectations formation in equilibrium and disequilibrium econometric commodity market models, Working Paper, Oxford University, Oxford.

Gondolfo, G. (1981) *Quantitative Analysis and Econometric Estimation of Continuous Time Series Dynamic Models*, North-Holland, Amsterdam.

Golombek, R., Hoel, M. and Vislie, J. (1987) *Natural Gas Markets and Contracts*, North-Holland, Amsterdam.

Gordon, R. B., Koopmans, T. C., Nordhaus, W. B. and Skinner, B. J. (1987) *Toward a New Iron Age*, Harvard University Press, Cambridge, MA.

Goreau, C. (1978) *Income and Food Consumption*, FAO Studies in Agricultural Economics and Statistics, UN Food and Agriculture Organization, Rome.

Gouriéroux, C., Laffont, J. J. and Monfort, A. (1980a) Disequilibrium econometrics in simultaneous equation systems. *Econometrica*, **48**, 75–96.

Gouriéroux, C., Laffont, J. J. and Monfort, A. (1980b) Coherency conditions in simultaneous linear equation models with endogenous switching regimes. *Econometrica*, **48**, 675–95.

Granger, C. W. J. (1969) Investigating causal relations by econometric models and cross spectral methods. *Econometrica*, **37**, 424–38.

Granger, C. W. J. (1986) Developments in the study of cointegrated economic variables. *Oxford Bulletin of Economics and Statistics*, **48**, 213–25.

Granger, C. W. J. and Newbold, P. (1986) *Forecasting Economic Time Series*, Academic Press, New York.

Greenberger, M., Crenson, M. A. and Crissey, B. L. (1976) *Models in the Policy Process: Public Decision Making in the Computer Era*, Russell Sage Foundation, New York.

Griffin, J. M. (1985) OPEC behaviour: a test of alternative hypotheses. *American Economic Review*, **75**, 954–63.

Griliches, Z. (1967) Distributed lags: a survey. *Econometrica*, **35**, 16–49.

Grilli, R. E. and Yang, C. M. (1981) Real and monetary determinants of non oil primary commodity price movement, Commodity Division Working Paper 1981-6, World Bank, Washington, DC.

Groncki, P. J. and Marcuse, W. (1980) The Brookhaven integrated energy/economy modelling system and its use in conservation policy analysis, in *Energy Modeling*

Studies and Conservation (ed. ECE), Pergamon Press for the United Nations, New York, pp. 535–56.

Guither, H. (1963) Factors influencing farm operators' decisions to leave farming. *Journal of Farm Economics*, **45**, 567–76

Haaland, J. I. and Norman, V. D. (1977) Introduction: modeling trade and trade policy. *Scandinavian Journal of Economics*, **89**, 217–26.

Haley, S. and Dixit, P. (1988) *Economic Welfare Analysis*, AGES 871215, Economic Research Service, US Department of Agriculture, Washington, DC.

Haltiwanger, J. and Waldman, M. (1985) Rational expectations and the limits of rationality: an analysis of heterogeneity. *American Economic Review*, **75**, 326–40.

Hannan, M. (1988) A factor analysis approach to modeling mineral market structures: an application to aluminium, copper, lead and zinc, Ph.D. Thesis, Department of Mineral and Energy Resource Economics, West Virginia University, Morgantown.

Hannan, M. and Labys, W. C. (1988a) Logistic function analysis of mineral substitution, Working Paper 131, Department of Mineral and Energy Resources Economics, West Virginia University, Morgantown, November.

Hannan, M. and Labys, W. C. (1988b) A quantitative approach to market structure in mineral markets, Working Paper 130, Department of Mineral and Energy Resources Economics, West Virginia University, Morgantown, November.

Hanson, K. and Robinson, S. (1989) *Data, Linkages, and Models*, AGES 89-5, Economic Research Service, US Department of Agriculture, Washington, DC.

Harlow, A. (1960) The hog cycle and the cobweb theorem. *Journal of Farm Economics*, **42**, 842–53.

Harrington, D. and Manchester, A. (1985) Profile of the U.S. farm sector, in *Agricultural-Food Policy Review*, AER 530, Economic Research Service, US Department of Agriculture, Washington, DC.

Harris, D. and Burrows, J. (1978) The economics of geology and mineral supply: an integrated framework for long run policy analysis, Report 327, Charles River Associates, Boston, MA.

Hashimoto, H. (1977) World food projection models, projections and policy evaluations, Ph.D. Thesis, Department of Economics, University of Illinois.

Hashimoto, H. and Sihsobhon, T. (1981) A world iron and steel economy model: WISE, in *World Bank Commodity Models*, Vols I and II, World Bank, Washington, DC.

Hathaway, D. E. (1963) *Government and Agriculture*, Macmillan, New York.

Hazell, P. B. (1983) Importance of risk in agricultural planning models, in *Programming Studies for Mexican Agriculture* (eds R. Norton and M. Leopoldo), Johns Hopkins University Press, Baltimore, MD.

Hazilla, M. and Kopp, R. J. (1984) A factor demand model for strategic nonfuel minerals in the primary metals sector. *Land Economics*, **60**, 328–39.

Herrara, A. O. and Scolnik, H. D. (1976) *Catastrophe or New Society: A Latin American World Model*, International Development Research Center, Ottawa.

Hnyilicza, E. (1975) An aggregate model of energy and economic growth, Energy Lab Paper MIT-E175010WP, Massachusetts Institute of Technology, Cambridge, MA.

Hnyilicza, E. and Pindyck, R. (1976) Pricing policies for a two-part exhaustible resource cartel: the case of OPEC. *European Economic Review*, **8**, 136–54.

Hoel, M., Holtsmark, B., and Vislie, J. (1986) Natural gas in Western Europe: a game theoretic approach, Communication, IAEE Meeting, Bergen.

Hogan, W. H. and Weyant, J. P. (1980) Combined energy models, Discussion Paper E80-82, Kennedy School of Government, Harvard University, Cambridge, MA.

Hogan, W. W. and Leiby, P. N. (1985) Oil market risk analysis, Report of the Harvard Energy Security Project, Harvard University, Cambridge, MA.
Hojman, D. (1984) An econometric model of the international bauxite–aluminium economy. *Resources Policy*, 7, 87–102.
Hotelling, H. (1931) The economics of exhaustible resources. *Journal of Political Economy*, **39**, 137–175.
Houck, J. (1986) *Elements of Agricultural Trade Policies*, Macmillan, New York.
Huang, K. (1985) *U.S. Demand for Food: A Complete System of Price and Income Effects*, TB 1714, Economic Research Service, US Department of Agriculture, Washington, DC.
Huang, K. and Haidacher, R. (1983) Estimation of a composite food demand system for the United States. *Journal of Business and Economic Statistics*, **1**, 285–91.
Huang, W. Y. and Uri, N. D. (1989) A regional model for the determination of an optimal policy for the protection of ground water quality. *Water Resources Bulletin*, **25** (4), 775–82.
Hudson, E. A. and Jorgenson, D. W. (1974) U.S. energy policy and economic growth, 1975–2000. *Bell Journal of Economics*, **5**, 461–514.
Hughes, B. B. and Mesarovic, M. D. (1978) Analysis of the WAES scenarios using the world integrated model. *Energy Policy*, **1**, 129–39.
Hwa, E. C. (1979) Price determination in several international primary commodity markets. *IMF Staff Papers*, **26**, 157–88.
Jacques, J. K., Lesourd, J. B. and Ruiz J. M. (eds) (1988) *Modern Applied Energy Conservation*, Ellis Horwood/Wiley, New York.
Johnson, D., Hemmi, K. and Lardinois, P. (1985) *Agricultural Policy and Trade*, New York University Press, New York.
Johnson, G. and Quance, L. (1982) *The Overproduction Trap in U.S. Agriculture*, Johns Hopkins University Press, Baltimore, MD.
Johnson, S., Womack, A., Meyers, W., Young, R. and Brandt, J. Options for the 1985 Farm Bill: an analysis and evaluation, in *U.S. Agricultural Policy: The 1985 Farm Legislation* (ed. B. Gardner), American Enterprise Institute, Washington, DC.
Johnson, S. R. (1981) *Alternative Designs for Policy Models of the Agricultural Sector*, Modeling Agriculture for Policy Analysis in the 1980s, Federal Reserve Bank of Kansas City, Kansas City, MO.
Johnson, S. R. (1986) Future challenges for modeling in agricultural economics. *American Journal of Agriculture Economics*, **68**, 387–94.
Kang, H. (1985) The effects of detrending in Granger causality test. *Journal of Business and Economic Statistics*, **3**, 344–9.
Kendrick, D. (1967) *Programming Investments in the Processing Industries*, MIT Press, Cambridge, MA.
Kendrick, D. and Stoutjesdijk, A. (1978) *The Planning of Industrial Investment Programs, A Methodology*, Johns Hopkins University Press for the World Bank, Baltimore, MD.
Kennedy, M. (1974) An econometric model of the world oil market. *Bell Journal of Economics*, **5**, 540–77.
Kennedy, M. and Miermayer, E. V. (1976) Energy supply and economic growth, Discussion Paper 4-76, Department of Economics, University of Texas at Austin.
Kim, S. (1984) Models of energy–economy interactions for developing countries: a survey, Energy Laboratory Working Paper MIT-EL 84 007 WA, Massachussetts Institute of Technology, Cambridge, MA.
Kislev, Y. and Peterson, W. (1982) Prices, technology and farm size. *Journal of Political Economy*, **90**, 578–95.

Klein, L. and Young, R. M. (1980) *An Introduction to Econometric Forecasting and Forecasting Models*, Heath Lexington Books, Lexington, MA.

Kolstad, C. (1982) Noncompetitive analysis of the world coal market, Working Paper, Energy Laboratory, Los Alamos, NM.

Kolstad, C. D. (1989) Computing dynamics spatial oligopolistic equilibrium in an exhaustible resource market, the case of coal, in *Quantitative Methods for Market-Oriented Economic Analysis Over Space and Time* (eds W. C. Labys, T. Takayama and N. Uri), Gower, London.

Kovisars, L. (1975) Copper trade flow model, in *World Minerals Availability*, SRI MED NO. 3742-74, Stanford Research Institute, Stanford, CA.

Kovisars, L. (1976) World production, consumption and trade in zinc, US BOM Report No. J-0166003, Stanford Research Institute, Stanford, CA.

Krueger, P. K. (1976) *Modeling Futures Requirements for Metals and Minerals*, Proceedings of the XIV Symposium of the Council for the Application of Computers and Mathematics in the Minerals Industry, Pennsylvania State University, University Park, PA.

Labys, W. C. (1973) *Dynamic Commodity Models: Specification, Estimation and Simulation*, Heath Lexington Books, Lexington, MA.

Labys, W. C. (1975) *Quantitative Models of Commodity Markets*, Ballinger, Cambridge, MA.

Labys, W. C. (1977) Minerals commodity modeling, in *Proceedings of the Mineral Economics Symposium on Minerals Policies in Transition*, Council of Economics of the AIME, Washingon, DC, pp. 80–106.

Labys, W. C. (1978) Commodity markets and models: the range of experiences, in *Stabilizing World Commodity Markets: Analysis, Practice and Policy* (ed. F. G. Adams), Heath Lexington Books, Lexington, MA.

Labys, W. C. (1980) *Market Structure, Bargaining Power and Resource Price Formation*, Heath Lexington Books, Lexington, MA.

Labys, W. C. (1981) *A General Disequilibrium Model of Commodity Market Adjustments*, NFS–NTIS Report, Projection Disequilibrium Adjustments and Stabilization in Mineral Markets.

Labys, W. C. (1982) A Critical Review of Energy Modeling Methodologies, Energy Laboratory Working Paper MIT-E6 82-034WP, M.I.T., Cambridge, MA.

Labys, W. C. (1988) Recent developments in commodity modeling: a World Bank focus, WPS 119, International Economics Department, World Bank, Washington, DC.

Labys, W. C. (1990) Model Validation, in *Handbook of Engineering-Economic Modeling* (eds H. Kuczmowski and J. Weyant), Praeger, New York.

Labys, W. C. and Granger, C. W. J. (1970) *Speculation, Hedging and Commodity Price Forecasts*, Heath Lexington Books, Lexington, MA.

Labys, W. C. and Pollak, P. K. (1984) *Commodity Models for Forecasting and Policy Analysis*, Croom Helm, London.

Labys, W. C. and Wood, D. O. (1985) Energy modeling, in *Economics of the Minerals Industry* (ed. W. A. Vogeley), American Institute of Mining Engineers, New York, pp. 303–66.

Labys, W. C. and Yang, C. W. (1980) A quadratic programming model of the Appalachian coal market. *Energy Economics*, 9, 86–95.

Labys, W. C., Paik, S. and Liebenthal, A. M. (1979) An econometric simulation model of the U.S. market for steam coal. *Energy Economics*, 1, 19–26.

Labys, W. C., Field, F. R. and Clark, J. (1985) Mineral models, in *Economics of the Mineral Industries* (ed. W. A. Vogely), American Institute of Mining Engineers, New York.

Labys, W. C., Takayama, T. and Uri, N. (eds) (1989) *Quantitative Methods for Market-Oriented Economic Analysis over Space and Time*, Gower, London.

Labys, W. C. and Thomas, H. (1975) Speculation, Hedging and Commodity Price Behaviour: An International Comparison, *Applied Economics*, 7, 287–301.

Labys, W. C., Thomas, H. and Gijsbers, D. (1989) Monetary and economic determinants of international commodity price behavior, in *A Reappraisal of the Efficiency of Financial Markets* (eds R. Guimares, B. Kingman and S. Taylor), Springer-Verlag, Heidelberg.

Langley, J., Reinsel, R., Craven, J., Zellner, J. and Nelson, F. (1987) Commodity price and income support policies in perspective, in *Agricultural-Food Policy Review*, AER 530, Economic Research Service, US Department of Agriculture, Washington, DC.

Leontief, W., Koo, Y., John, I. and Fisher, S. (1983) *The Future of Nonfuel Minerals in the U.S. and World Economy*, Heath Lexington Books, Lexington, MA.

Lesourd, J. B., Percebois, J. and Ruiz, J. M. (1987) Equilibre et Déséquilbre sur le marché international du gaz naturel, in *Pétrole: Marchés et Strategies* (eds A. Ayoub and J. Percebois), Economica, Paris.

Lesourd, J. B., Percebois, J. and Ruiz, J. M. (1988) Modelling the international natural gas market: the case of the Western European natural gas market, Chapter 10, this volume.

Linneman, H. (1976) MOIRA: a model of international relations in agriculture—the energy sector, Working Paper, Institute for Economic and Social Research, Free University, Amsterdam.

Lucas, R. E. (1967) Adjustment cost in the theory of supply. *Journal of Political Economy*, 75, 321–34.

MacDonald, R. and Taylor, M. (1988) Metal prices efficiency and cointegration: some evidence from the London Metal Exchange. *Bulletin of Economic Research*, **40**, 235–9.

Maddala, G. S. (1983) *Disequilibrium Models*, in *Limited Dependent and Qualitative Variables in Econometrics*, Cambridge University Press, Cambridge, Chapter 10.

Maki, W. (1962) Decomposition of the beef and pork cycles. *Journal of Farm Economics*, **44**, 731–43.

Manne, A. S. and Markowitz, H. M. (eds) (1963) *Studies in Process Analysis*, Wiley, New York.

Marks, S. and Yetley, M. (1987) *Global Food Demand Patterns Over Changing Levels of Economic Development*, AGES 870910, Economic Research Service, US Department of Agriculture, Washington, DC.

Mathiesen, L. (1985a) Computational experience in solving equilibrium models by a sequence of linear complementary problems. *Operations Research*, **33**, 1225–50.

Mathiesen, L. (1985b) Computation of economic equilibrium by a sequence of linear complementary problems. *Mathematical Programming Studies*, **23**, 223–47.

McAvoy, P. W. and Pindyck, R. S. (1972) Alternative regulatory policies for dealing with the natural gas shortage. *Bell Journal of Economics*, **2**, 454–98.

McAvoy, P. W. and Pindyck, R. S. (1975) *The Economics of the Natural Gas Shortage, 1960–1980*, North-Holland, Amsterdam.

Minguet, L. (1985) The MORE model, in *Energie: Modélisation et Econométrie* (eds J. Fericelli and J. B. Lesourd), Economica, Paris, pp. 83–7.

Monke, E. A. and Taylor, L. D. (1984) International trade constraints and commodity market models: an application to the cotton market, Working Paper, University of Arizona, Tucson.

Moroney, J. R. (1986) *Advances in the Economics of Energy and Resources*, JAI Press, Greenwich, CN.

Moroney, J. R. and Trapani, J. M. (1981) Alternative models of substitution and technical change in natural resource intensive industries, in *Modeling and Measuring Resource Substitution* (eds E. R. Berndt and B. C. Fields), MIT Press, Cambridge, MA.

Mueller, R. and Jansen, H. (1988) Farmer and farm concepts in measuring adoption lags. *Journal of Agricultural Economics*, **39**, 121–4.

Murata, T. (1984) Modeling and analysis of concurrent systems, in *Handbook of Software Engineering* (eds C. Vick and C. Ramamoorthy), Van Nostrand-Reinhold, New York.

Muth, J. F. (1961) Rational expectations and the theory of price movements, *Econometrica,* **29**, 315–35.

Nash, T. F. (1953) Two person competitive games. *Econometrica*, **21**, 128–40.

Nehring, R. (1982) Prospects for conventional oil resources. *Annual Review of Energy*, 7, 175–200.

Newbery, D. M. and Stiglitz, E. (1981) *The Theory of Commodity Price Stabilization*, Oxford University Press, Oxford.

Nguyen, H. D. (1977) World food projection models and short run world trade and reserve policy evaluation, Ph.D Thesis, University of Illinois.

Nielsen, E., Moranowski, J. and Morehart, M. (1989) *Investment in Soil Conservation and Land Improvements*, AER 601, Economic Research Service, Washington, DC.

Orcutt, G., Merz, J. and Quinke, H. (1986) *Microanalytic Simulation Models to Support Social and Financial Policy*, North-Holland, Amsterdam.

Osteen, C. and Kuchler, F. (1986) *Potential Bans of Corn and Soybean Pesticides: Economic Implications for Farmers and Consumers*, AER 546, Economic Research Service, US Department of Agriculture, Washington, DC.

Pepper, M. P. C. (1985) Multivariate Box–Jenkins analysis, A case study in UK energy demand forecasting. *Energy Economics*, 7, 168–78.

Percebois, J. (1989) *Economie de l'Energic*, Economica, Paris.

Pierce, F., Larson, W., Dowdy, R. and Graham, W. (1983) Productivity of soils: assessing long-term changes due to erosion, *Journal of Soil and Water Conservation*, **39**, 39–44.

Pindyck, R. S. (1978) Gains to producers from the cartelization of exhaustible resources. *Review of Economics and Statistics*, **60**, 238–51.

Pinstrup-Anderson, P. and Caicedo, E. (1978) The potential impact of changes in income distribution on food demand and human nutrition. *American Journal of Agricultural Economics*, **60**, 321–32.

Priovolos, T. and Dunietz, T. (1987) Substitutability of metals in US industry, Working Paper 1987-11, International Economic Department, World Bank, Washington, DC.

Puri, R. G., Meilke, K. D. and MacAuby, T. G. (1977) North-American Japanese Pork Trade. The Application of Quadrated Planning. *Canadian Journal of Agricultural Economics*, **25**, 61–79.

Quance, L. and Lu, Y. C. (1978) Applications of system science in projecting alternative food and agricultural futures, in *System Theory Applications to Agricultural Modeling* (eds A. Levis and L. Quance), Economics, Statistics and Cooperative Service, US Department of Agriculture, Washington, DC.

Quandt, R. (1985) The estimation of the parameters of a linear regression system obeying two separate regimes. *Journal of American Statistical Association*, **68**, 873–80.

Rausser, G. C. (1982) *New Directions in Econometric Modeling and Forecasting in the U.S. Agriculture*, North-Holland, Amsterdam.

Rausser, G. and Hochman, E. (1979) *Dynamic Agricultural Systems: Economic*

Prediction and Control, North-Holland, Amsterdam.
Ray, W. H. and Szekely, J. (1973) *Process Optimization: With Application in Metallurgy and Chemical Engineering*, Wiley, New York.
Regier, D. and Goolsby, H. (1980) *Growth in World Demand from Feed Grains: Related to Meat and Livestock Products and Human Consumption of Grain in 1980*, FAER 63, Economic Research Service, US Department of Agriculture, Washington, DC.
Reilly, J. (1988) *Cost-Reducing and Output Enhancing Technologies*, TB 1740, Economic Research Service, US Department of Agriculture, Washington, DC.
Ridker, R. G. and Watson, W. D. (1980) *To Choose a Future*, Johns Hopkins University Press, Baltimore, MD.
Roe, T., Shane, M. and Huu Vo, D. (1986) *Price Responsiveness of World Grain Markets*, TB 1720, Economic Research Service, US Department of Agriculture, Washington, DC.
Rose, A., Labys, W. C. and Witt, T. (1986) Estimation and analysis of a state natural gas econometric model. *Proceedings of the American Statistical Association*. pp. 380–5.
Rosenberg, N. (1976) *Perspectives on Technology*, Cambridge University Press, Cambridge.
Ryan, T. (1985) A beef feedlot simulation model. *Journal of Agricultural Economics*, 25, 265–76.
Salant, S. (1976) Exhaustible resources and industrial structure: a Nash–Cournot approach to the world oil market. *Journal of Political Economy*, 8, 1049–94.
Salant, S., Miracort, F., Sanghri, A. and Wagner, M. (1981) *Imperfect Competition in the International Oil Market*, Heath Lexington Books, Lexington, MA.
Scott, M. J. and Goldsmith, O. S. (1987) Assessing regional econometric models: a discussion and application, US Department of the Interior, Mineral Management Service, Washington, DC.
Seeley, R. (1985) *Price Elasticities from the IIASA World Agriculture Model*, AGES850418, Economic Research Service, US Department of Agriculture, Washington, DC.
Setia, P., Magleby, R. and Carvey, D. (1988) *Illinois Rural Clean Water Project*, AGES880617, Economic Research Service, US Department of Agriculture, Washington, DC.
Shei, S. Y. and Thompson, R. L. (1977) The impact of Trade Restrictions on Price Stability in the World Wheat Market. *American Journal of Agricultural Economics*, 59, 628–38.
Shoemaker, R. (1986) *Effects of Changes in U.S. Agricultural Production on Demand for Farm Inputs*, TB 1722, Economic Research Service, US Department of Agriculture, Washington, DC.
Shoven, J. and Whalley, J. (1984) Applied general equilibrium models of taxation and trade. *Journal of Economic Literature*, 22, 1007–51.
Smallwood, D., Blaylock, J. and Harris, J. (1987) *Food Spending in American Households*, SB 753, Economic Research Service, US Department of Agriculture, Washington, DC.
Solnick, B. (1975) The advantages of domestic and international diversification, in *International Capital Markets* (eds E. Elton and M. Gruber), North-Holland, Amsterdam.
Soyster, A. and Sherali, H. D. (1981) On the influence of market stucture in modeling the U.S. copper industry. *International Journal of Management Science*, 9, 381–8.
Sparrow, F. and Soyster, A. (1980) Process models of minerals industries, in

Proceedings of the Council of Economics of the AIME, American Institute of Mining, Metallurgical, and Petroleum Engineers, New York, pp. 93–101.

Steigum, E. (1987) ARMOD: a small numerical macroeconomic model of world non-clearing markets. *Scandinavian Journal of Economics*, **89**, 227–46.

Stern, R., Francis, J. and Schumacher, B. (1976) *Price Elasticities in International Trade*, Macmillan, New York.

Stillman, R. (1985) *A Quarterly Model of the Livestock Industry*, TB 1711, Economic Research Service, US Department of Agriculture, Washington, DC.

Stillman, R. (1987) *A Quarterly Forecasting Model of the U.S. Egg Sector*, TB 1729, Economic Research Service, US Department of Agriculture, Washington, DC.

Sullivan, J., Wainlio, J. and Roningen, V. (1989) *A Database for Trade Liberalization Studies*, AGES89-12, Economic Research Service, US Department of Agriculture, Washington, DC.

Sun, T. (1987) *Quality Demand and Policy Implications for Florida Green Tomatoes*, TB 1728, Economic Research Service, US Department of Agriculture, Washington, DC.

Sutton, J. (1988) *Agricultural Trade and Natural Resources*, Lynne Rienner Publishers, Boulder, CO.

Takayama, T. and Labys, W. C. (1985) Spatial equilibrium analysis: mathematical and programming model formation of agricultural, energy and mineral models, in *Handbook of Regional Economics* (ed. P. Nijkamp), North-Holland, Amsterdam.

Taylor, L. (1975) Theoretical Foundations and Technical Implications, in *Economy-Wide Models and Development Planning* (eds C. Blitzer, P. Clark and L. Taylor), Oxford University Press, Oxford.

Teigen, L. (1988) *Agricultural Policy, Technology Adoption and Farm Structure*, AGES 880810, Economic Research Service, US Department of Agriculture, Washington, DC.

Trede, L. Boshlje, M. and Geasler, M. (1977) Systems approach to management for the cattle feeder. *Journal of Animal Science*, **45**, 1213–21.

Tsao, C. S. and Day, R. H. (1971) A process analysis model of the U.S. steel industry. *Management Science*, **17**, 588–608.

Tweeten, L. (1967) The demand for United States farm output. *Food Research Institute Studies*, **7**, 343–69.

Ulph, A. and Follie, M. (1980) Exhaustible resources and cartels: an intertemporal Nash–Cournot model. *Canadian Journal of Economics*, **13**, 645–58.

Uri, N. (1976) *Toward an Efficient Allocation of Electric Energy*, Heath Lexington Books, Lexington, MA.

Uri, N. (1977) Planning in the public utilities. *Regional Science and Urban Economics*, **6**, 105–25.

Uri, N. (1990) Delineating the nature and extent of the market for agricultural commodities. *International Trade Journal*, forthcoming.

Uri, N. (1989) Target prices, market prices and economic efficiency. *Journal of Consumer Policy*, **12**, 1–17.

Uri, N. (1989) Delineating the nature and extent of the market for agricultural commodities. *Journal of Economic Development*, **14** (1), 47–64.

Uri, N. and Jones, J. D. (1988) The price elasticity of export demand for U.S. agricultural commodities reconsidered. *Agricultural Systems*, **28**, 273–97.

Uri, N. and Rifkin, E. (1985) Geographic markets, railroad deregulation and causality. *Review of Economics and Statistics*, **67**, 422–8.

US Congress (1980) The world oil market in an interactive computer model called OILMAR. *Energy and Power*, Washington, DC.

US Department of Agriculture (1965) *Elasticity of Food Consumption Associated*

with Changes in Income in Developing Countries, Economic Research Service, US Department of Agriculture, Washington, DC..
US Department of Agriculture (1988) *Global Review of Agricultural Policies*, AGES 880304, US Department of Agriculture, Washington, DC.
US Department of Energy (1984) International coal trade model, DOE/EIA-0444, US Department of Energy, Washington, DC.
US Federal Power Commission (1972) Natural gas supply and demand: 1971 to 1990, Staff Report 2, Bureau of Natural Gas, Federal Power Commission, Washington, DC.
US Federal Reserve Bank of New York (1985) Commodity prices in the current recovery. *FRBNY Quarterly Review*, **1**, 11–19.
US National Academy of Sciences (1982) *Mineral Demand Modeling*, Committee on Nonfuel Mineral Demand Relationships, National Research Council, Washington, DC.
Vansickle, J. and Alvarado, G. (1983) Florida tomato market order restrictions. *Southern Journal of Agricultural Economics*, **15**, 109–13.
Vasavada, U. and Ball, V. E. (1988) *Modelling Dynamic Adjustment in a Multi-Output Framework*, SRAGES 880205, Economic Research Service, US Department of Agriculture, Washington, DC.
Von Stackelberg, H. (1952) *The Theory of Market Economy*, Oxford University Press, Oxford.
Waddell, L. and Labys, W. C. (1988) Transmaterialization: technology and materials demand cycles. *Materials and Society*, **12**, 59–86.
Waugh, F. (1966) Consumer aspects of price instability. *Econometrica*, **34**, 504–8.
Weaver, R. (1983) Multiple input, multiple output production choices and technology in the U.S. wheat region. *American Journal of Agricultural Economics*, **65**, 45–56.
Webb, A., Parrlberg, P., Dunmore, J. and Deaton, J. (1985) The U.S. competitive position in world commodity trade, in *Agricultural-Food Policy Review*, AER 530, Economic Research Service, US Department of Agriculture, Washington, DC.
Webster, J. and Williams N. (1988) Changes in cereal production and yield variability on farms in south east England. *Journal of Agricultural Economics*, **39**, 255–62.
Whitacre, R. (1979) An evaluation of Japanese agricultural policies with a multiregion–multicommodity model, Ph.D. Thesis, Department of Agricultural Economics, University of Illinois.
Williams, J., Jones, C. and Dyke P. (1984) A modeling approach to determining the relationship between erosion and productivity. *Transactions of the American Society of Agricultural Engineers*, **27**, 129–84.
Wood, D. O. and Spierer, C. (1984) Modeling Swiss industry interfuel substitution in the presence of natural gas supply constraints, Energy Laboratory Working Paper WP84-011, Massachusetts Institute of Technology, Cambridge, MA.
World Bank (1981) *World Bank Commodity Models*, Papers presented at the Workshop on Commodity Models and Policies at Aarhus, 14–17 December 1979, Staff Commodity Working Paper 6, Washington, DC.
Yang, C. W. and Labys, W. C. (1985) Sensitivity analysis of the linear complementarity programming model of the U.S. coal market. *Energy Economics*, **7**, 145–52.
Yang, C. W. and Labys, W. C. (1989) A sensitivity analysis of the linear complementarity programming model: Appalachian steam coal and natural gas, in *Quantitative Methods for Market-Oriented Economic Analysis Over Space and Time* (eds W. C. Labys, T. Takayama and N. Uri), Gower, London.
Zangwill, W. and Garcia, C. (1981) *Pathways to Solutions, Fixed Points and Equilibria*, Prentice-Hall, Englewood Cliffs, NJ.

2

Computing equilibria in imperfectly competitive commodity markets

CHARLES D. KOLSTAD

AND LARS MATHIESEN

2.1 INTRODUCTION

The computation of partial and general competitive equilibria has been a field of enormous empirical importance as well as a source of diverse and fundamental theoretical research questions. However, aside from the case of a pure monopolist, until recently the question of the computation of equilibria under conditions of imperfect competition has been largely neglected.[1]

The purpose of Chapter 1 is to present a modelling format and an algorithm for the computation of Cournot–Nash (or, more simply, Cournot) equilibria. The method is based on formulating the equilibrium problem as that of finding a solution to a non-linear complementarity problem (CP). We solve this CP by Newton's method whereby the CP is sequentially linearized and the resulting linear complementarity problems (LCPs) are solved using Lemke's algorithm.

This sequential LCP (SLCP) algorithm has been used before in contexts other than the computation of imperfectly competitive equilibria. Specifically, Josephy (1979) and Friesz *et al.* (1983) have solved competitive partial equilibrium models and Mathiesen (1985) has solved a range of partial and general equilibrium models using this method.

[1]Exceptions include the work of Murphy *et al.* (1982) and Harker (1984) for computing Cournot equilibria and that of Salant (1982) for computing Cournot equilibria in the international oil market. Recent algorithmic contributions include Harker (1986*a*, *b*), Hashimoto (1984), Sherali *et al.* (1983), Dafermos and Nagurney (1987), Nagurney (1988), Hobbs (1986) and Haurie and Marcotte (1985), all for Cournot equilibria.

In an earlier paper (Kolstad and Mathiesen, 1989), we developed conditions for and demonstrated the convergence of SLCPs to a unique equilibrium of the Cournot–Nash model. In this paper we focus on the application to two variants of the Cournot–Nash model and review some empirical analyses that have employed this approach.

In the next section of the paper we provide background on the relevance of imperfect competition in trade models. We then review briefly the modelling format and the relevant results for the existence of a unique equilibrium and convergence of the algorithm. The format is next applied to the homogeneous product, segmented market problem and to the differentiated product problem where in both cases producers adhere to Cournot behaviour.

2.2 IMPERFECT COMPETITION IN COMMODITY TRADE MODELS[2]

As in much applied economics, the paradigm of perfect competition is widely used in evaluating trade in agricultural and other bulk commodities. In a thorough review of US developments in agricultural trade models, Thompson (1981) states that spatial price equilibrium models are 'the most common class of agricultural trade models, particularly for comparative statics analysis of the effect of a change in policy'. Thompson cites nearly three dozen spatial equilibrium models of international markets for individual agricultural commodities.

Despite its popularity, it has been recognized for some time that the paradigm of spatial competitive equilibrium suffers from significant deficiencies, most notably poor performance in explaining trade patterns. Characteristically, such models predict fewer bilateral trades than actually occur,[3] although many actual transactions are quite small.[4] The class of models also exhibits a high degree of sensitivity of equilibrium trade levels to small parameter changes. This has been recognized in agricultural trade models as well as models of other bulk commodities such as coal (Kolstad and Abbey, 1984). In essence, the simple model of spatial perfect competition does not seem consistent with how many international commodity markets operate.

A major thrust of the literature in correcting the predictive deficiency of spatial equilibrium models is to examine more closely the role of govern-

[2]Parts of this discussion are drawn from Kolstad and Burris (1986).

[3]This is the 'no cross-hauling' result of international trade and the insight from the 'transportation problem' of operations research, where at most $n + m - 1$ flows are positive, n and m being the numbers of suppliers and demanders.

[4]In partial equilibrium models, researchers have circumvented the problem by putting additional constraints (interpreted as import or export quotas) on flows; in other words, a particular trade pattern is assumed rather than predicted.

ments in markets. Endogenizing government policy in trade analysis has taken two directions. One direction has been to assume that government policy is determined by domestic political factors and not by market power considerations.[5] The focus of such work is on government objectives in the policy-setting process—non-economic factors are used to explain trade.

A second approach to analysis with endogenous government policy is to assume that such policy is motivated by economic factors; in essence, policy serves to coordinate producers and consumers so that they may exercise power in the international market (e.g. Brock and Magee, 1978). Thus, the international trade would be expected to operate as any oligopoly/oligopsony with countries as participants.[6] Although not in the context of commodity trade models, McCalla (1966) and Alaouze *et al.* (1978) suggested just this sort of oligopoly conduct in international wheat markets. Carter and Schmitz (1979) suggested that a duopsony was more appropriate for wheat.

A problem with imperfect competition is that it is difficult to implement empirically. Jeon (1981) has developed an empirical but non-spatial model of the international wheat market; Kolstad and Burris (1986) have done the same in a spatial context. Karp and McCalla (1983) have modelled the international corn market as a differential game. In non-agricultural markets, Kolstad and Abbey (1984) have modelled the international steam coal market as a Nash n-person game, Salant (1982) has modelled the international oil market and Mathiesen *et al.* (1987) have analysed the European natural gas market.

Before proceeding to present our algorithm for computing imperfectly competitive equilibria, it is useful to connect the traditional Takayama–Judge spatial equilibrium models with our imperfectly competitive version. Consider a static market for a single homogeneous commodity where a set of producers is separated by a transportation network from a set of consumers. Let $i = 1, \ldots, I$ index producers and $j = 1, \ldots, J$ index consumers and, further, let q_{ij} denote the quantity shipped from producer i to consumer j, t_{ij} denote the unit transport cost from producer i to consumer j, $P_j(\Sigma_i q_{ij})$ denote the inverse demand function for consumer j, and $C_i(\Sigma_j q_{ij})$ denote the cost function for producer i. The first of these variables (q_{ij}) is endogenous, to be computed. The unit transport cost, marginal cost and the inverse demand functions are exogenous. In the model, we shall permit producers with market power to price discriminate and further we preclude arbitrage among consumers.

In our model, we have some producers exercising market power and all

[5]See, for example, the work of Rausser *et al.* (1982), Sarris and Freebairn (1983) and Meilke and Griffith (1983).

[6]Cartels are the simplest example of international market power. However, since cartels are effectively monopolies with a competitive fringe, they are not difficult to model.

others in a competitive fringe. All consumers can be totally represented by their inverse demand curves. Thus there are two types of producers in the market: oligopolists and the competitive fringe. Denote the set of oligopolists by $M = \{i|i \text{ is an oligopolist}\}$. Each producer has the same objective, to maximize profits over $q_{ij} \geqslant 0$:

$$\pi_i = \sum_j \left[P_j\left(\sum_{i'} q_{i'j}\right) - t_{ij}\right] q_{ij} - C_i\left(\sum_j q_{ij}\right) \tag{2.1}$$

where i' is an index of summation over all producers. The price the producer faces is the consumer price P_j less the transport cost t_{ij}. Producers maximize profits by choosing the quantity q_{ij} sold to each consumer. The first-order conditions for a maximum of 2.1 are

$$w_{ij} \equiv \left\{P_j\left(\sum_{i'} q_{i'j}\right) - t_{ij} + q_{ij}\frac{\mathrm{d}P_j}{\mathrm{d}q_{ij}} - C_i'\left(\sum_{j'} q_{ij'}\right)\right\} \leqslant 0 \tag{2.2}$$

$$q_{ij} \geqslant 0 \qquad w_{ij}q_{ij} = 0 \qquad \text{for all } j$$

In this expression, the dummy variables w_{ij} is used to reveal easily that the term in braces must be non-positive and that its product with q_{ij} must be zero. The complementarity condition is necessary because we would expect some flows q_{ij} to be zero and thus the corresponding w_{ij} perhaps to be negative. The first three terms of the expression in braces represent marginal revenue. A fundamental distinguishing characteristic of a spatial model of trade is that we cannot determine *a priori* which producer–consumer pairs will have no trade ($q_{ij} = 0$)—thus the necessity for this condition which is more complex than just marginal revenue equals marginal cost.

For a price taker (the competitive fringe), $\mathrm{d}P_j/\mathrm{d}q_{ij}$ is considered to be zero since the price taker assumes that the market price is insensitive to changes in his output. Thus for the competitive fringe the first-order conditions 2.1 are quite simple:

$$w_{ij} \equiv \left\{P_j\left(\sum_{i'} q_{i'j}\right) - t_{ij} - C_i'\left(\sum_{j'} q_{ij'}\right)\right\} \leqslant 0 \tag{2.3}$$

$$q_{ij} \geqslant 0 \qquad w_{ij}q_{ij} = 0 \qquad \text{for all } j,\ i \notin M$$

This expression merely states that the price net of transport cost is equal to marginal cost unless no transactions take place ($q_{ij} = 0$), in which case price can be less than marginal cost.

The entire imperfectly competitive spatial equilibrium model consists of $I \times J$ variables (the q_{ij}) and a vector-valued function w of dimension $I \times J$ defined by Equations 2.2 and 2.3. In the next section we consider an algorithm for solving this problem where all producers are oligopolists. The standard Takayama–Judge competitive spatial equilibrium model finds an equilibrium set of prices and quantities in a competitive market by maximizing consumer and producer surplus. First-order conditions for such a maximization are identical with Equation 2.3. However, except when demand is linear, there is generally no single maximization problem whose first-order conditions are as in Equation 2.2 (Slade, 1988).

Before considering the algorithm for solving this problem, we present two analyses where this approach is applied. The first is from Kolstad and Burris (1986) concerning the international wheat market. In that paper, a competitive equilibrium model of Shei and Thompson (1977) was modified to include Cournot behaviour on the part of the USA and Canada (as suggested by McCalla, 1966). Table 2.1 presents actual wheat trade in 1972–3, Table 2.2 presents the results of a competitive equilibrium model of trade and Table 2.3 presents the results of our Cournot model. Although Table 2.3 does not replicate actual trade, it appears that the duopoly model is better than the competitive model. Statistical tests reported by Kolstad and Burris (1986) confirm this.

The second application, by Mathiesen *et al.* (1987), concerns the market for natural gas in Western Europe. The UK, the Netherlands, Norway, the USSR and Algeria are recognized as the major sellers. The static competitive, collusive and Nash–Cournot equilibria are computed and compared with observed trade for 1983. The Nash–Cournot equilibrium is found to provide the best approximation. In these equilibria, producing countries are treated symmetrically. The UK and the Netherlands, however, enter both the supply and the demand side and it seems reasonable to model their behaviour differently.

The equilibrium of a slightly modified model, where the UK and the Netherlands are assumed to behave as price takers in their respective home-markets but oligopolists elsewhere, matches observations remarkably well (Table 2.4). British and Dutch producers drive domestic prices down and consumption up compared with the Nash–Cournot equilibrium. Total computed consumption is 2% too high, and except for the Netherlands consumption in any region deviates less than 5% from observed figures. Also regional production levels are reasonably similar to observed quantities, although the Algerian supply is definitely too high.

The remaining deviations between observed and computed regional consumption and production quantities may stem from several sources not incorporated in the model, in particular the lack of an intertemporal dimension and uncertainty. There are also ample opportunities for errors in the estimated functions, cost parameters and behavioural assumptions. The Dutch home supply and the Algerian sales and production can probably be ascribed to misrepresented behaviour.

We now turn to our algorithm for computing imperfectly competitive equilibria. First we consider a spaceless homogeneous market. After developing the algorithm in this notationally more tractable context, we expand the treatment to include spatial markets and product-differentiated markets.

Table 2.1 Actual wheat trade shares by source and primary destination, 1972–3

Destination	*Source*					*Total imports (10^6 tonnes)*
	USA	*Canada*	*Australia*	*Argentina*	*EEC*	
Japan	0.057	0.023	0.012	0.001		5.5
PRC	0.010	0.074	0.005			5.3
Other Asia	0.123	0.024	0.024	0.017	0.023	12.5
USSR	0.160	0.070	0.015		0.012	15.2
E. Europe	0.014	0.002	0.002		0.001	1.1
EEC	0.047	0.038	0.008	0.006		5.9
Other W. Europe	0.015	0.002	0.001		0.005	1.2
Africa	0.029	0.009	0.014	0.001	0.051	6.1
South America	0.052	0.009	0.010	0.034		6.2
Total exports (10^6 tonnes)	29.9	14.9	5.5	3.5	5.4	59.2
Export price (US$/tonne)	92.35	90.94	90.90	85.98	105.93	

Shares as fraction of total world trade.
Source: Shei and Thompson (1977), Kolstad and Burris (1986).

Table 2.2 Simulated wheat trade shares by source and primary destination 1972–3, perfectly competitive market

	Source					*Total imports*	*Import price*
Destination	*USA*	*Canada*	*Australia*	*Argentina*	*EEC*	*(10^6 tonnes)*	*(US$/tonne)*
Japan	0.097					5.6	99
PRC	0.091					5.3	105
Other Asia		0.067	0.142			12.1	108
USSR	0.058	0.198				14.8	101
E. Europe		0.015				0.9	101
EEC	0.100					5.8	96
Other W. Europe	0.022					1.3	98
Africa				0.032	0.070	5.9	105
South America	0.108					6.3	98
Total exports (10^6 tonnes)	27.5	16.2	8.2	1.8	4.1	57.8	
Export price (US$/tonne)	88	91	93	86	95		

Shares as fraction of total world trade.
Source: Kolstad and Burris (1986).

Table 2.3 Simulated wheat trade shares by source and primary destination 1972–3, US–Canada duopoly

Destination	*Source*					*Total imports* (10^6 *tonnes*)	*Import price* (*US$/tonne*)
	USA	*Canada*	*Australia*	*Argentina*	*EEC*		
Japan	0.016	0.011	0.068			5.4	113
PRC	0.024	0.015	0.048			4.9	116
Other Asia	0.097	0.070	0.030			11.1	115
USSR	0.142	0.101				13.7	107
E. Europe	0.020	0.009				1.6	88
EEC	0.077	0.038				6.5	91
Other W. Europe	0.022	0.008				1.7	89
Africa	0.030	0.021			0.049	5.6	110
South America	0.022	0.013		0.035	0.035	5.9	113
Total exports (10^6 tonnes)	25.3	16.1	8.2	2.0	4.7	56.4	
Export price (US$/tonne)	91	93	100	99	100		

Shares as fraction of total world trade.
Source: Kolstad and Burris (1986).

Table 2.4 Computed gas trade patterns, consumption and supplies billions per cubic centimetre (bcm) for modified Nash–Cournot equilibrium, 1983

	Netherlands	*UK*	*FRG*	*France*	*Delux*	*Italy*	*Austria*	*Computed supplies A*	*Actual supplies B*	*A – B*
Netherlands	43.2		8.9	5.3	3.3	4.1	0.9	65.7	67.4	−1.7
UK		39.7						39.7	38.9	0.8
USSR			8.0	4.7	2.7	3.6	0.9	19.9	22.5	−2.6
Algeria			6.1	5.0	2.2	4.1	0.5	17.9	12.1	5.8
Norway		7.4	5.0	3.7	1.3	1.7	0.5	19.6	22.8	−3.2
Indigenous production[a]			14.7	6.8		12.3	1.3	35.1	35.1	—
Computed consumption *C*	43.2	47.1	42.7	25.5	9.5	25.8	4.1	197.9	198.8	−0.9
Actual consumption *D*	35.9	49.4	43.1	27.6	9.1	25.1	4.3	194.5		
C – D	7.7	−2.3	−0.4	−2.1	0.4	0.7	−0.2	3.4		

[a]Indigenous production in FRG, France, Italy and Austria in 1983 is exogenously given in the model.
Source: Mathiesen *et al.* (1987).

2.3 AN ALGORITHM FOR COMPUTING COURNOT EQUILIBRIA

2.3.1 Cournot equilibrium: existence and uniqueness

Let there be N firms, each providing the same good. Denote the output of the ith firm by q_i, produced at cost $C_i(q_i)$. The firms face a market inverse demand function $P(Q)$. Profits for the ith firm are then given by

$$\pi_i(q_i) = P(Q)q_i - C_i(q_i) \tag{2.4}$$

where $Q = \Sigma_{i=1}^{N} q_i$. The Cournot equilibrium is defined in the conventional manner.

Definition. A Cournot equilibrium is a vector of outputs q^* such that

$$\pi_i(q^*) = \max_{q_i} \pi_i(q_1^*, \ldots, q_{i-1}^*, q_i, q_{i+1}^*, \ldots, q_N^*) \qquad i = 1, \ldots, n$$

In other words, holding output of other firms constant, each firm can do no better.

We shall assume inverse demand and costs are twice continuously differentiable and that profits are concave (pseudoconcavity[7] would suffice) with respect to own output. Thus, first-order stationarity conditions for a profit maximum are necessary and sufficient for a global (not necessarily unique) maximum of profit. Consequently, the set of Cournot equilibria is the same as the set of solutions to the following first-order conditions for a profit maximum:

$$f_i \equiv \frac{\partial \pi_i}{\partial q_i} = C_i'(q_i) - P\left(\sum_{j=1}^{N} q_j\right) - q_i P'\left(\sum_{j=1}^{N} q_j\right) \geqslant 0 \tag{2.5a}$$

$$\frac{\partial \pi_i}{\partial q_i} q_i = 0 \tag{2.5b}$$

$$q_i \geqslant 0 \qquad \text{for } i = 1, \ldots, N \tag{2.5c}$$

Equations 2.5 are a CP which can be more compactly written as

$$\begin{aligned} &\text{CP}(f)\text{: find } q \in R^N \text{ such that} \\ &\qquad f(q) \geqslant 0,\ q \geqslant 0 \text{ and } q^{\mathrm{T}} f(q) = 0 \end{aligned} \tag{2.6}$$

where f: $R_+^N \to R^N$ and T denotes transpose.

In order to ensure that a Cournot equilibrium exists, we introduce the notion of a bound on industry output.

Definition. Industry output is said to be bounded by $Q > 0$ if output in

[7] Recall that f is pseudoconcave on $X \subset R^n$ if $(x_1 - x_2)\nabla f(x_2) \geqslant 0$ implies $f(x_1) \leqslant f(x_2)$ for any $x_1, x_2 \in X$.

excess of Q from any producer implies that the marginal profits of all producers are negative.

The definition merely states that for some industry output level no firm is at a profit maximum and that further expanding the industry will not improve profits. That is, marginal profits become negative and the functions $f(q)$ become positive.

(a) *Existence of a unique equilibrium (Kolstad and Mathiesen, 1989)*

Assume that

1. inverse demand and costs are C^2,
2. industry output is bounded and
3. profits are strictly concave for all $q \geqslant 0$.

Then there exists a unique Cournot equilibrium.

A special case is where the Jacobian matrix of f has a positive dominant diagonal[8] and thus is a P-matrix.[9] In this case the interpretation is that if own-effects on marginal profits dominate cross-effects, an equilibrium will be unique. It should be noted that the above result is a sufficient condition for uniqueness and can be considerably relaxed (Kolstad and Mathiesen, 1987). One consequence of the result is the following. Let demand have constant price elasticity e and let the marginal cost function be non-decreasing. Then if n (the number of firms with $q_i^* > 0$) satisfies $n > -1/e$, the equilibrium is unique (Mathiesen, 1989).

2.3.2 The algorithm

The SLCP algorithm applied to the function f involves linearizing f at some initial point x^0 and solving the resulting LCP. f is then re-linearized at this solution and the process is continued until convergence is achieved. Eaves (1978) and Josephy (1979) were the first to suggest this method and prove local convergence based on a norm-contraction approach.[10]

Let us define the linearization of f at y as the first-order Taylor expansion:

$$\mathrm{L}f(x|y) = f(y) + \nabla f(y)(x - y) \tag{2.7}$$

The LCP $\mathrm{CP}(\mathrm{L}f(x|y))$ is

[8] A is said to have a positive dominant diagonal if $q_{ii} > 0$ for all i and $a_{ii} > \Sigma_{j \neq i} |a_{ij}|$ for all i. Frequently the definition of diagonal dominance allows scaling of the matrix before imposition of this inequality.

[9] A is a P-matrix if all of the principal minors of A are positive definite.

[10] See also Pang and Chan (1982) who establish both local and global results for this and other algorithms using norm-contraction as well as other approaches.

find $x \in R_+^N$ such that

$$\mathrm{L}f(x|y) = q + Mx \geqslant 0, x \geqslant 0 \tag{2.8a}$$

$$x^{\mathrm{T}}(q + Mx) = 0 \tag{2.8b}$$

where $q = [f(y) - \nabla f(y)y]$ and $M = \nabla f(y)$. A vector x is said to be feasible when it satisfies 2.8a.

The sequential linear complementarity problem algorithm

Step 0 Stipulate $x^0 \in R_+^N$.

Step k ($k = 1, 2, \ldots$) Compute x^k, the solution to LCP($f|x^{k-1}$).

Thus the solution to the LCP for f at y is the solution to CP(L$f(x|y)$). It should be pointed out that, even if there is a unique solution to CP(f), there may be some y for which there is no solution or multiple solutions to CP(L$f(x|y)$).[11] If the Jacobian matrix of f is everywhere a P-matrix, however, then the Jacobian of each linearization of f is also a P-matrix. Thus there will never be more than one solution to CP(L$f(x|y)$). Furthermore, since in our case the Jacobian matrix of f is positive definite, Dorn's theorem (Karamardian, 1969) implies existence (as well as uniqueness) of solutions to CP(L$f(x|y)$). It is also known that Lemke's algorithm computes the solution to the LCP in this case (see Lemke (1965) or Cottle and Dantzig (1968)).

Based on Pang and Chan (1982) we can state our local convergence result for the SLCP applied to the Cournot model.

Local convergence (Kolstad and Mathiesen, 1989). Let q^* solve 2.5 and let f: $R_+^N \rightarrow R^N$ be continuously differentiable. Assume that $\nabla f(q^*)$ is positive definite. Then there exists a neighbourhood of q^* such that, if the initial iterate q^0 is chosen there, the SLCP computes a sequence $\{q^k\}$ converging to q^*.

Remark. Concavity of profits π in a neighbourhood of q^* implies that $\nabla^2\pi(q^*)$ is negative definite; thus $\nabla f = -\nabla^2\pi$ is positive definite.

In order to prove global convergence for the Newton process more stringent assumptions are required. Several of the available results that employ norm-contraction (as above) are based on overly strong conditions (e.g. Pang and Chan (1982), Theorem 2.9 and Corollary 2.10). To appreciate this statement consider the function

$$f(x) = x^2 - 1 \tag{2.9}$$

whose solution is $x^* = 1$. The SLCP converges to this solution for $x^0 > 0$, yet the conditions of Pang and Chan's theorem are not satisfied.

[11]See Mathiesen (1987).

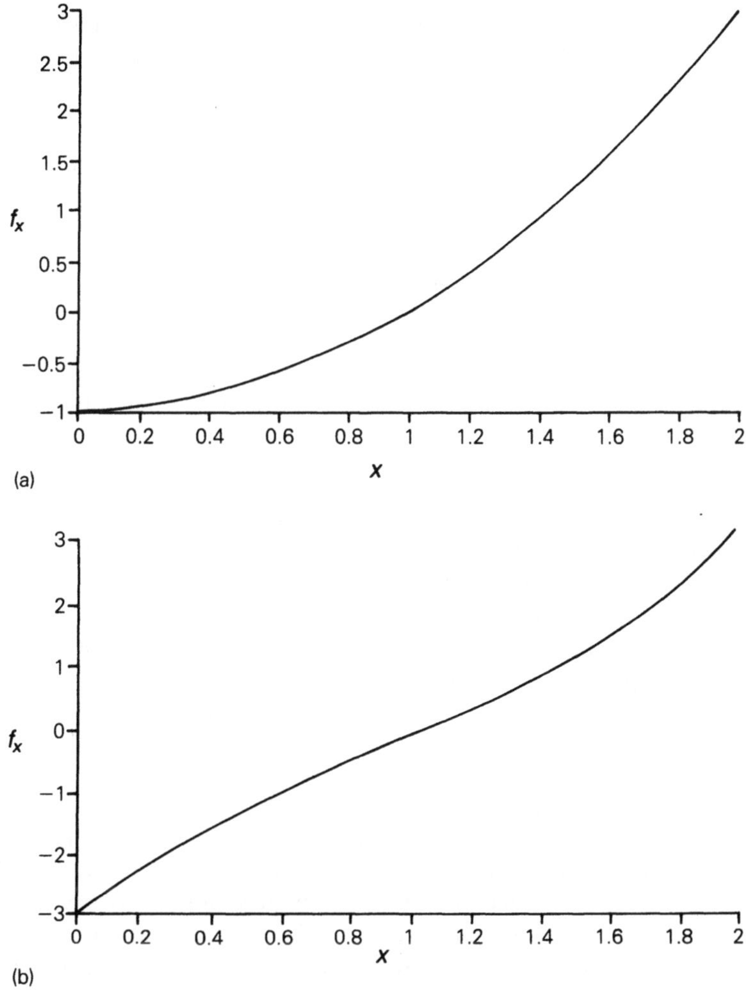

Fig. 2.1 Plots of (a) Equation 2.9 and (b) Equation 2.10.

Following Pang and Chan's monotone approach (essentially assuming f to be convex) Kolstad and Mathiesen (1989) established global convergence. Observe, however, that even this assumption is too strong. For example consider the function

$$f(x) = \begin{cases} f_1(x) = (x-1)(3-x) & 0 < x \leqslant 1 \\ f_2(x) = x^2 - 1 & 1 \leqslant x \end{cases} \tag{2.10}$$

whose solution is $x^* = 1$. f is continuously differentiable but not convex, yet the SLCP converges for $x^0 \geqslant 0$. Thus there is ample room for less stringent assumptions on f (and marginal profits).

Global convergence (Kolstad and Mathiesen, 1989). Assume that

1. profits π_i are twice continuously differentiable,
2. industry output is bounded,
3. marginal profits (defined in equation 2.5.) are concave, and
4. the Jacobian of marginal profits has a negative dominant diagonal.

Then there exists a well-defined sequence $\{q^k\}$, where each q^k solves CP(L$f(q|q^{k-1})$, such that $\{q^k\}$ converges to the unique solution of the CP for any feasible q^0.

This result gives a restricted form of global convergence: starting at any *feasible* point, global convergence is assured. The assumption that industry output is bounded gives a natural feasible starting point: a vector of outputs for which all marginal profits are non-positive. Two other significant restrictions of the theorem are that marginal profits are concave and the Jacobian of marginal profits has a negative dominant diagonal. Diagonal dominance implies that own-effects on marginal profit dominate cross-effects. Thus if everyone raises output equally the change in marginal profits from increasing own-output dominates the change in marginal profits from the actions of other producers.

The concavity of marginal profits is probably the most severe restriction. Recall that marginal profits are the sum of marginal revenue and negative marginal costs. Negative marginal costs will typically be concave, even with increasing returns. Marginal revenue may or may not be concave. Concave demand yields concave marginal revenue. A constant elasticity of demand yields convex marginal revenue and thus possible non-concave marginal profits.

2.4 VARIANTS ON THE BASIC MODEL

The homogeneous product Cournot model where quantities are decision variables is a standard textbook model that can be extended in several directions. We shall consider two such extensions: a spatial or segmented homogeneous product market and a differentiated product market. These models can be solved by the algorithm of the previous section under essentially similar conditions. In this section we describe these models and develop an economic interpretation of their assumptions. Proofs of the results presented here may be found in Kolstad and Mathiesen (1989).

2.4.1 The homogeneous product segmented market Cournot model

In this application the N firms sell their product in M different segments of the market. The segmentation may be on the basis of a number of criteria, such as geographic location or consumer group. It is assumed that these segments are completely separated from each other and that there are no possibilities of arbitrage among segments. Hence the Cournot equilibrium will typically imply discriminatory pricing. This is the model presented in Equations 2.1–2.3 with the exception that there is no competitive fringe here. Including the competitive fringe is a trivial extension.

Major features of this model date back to Enke (1951) and Samuelson (1952) and are perhaps best known through Takayama and Judge (1971) and a long series of applications to a wide range of industries. All these models of competitive equilibrium are cast as optimization models where generally supply and demand are either fixed, step functions (LP models) or described by linear functions (QP models). MacKinnon (1976) allows for non-linear supply and demand functions and uses a fixed point algorithm to compute the equilibrium.

Recently the behavioural assumption has been modified to that of a Cournot–Nash strategy. In these applications (Okuguchi, 1983; Harker, 1984; Kolstad and Abbey, 1984; Kolstad and Burris, 1986; Mathiesen *et al.*, 1987) the complementarity approach is generally followed by using different iterative processes. Salant (1982) and Murphy *et al.* (1982) find an equilibrium through a sequence of optimization problems.

Assume that there are N firms and M segments. For firm i and segment j let q_{ij} denote the quantity sold, $C_i(\Sigma_{j=1}^{M} q_{ij})$ denote the cost function for firm i and $P_j(Q_j)$ denotes the inverse demand function for segment j, where $Q_j = \Sigma_{i=1}^{N} q_{ij}$. Finally, let q denote the vector of all quantities, i.e. $q = (q_{11}, q_{21}, \ldots, q_{N1}, q_{12}, \ldots, q_{NM})$. Profits for the ith firm are then

$$\pi_i(q) = \sum_{j=1}^{M} P_j(Q_j) q_{ij} - C_i\left(\sum_{j=1}^{M} q_{ij}\right) \tag{2.11}$$

and the first-order conditions for a profit maximum are

$$-\frac{\partial \pi_i}{\partial q_{ij}} = C_i'\left(\sum_{j=1}^{M} q_{ij}\right) - P_j\left(\sum_{i=1}^{N} q_{ij}\right) - q_{ij} P_j'\left(\sum_{i=1}^{N} q_{ij}\right) \geqslant 0 \tag{2.12a}$$

$$-\frac{\partial \pi_i}{\partial q_{ij}} q_{ij} = 0 \tag{2.12b}$$

$$q_{ij} \geqslant 0 \tag{2.12c}$$

We make several assumptions about the market.

Assumption S1. Inverse demand functions $P_j(\cdot)$, $j = 1, \ldots, M$, and cost functions $C_i(\cdot)$, $i = 1, \ldots, N$, are C^2.

Assumption S2. Industry output is bounded.

Assumption S3. Marginal profits are concave.

Finally, consider the Jacobian of the negative marginal profit functions:

$$J = \nabla f = -\nabla^2 \pi = \begin{bmatrix} A_1 & D & \dots & D \\ D & A_2 & \dots & D \\ \vdots & \vdots & \vdots & \vdots \\ D & D & \dots & A_M \end{bmatrix} \tag{2.13}$$

where the matrix A_j has diagonal elements $C_i'' - 2P_j' - q_{ij}P_j''$ and off-diagonal elements $-(P_j' + q_{ij} + q_{ij}P_j'')$ and where

$$D = \begin{bmatrix} C_1'' & 0 & \dots & 0 \\ 0 & C_2'' & \dots & 0 \\ \vdots & \vdots & & \vdots \\ 0 & 0 & \dots & C_N'' \end{bmatrix}$$

Observe that each A_j, ignoring the index j, is equivalent to the Jacobian of the negative marginal profit functions of the non-segmented market model.

Assumption S4. The $(N \times M) \times (N \times M)$ matrix J given by 2.13 has a positive dominant diagonal:

$$C_i'' - 2P_j' - q_{ij}P_j'' > (N - 1)|(P_j' + q_{ij}P_j'')| + (M - 1)|C_i''| \text{ for all } i$$

In comparison with the non-segmented market model, there is an additional term $(M - 1)|C_i''|$. This is a stricter, through natural, extension. The feedback effects on a firm's marginal profits from selling in segment j are now both from its $N - 1$ competitors in segment j and from its own selling in the $M - 1$ other segments.

Under assumptions S1–S4, the sequence $\{q^k\}$ generated by the SLCP converges to a unique equilibrium q^* for any feasible q^0.

2.4.2 The differentiated product model

In this application we consider N firms, each selling a similar though not identical product; i.e. each firm has its own product (or 'brand' of a product). As the products are not perfect substitutes for one another, it makes sense to talk about demand for a given firm's product, and each firm may charge a different price. Let $p = (p_1, \dots, p_N)$ be the vector of prices obtained by each firm and $q = (q_1, \dots, q_N)$ the quantities.

The demand function of the ith firm is $d_i(p)$. It is assumed to be C^2 and to satisfy gross substitutability, i.e. its first derivatives satisfy $d_{ii} < 0$, $d_{ij} > 0$ and $|d_{ii}| > \Sigma_{j \neq i}|d_{ij}|$. The interpretation is that demand for product i decreases when its price is increased; it increases when some other price is increased;

and when all firms increase their prices by an equal amount, demand decreases.

Let the demand function be $d(p) = (d^1(p), \ldots, d^N(p))$ with Jacobian $D(p)$. Assume $D(p)$ has a dominant negative diagonal and that $-D(p)$ is a K-matrix. Then the inverse demand function $p = d^{-1}(q) = P(q)$ exists and its first derivatives are all non-positive.

As before, let $C_i(q_i)$ denote the cost function of the ith firm. The profit function of this firm is

$$\pi_i = P_i(q)q_i - C_i(q_i) \tag{2.14}$$

and the first-order conditions for a profit maximum are

$$-\frac{\partial \pi_i}{\partial q_i} = C_i'(q_i) - (p_i^i(q)q_i + P_i) \geqslant 0 \tag{2.15a}$$

$$-\frac{\partial \pi_i}{\partial q_i} q_i = 0 \tag{2.15b}$$

$$q_i \geqslant 0 \tag{2.15c}$$

where superscripts on P_i indicate differentiation.

The Jacobian of the negative marginal profit functions $\nabla f(q)$ has in this case diagonal elements $C_i'' - 2P_i^i - P_i^{ii} - q_i$ and off-diagonal (i, j) elements $-P_i^j - P_i^{ij} q_i$.

Assumption D1. Inverse demand functions $P_i(\cdot)$, $i = 1, \ldots, N$, and cost functions $C_i(\cdot)$, $i = 1, \ldots, N$ are C^2.

Assumption D2. Industry output is bounded.

Assumption D3. Marginal profits are concave.

Assumption D4. The $N \times N$ matrix $\nabla f(q)$ of negative marginal profits has a positive dominant diagonal. That is, for $i = 1, \ldots, N$,

1. $C_i'' - 2P_i^i - P_i^{ii} q_i > 0$ and
2. $|C_i'' - 2P_i^i - P_i^{ii} q_i| > \sum_{j \neq i} |P_i^j + P_i^{ij} q_i|$.

Under assumptions D1–D4, the SLCP algorithm converges to a unique equilibrium q^* for any sequence $\{q^k\}$ generated by the feasible q^0.

CONCLUSIONS

The basic purpose of this chapter has been to present a modelling format and algorithm for finding Cournot–Nash equilibria and to demonstrate conditions for global convergence of this algorithm. Within this context, we examined three generic classes of markets: homogeneous products; homogeneous product, segmented markets; differentiated products. Clearly, there are other types of markets to which the algorithm can be applied but which are not considered here. We have also demonstrated the utility of using empirical models of imperfectly competitive commodity markets by briefly reviewing experience with the European gas market and the international wheat market.

REFERENCES

Alaouze, C. M., Watson, A. S. and Sturgess, N. H. (1978) Oligopoly pricing in the world wheat market. *American Journal of Agricultural Economics*, **60**, 17–85.

Brock, W. A. and Magee, S. P. (1978) The economics of special interest politics: the case of a tariff. *American Economic Review*, **68**, 246–50.

Carter, C. and Schmitz, A. (1979) Import tariffs and price formation in the world wheat market. *American Journal of Agricultural Economics* **61**, 517–22.

Cottle, R. W. and Dantzig, G. B. (1968) Complementary pivot theory of mathematical programming. *Linear Algebra and Its Applications*, **1**, 103–25.

Dafermos, S. and Nagurney, A. (1987) Oligopolistic and competitive behavior of spatially separated markets. *Regional Science and Urban Economics*, **17**, 245–54.

Eaves, B. C. (1978) A locally quadratically convergent algorithm for computing stationary points, Technical Report 78-13, Systems Optimization Laboratory, Department of Operations Research, Stanford University, Stanford, CA.

Enke, S. (1951) Equilibrium among spatially separated markets: solution by electric analog. *Econometrica*, **19**, 40–7.

Freisz, T. L., Tobin, R. L., Smith, T. E. and Harker, P. T. (1983) A nonlinear complementary formulation and solution procedure for the general procedure demand network equilibrium problem. *Journal of Regional Science*, **23** (3), 337–59.

Harker, P. T. (1984) A variational inequality approach for the determination of Oligopolistic Market Equilibrium. *Mathematical Programming*, **30**, 105–11.

Harker, P. T. (1986a) Alternative models of spatial competition. *Operations Research*, **34**, 410–25.

Harker, P. T. (1986b) Generalized Nash games and quasi-variational inequalities, Wharton School Working Paper 86-05-05, Philadelphia, PA, May.

Hashimoto, H. (1984) A spatial Nash equilibrium model, in *Spatial Price Equilibrium: Advances in Theory, Computation and Application* (ed. P. T. Harker), Springer-Verlag, Berlin.

Haurie, A. and Marcotte, P. (1985) On the relationship between Nash–Cournot and Wardrop equilibrium. *Networks*, **15**, 295–308.

Hobbs, B. F. (1986) Network models of spatial oligopoly with an application to deregulation of electricity generation. *Operations Research*, **34**, 395–409.

Jeon, D. (1981) A study of stability conditions for the imperfect world wheat market: a duopoly simulation model. *Journal of Rural Development*, **4**, 37–53.

Josephy, N. H. (1979) Newton's method for generalized equations, Technical Report 1965, Mathematics Research Center, University of Wisconsin, Madison.

Karamardian, S. (1969) The nonlinear complementary problem with applications, Part I. *Journal of Optimization Theory and Applications*, **4** (2), 87–98.

Karp, L. S. and McCalla, A. F. (1983) Dynamic games and international trade: an application to the world corn market. *American Journal of Agricultural Economics*, **65**, 641–56.

Kolstad, C. D. and Abbey, D. S. (1984) The effect of market conduct on international steam coal trade. *European Economic Review*, **24**, 39–59.

Kolstad, C. D. and Burris, A. E. (1986) Imperfectly competitive equilibria in international commodity markets. *American Journal of Agricultural Economics*, **68**, 27–36.

Kolstad, C. D. and Mathiesen, L. (1987) Necessary and sufficient conditions for uniqueness of a Cournot equilibrium. *Review of Economic Studies*, **54**, 681–90.

Kolstad, C. D. and Mathiesen, L. (1989) Computing Cournot–Nash equilibria, unpublished manuscript.

Lemke, C. E. (1965) Bimatrix equilibrium points and mathematical programming. *Management Science*, **11** 681–9.

MacKinnon, J. (1976) A technique for the solution of spatial equilibrium models. *Journal of Regional Science*, **16**, 293–307.

Mathiesen, L. (1985) Computational experience in solving equilibrium model by a sequence of linear complemetary problems. *Operations Research*, **33**(6), 1225–50.

Mathiesen, L. (1987) An algorithm based on a sequence of linear complementarity problems applied to Walrasian equilibrium model: an example. *Mathematical Programming*, **37**, 1–18.

Mathiesen, L. (1989) On the uniqueness of a Cournot equilibrium, Working Paper 53, Center for Applied Research, Norwegian School of Economics and Business Adminstration, Bergen.

Mathiesen, L., Roland, K. and Thonstad, K. (1987) The European natural gas market. Degrees of market power on the selling side, in *Natural Gas Markets and Contracts,* (eds. E. Golombet, M. Hoel and J. Vislie), North-Holland, Amsterdam.

McCalla, A. F. (1966) A duopoly market of world wheat pricing. *Journal of Farm Economics*, **65**, 65–73.

Mielke, K. D. and Griffith, G. R. (1983) Incorporating policy variables in a model of the world soybean/rapeseed market. *American Journal of Agricultural Economics*, **65**, 65–73.

Murphy, F. H., Sherali, H. D. and Soyster, A. L. (1982) A mathematical programming approach for determining oligopolistic market equilibria. *Mathematical Programming*, **24**, 92–106.

Nagurney, A. (1988) Algorithms for oligopolistic market equilibrium problems. *Regional Science and Urban Economics*. **18**, 425–45.

Okuguchi, K. (1983) The Cournot oligopoly and competitive equilibria as solutions to nonlinear complementarity problems. *Economics Letters,* **12**, 127–33.

Pang, J. S. and Chan, D. (1982) Iterative methods for variational and complementary problems. *Mathematical Programming*, **24**, 284–313.

Rausser, G. C., Lichtenberg, E. and Lattimore, R. (1982) Developments in theory and empirical application of endogenous government behaviour, *New Directions in Econometric Modeling and Forecasting in U. S. Agriculture* (ed. G. C. Rausser), North-Holland, Amsterdam, Chapter 18.

Salant, S. W. (1982) Imperfect competition in the international energy market: a computerized Nash–Cournot model. *Operations Research,* **30**, 252–80.

Samuelson, P. (1952) Spatial price equilibrium and linear programming. *American Economic Review*, **42**, 283–303.

Sarris, A. H. and Freebairn, J. (1983) Endogenous price policies and international wheat prices. *American Journal of Agricultural Economics*, **65**, 214–24.

Shei, S. Y. and Thompson, R. L. (1977) The impact of trade restrictions on price stability in the world wheat market. *American Journal of Agricultural Economics*, **59**, 628–38.

Sherali, H. D., Soyster, A. L. and Murphy, F. H. (1983) Stackelberg–Nash–Cournot equilibria: characterizations and computations *Operations Research*, **31**, 253–76.

Slade, M. E. (1988) What does an oligopoly maximize? Necessary and sufficient conditions for the equivalence between a Nash equilibrium and an optimization problem, Discussion Paper 88-35, Department of Economics, University of British Columbia, Vancouver.

Takayama, T. and Judge, G. G. (1971) *Spatial and Temporal Price and Allocation Models*, North-Holland, Amsterdam.

Thompson, R. L. (1981) A survey of recent U. S. developments in international agricultural trade models, US Department of Agriculture, Economics Research Service, Bibliography and Literature of Agriculture Report 21, Washington, DC, September.

3

Recent developments in spatial (temporal) equilibrium models: non-linearity, existence and other issues

TAKASHI TAKAYAMA AND T. GORDON MACAULAY

3.1 INTRODUCTION

Following the work by Paul A. Samuelson in 1952, Takayama and Judge (1964) successfully formulated the spatial equilibrium problem as a quadratic programming model and used the Wolfe (1959) simplex method to obtain solutions for a variety of standard spatial equilibrium models. Since then, a large number of applied research studies have been produced on the basis of this model. A clear advantage of the quadratic programming approach is that a solution can be obtained within a finite number of iterations. Because of this advantage, and others related to applied research requirements, the approach has been further developed by the use of linear complementarity programming in the past decade (Hashimoto, 1977; Nguyen, 1977; Whitacre, 1979; Takayama and Labys, 1986; Takayama and Hashimoto, 1988; Takayama *et al.*, 1988 etc.). For each of these models the existence of linear demand and/or linear supply functions for each region or nation has been maintained. Others including King and Ho (1974), MacKinnon (1976), Rowse (1981), Harker (1983), Dafermos and Nagurney (1984), Rathburn and Zwart (1985), Fox (1986) and Nagurney (1987), have used non-linear methods to solve various forms of the spatial equilibrium model.

Recently, MacAulay *et al.* (1988) presented a paper in which the formulation of a spatial equilibrium model with quadratic average costs of transformation was solved using the non-linear programming solver MINOS (Murtagh and Saunders, 1986). The formulation was the so-called 'net social revenue' maximization problem subject to a set of inequalities relating to the balance of supply and demand quantities and a set of spatial price equilibrium conditions.

In the first section of this paper a simplified spatial equilibrium model is presented which follows the work of Takayama and Judge (1971) and uses a primal–dual format and a net social revenue objective function. It is used to examine the special nature of the primal and dual solutions. Then, following the approach of MacAulay *et al.* (1988), the model structure is clarified in section 3.3. Characteristics of the solution, following the logic of section 3.2 are then presented. The results of section 3.3 indicate that the transport or transformation functions can be of any functional form as long as, at a solution set, the total transformation functions are concave with respect to X (vector of quantities shipped) and ρ (a vector of prices) in the neighbourhood of $\bar{X}$ and $\bar{\rho}$, at a solution set $(\bar{X}'\bar{\rho}') \geqslant 0$ if it exists.

In section 3.4 the condition of linearity on the demand and/or supply functions in the model is relaxed, in addition to the assumption, already made in section 3.3, that $T(X)$ (transformation cost function) is a function of X. This leads to an ultimate generalization of the standard Samuelson–Takayama–Judge spatial equilibrium model. The net social revenue maximization (primal–dual) formulation has two remarkable properties that have been anticipated but never shown concretely in the past. In sections 3.3 and 3.4 an example model and solution are presented to clarify these interesting and unique properties.

In section 3.5 an attempt is made to prove existence and uniqueness of the solutions for problems SM1, SM2, and GM1, GM2 (standard models and general models). The proof in one sense justifies the conclusion reached by Takayama in 1971 (Takayama and Judge, 1971) but more localized and sharper results are presented for problems SM2, GM1 and GM2.

In section 3.6 it is concluded that advancement in non-linear programming software and efficient solution procedures for non-linear problems is a crucial contributing factor for the development of bolder modelling efforts in the spatial and temporal equilibrium modelling field. We bring together in a consistent framework a number of non-linear spatial model formulations which should contribute to this modelling effort.

3.2 THE STANDARD SPATIAL EQUILIBRIUM MODEL AND A PRIMAL–DUAL FORMULATION

A single-commodity spatial equilibrium model takes the following format (standard model):

Problem SM1 Find $(\bar{y}'\bar{x}'\bar{X}') \geqslant 0'$ that maximizes

$$Z = \lambda' y - \tfrac{1}{2} y'\Omega y - \upsilon' x - \tfrac{1}{2} xHx - T'X \tag{3.1}$$

subject to

$$\begin{bmatrix} -I & 0 & G_y \\ 0 & I & G_x \end{bmatrix} \begin{bmatrix} y \\ x \\ X \end{bmatrix} \geqslant \begin{bmatrix} 0 \\ 0 \end{bmatrix} \tag{3.2}$$

and

$$(y'x'X') \geqslant 0'$$

where

$$\lambda = \begin{bmatrix} \lambda_1 \\ \lambda_2 \\ \vdots \\ \lambda_n \end{bmatrix} \quad v = \begin{bmatrix} v_1 \\ v_2 \\ \vdots \\ v_n \end{bmatrix} \quad T = \begin{bmatrix} t_{11} \\ t_{12} \\ \vdots \\ t_{nn} \end{bmatrix}$$

$$(n \times 1) \qquad (n \times 1) \qquad (n^2 \times 1)$$

$$\Omega = \begin{bmatrix} \omega_1 & & & 0 \\ & \omega_2 & & \\ & & \ddots & \\ & 0 & & \omega_n \end{bmatrix} \quad H = \begin{bmatrix} \eta_1 & & & 0 \\ & \eta_2 & & \\ & & \ddots & \\ & 0 & & \eta_n \end{bmatrix}$$

$$(n \times n) \qquad\qquad (n \times n)$$

$$G_y = \begin{bmatrix} 1 & & & 1 & & & & 1 & & \\ & 1 & & & 1 & & & & 1 & \\ & & \ddots & & & \ddots & \cdots & & & \ddots \\ & & 1 & & & 1 & & & & 1 \end{bmatrix}$$

$$(n \times n^2)$$

$$G_x = \begin{bmatrix} -1-1\ldots-1 & & & \\ & -1-1\ldots-1 & & \\ & & \ddots & \\ & & & -1-1\ldots-1 \end{bmatrix}$$

$$(n \times n^2)$$

$$y = \begin{bmatrix} y_1 \\ y_2 \\ \vdots \\ y_n \end{bmatrix} \quad x = \begin{bmatrix} x_1 \\ x_2 \\ \vdots \\ x_n \end{bmatrix} \quad X = \begin{bmatrix} x_{11} \\ x_{12} \\ \vdots \\ x_{nn} \end{bmatrix}$$

$$(n \times 1) \qquad (n \times 1) \qquad (n^2 \times 1)$$

If both Ω and H are positive definite ($\omega_1 > 0$ and $\eta_i > 0$ for all i), then Z is strictly concave with respect to y and x and just concave with respect to X, and therefore concave with respect to all the primal variables. It is easy to see that the Kuhn–Tucker (Slater or Arrow–Enthoven) constraint qualification is met for 3.2 (Mangasarian, 1969). Therefore, we can conclude that Problem SM1 is solvable for $(\bar{y}'\bar{x}'\bar{X}') \geqslant 0'$.

Defining the Lagrangian function for Problem SM1 as

$$\phi(y, x, X, \rho_y, \rho_x) = \lambda_y' - \tfrac{1}{2}y'\Omega y - v_x' - \tfrac{1}{2}x'Hx - T'X + (\rho_y'\rho_x')\begin{bmatrix} -I & 0 & G_y \\ 0 & I & G_x \end{bmatrix}\begin{bmatrix} y \\ x \\ X \end{bmatrix} \geqslant \begin{bmatrix} 0 \\ 0 \end{bmatrix} \tag{3.3}$$

we can derive the following optimality (necessary) conditions:

$$\frac{\partial\bar{\phi}}{\partial Q} = \begin{bmatrix} \lambda \\ -v \\ -T \end{bmatrix} + \frac{1}{2}\begin{bmatrix} \Omega + \Omega' & & \\ & H + H' & \\ & & 0 \end{bmatrix}\begin{bmatrix} \bar{y} \\ \bar{x} \\ \bar{X} \end{bmatrix} + \begin{bmatrix} -I & 0 \\ 0 & I \\ G_y' & G_x' \end{bmatrix}\begin{bmatrix} \bar{\rho}_y \\ \bar{\rho}_x \end{bmatrix} \leqslant \begin{bmatrix} 0 \\ 0 \\ 0 \end{bmatrix} \tag{3.4}$$

and

$$\left(\frac{\partial\bar{\phi}}{\partial Q}\right)'\bar{Q} = 0$$

$$\frac{\partial\bar{\phi}}{\partial Q} = \begin{bmatrix} -I & 0 & G_y' \\ 0 & I & G_x' \end{bmatrix}\begin{bmatrix} \bar{y} \\ \bar{x} \\ \bar{X} \end{bmatrix} \geqslant \begin{bmatrix} 0 \\ 0 \end{bmatrix} \tag{3.5}$$

and

$$\left(\frac{\partial\bar{\phi}}{\partial U}\right)'\bar{U} = 0$$

where

$$Q = \begin{bmatrix} y \\ x \\ X \end{bmatrix} \qquad U = \begin{bmatrix} \rho_y \\ \rho_x \end{bmatrix}$$

are both non-negative quantity and shadow price vectors such that

$$\rho_y = \begin{bmatrix} \rho_1 \\ \rho_2 \\ \vdots \\ \rho_n \end{bmatrix} \geqslant 0 \quad \text{and} \quad \rho_x = \begin{bmatrix} \rho^1 \\ \rho^2 \\ \vdots \\ \rho^n \end{bmatrix} \geqslant 0$$

which are shadow demand and supply price vectors, respectively.

An alternative formulation for Problem SM1 (presented by Plessner and Heady, 1965) takes the following net social revenue maximization format.

Problem SM2 Find $(\bar{y}'\bar{x}'\bar{X}'\bar{\rho}_y'\bar{\rho}_x') \geqslant 0'$ that maximizes

$$Z_a = (y'x'X'\rho_y'\rho_x')\left\{\begin{bmatrix}\lambda\\ -v\\ -T\\ 0\\ 0\end{bmatrix} - \begin{bmatrix}\Omega & 0 & 0 & I & 0\\ 0 & H & 0 & 0 & -I\\ 0 & 0 & 0 & -G_y' & -G_x'\\ -I & 0 & G_y & 0 & 0\\ 0 & I & G_x & 0 & 0\end{bmatrix}\begin{bmatrix}y\\ x\\ X\\ \rho_y\\ \rho_x\end{bmatrix}\right\} \tag{3.6}$$

subject to

$$\begin{bmatrix}-\lambda\\ v\\ T\\ 0\\ 0\end{bmatrix} + \begin{bmatrix}\Omega & 0 & 0 & I & 0\\ 0 & H & 0 & 0 & -I\\ 0 & 0 & 0 & -G_y' & -G_x'\\ -I & 0 & G_y & 0 & 0\\ 0 & I & G_x & 0 & 0\end{bmatrix}\begin{bmatrix}y\\ x\\ X\\ \rho_y\\ \rho_x\end{bmatrix} \geqslant \begin{bmatrix}0\\ 0\\ 0\\ 0\\ 0\end{bmatrix} \tag{3.7}$$

and

$$(y'x'X'\rho_y'\rho_x') \geqslant 0'.$$

By simple inspection, it is clear that 3.3 is the net social revenue, or total revenue $(\lambda - \Omega y)'y$ minus total production costs $(v + Hx)'x$ minus total transport costs TX.

It is clear, by reference to 3.4 and 3.5, that 3.7 is the combined optimality constraints for a set of perfectly competitive spatial equilibrium market conditions. In other words, problem SM2 provides the optimal quantity and price solutions that maximize net profit for the society as a whole (monopolistic behaviour). However, competitive market consumption–production–allocation (distribution) rules must be strictly adhered to in reaching the solution.

It is interesting to observe that if there is a set of solutions for Problem SM2 it should satisfy the following condition:

$$(\bar{y}', \bar{x}', \bar{X}', \bar{\rho}_y', \bar{\rho}_x', \bar{R}_y', \bar{R}_x', \bar{R}_X', \bar{S}_y', \bar{S}_x') = (\bar{y}', \bar{x}', \bar{X}', \bar{\rho}_y', \bar{\rho}_x'\bar{y}', \bar{x}', \bar{X}', \bar{\rho}_y', \bar{\rho}_x') \geqslant 0' \tag{3.8}$$

where $(\bar{R}_y', \bar{R}_x', \bar{R}_X', \bar{S}_y', \bar{S}_x') \geqslant 0'$ is the Lagrangian vector for 3.7, i.e. the primal solution vectors and the dual solution vectors are exactly the same.

Another interesting property of the primal solution vector is that the optimal net social revenue Z_a, is exactly zero. The economic implication of this result is that under a spatially competitive equilibrium any spatial rents are bid to zero. By inspection of the objective function, it is also true that the linear part and the non-linear (quadratic) part of the objective function are exactly the same. But the 'economic' implication of this solution characteristic is almost meaningless.

3.3 THE STANDARD SPATIAL EQUILIBRIUM MODEL WITH NON-LINEAR TRANSPORT COST

MacAulay *et al.* (1988) successfully introduced non-linearity into the transportation cost function of a spatial equilibrium model and showed that the resulting cubic programming problem could actually be solved by using the non-linear programming software known as MINOS (Murtagh and Saunders, 1986). In this section the approach of MacAulay *et al.* is followed, using a simplified model to show that exactly the same two characteristics that were discussed in the previous section are present in this model. A numerical solution for an example problem is provided.

The model will be referred to as the generalized model 1 (GM1) and takes the following form:

Problem GM1 Find $(\bar{y}', \bar{x}', \bar{X}', \bar{\rho}_y \bar{\rho}_x) \geqslant 0'$ that maximizes

$$Z_1 = (y'x'X'\rho_y'\rho_x') \left\{ \begin{bmatrix} \lambda \\ -v \\ -T(X) \\ 0 \\ 0 \end{bmatrix} - \begin{bmatrix} \Omega & 0 & 0 & I & 0 \\ 0 & H & 0 & 0 & -I \\ 0 & 0 & 0 & -G_y' & -G_x' \\ -I & 0 & G_y & 0 & 0 \\ 0 & I & G_x & 0 & 0 \end{bmatrix} \begin{bmatrix} y \\ x \\ X \\ \rho_y \\ \rho_x \end{bmatrix} \right\} \tag{3.9}$$

subject to

$$\begin{bmatrix} -\lambda \\ v \\ T(X) \\ 0 \\ 0 \end{bmatrix} + \begin{bmatrix} \Omega & 0 & 0 & I & 0 \\ 0 & H & 0 & 0 & -I \\ 0 & 0 & 0 & -G_y' & -G_x' \\ -I & 0 & G_y & 0 & 0 \\ 0 & I & G_x & 0 & 0 \end{bmatrix} \begin{bmatrix} y \\ x \\ X \\ \rho_y \\ \rho_x \end{bmatrix} \geqslant \begin{bmatrix} 0 \\ 0 \\ 0 \\ 0 \\ 0 \end{bmatrix} \tag{3.10}$$

and

$$(y'x'X'\rho_y'\rho_x') \geqslant 0'$$

where $T(X)$ is a function of X and in the case of MacAulay, *et al.* (1988) is given as a quadratic function (i.e. the total transport cost function, $T(X)'X$ is a cubic function, as is readily confirmed).

Defining the Lagrangian function of GM1 as ϕ_1, and deriving the Kuhn–Tucker necessary conditions, we have[1]

[1]Note the structure of the quadratic matrix in Z_1 Equation 3.9, for the structure of the second matrix (off-diagonal terms cancel). A more remarkable skew-symmetric form of this matrix is found in Equation 3.13 later in the chapter.

$$
\frac{\partial \bar{\phi}}{\partial Q} = \begin{bmatrix} \lambda \\ -v \\ -T(\bar{X}) - \left(\dfrac{\partial T(\bar{X})}{\partial X}\right)\bar{X} \\ 0 \\ 0 \end{bmatrix}
- \begin{bmatrix} \Omega + \Omega' & & & & \\ & H + H' & & & \\ & & 0 & & \\ & & & 0 & \\ & & & & 0 \end{bmatrix}
\times \begin{bmatrix} \bar{y} \\ \bar{x} \\ \bar{X} \\ \bar{\rho}_y \\ \bar{\rho}_x \end{bmatrix}
+ \begin{bmatrix} 0 \\ 0 \\ \left(\dfrac{\partial T(\bar{X})}{\partial X}\right)' \bar{R}_X \\ 0 \\ 0 \end{bmatrix}
+ \begin{bmatrix} \Omega' & 0 & 0 & -I & 0 \\ 0 & H' & 0 & 0 & I \\ 0 & 0 & 0 & G_y' & G_x' \\ I & 0 & -G_y & 0 & 0 \\ 0 & -I & -G_x & 0 & 0 \end{bmatrix}
\begin{bmatrix} \bar{R}_y \\ \bar{R}_x \\ \bar{R}_X \\ \bar{S}_y \\ \bar{S}_x \end{bmatrix} \leqslant \begin{bmatrix} 0 \\ 0 \\ 0 \\ 0 \\ 0 \end{bmatrix}
\tag{3.11}
$$

and $(\partial \bar{\phi}_1/\partial P)'P = 0$;

$$
\frac{\partial \bar{\phi}_1}{\partial D} = \begin{bmatrix} -\lambda \\ v \\ T(X) \\ 0 \\ 0 \end{bmatrix} + \begin{bmatrix} \Omega & 0 & 0 & I & 0 \\ 0 & H & 0 & 0 & -I \\ 0 & 0 & 0 & -G_y' & -G_x' \\ -I & 0 & G_y & 0 & 0 \\ 0 & I & G_x & 0 & 0 \end{bmatrix} \begin{bmatrix} y \\ x \\ X \\ \rho_y \\ \rho_x \end{bmatrix} \geqslant \begin{bmatrix} 0 \\ 0 \\ 0 \\ 0 \\ 0 \end{bmatrix}
\tag{3.12}
$$

and $(\partial \bar{\phi}_1/\partial D)'D = 0$; where

$$
P = \begin{bmatrix} y \\ x \\ X \\ \rho_y \\ \rho_x \end{bmatrix} \geqslant 0 \qquad D = \begin{bmatrix} R_y \\ R_x \\ R_X \\ S_y \\ S_x \end{bmatrix} \geqslant 0
$$

are the primal and dual variable vectors respectively (both of the same dimensions).

It is easy to confirm that

1. If the primal solution vector P turns out to be the same as the dual solution vector D, then 3.11 and 3.12 are exact replicates of each other, and
2. the objective function value turns out to be zero as expected.

Thus it has been shown that the above two characteristics that held for Problem SM2 hold for Problem GM1, even though the transport cost function in the objective function is not a linear function of X.

The example of Problem GM1 given in the Appendix confirms that a cubic programming spatial equilibrium problem is now easily solvable as a result of the recent advances in non-linear programming computer software.

3.4 NON-LINEAR DEMAND, SUPPLY AND TRANSPORT COST FUNCTIONS

In much applied work in international or inter-regional trade where market demand and/or supply functions and transport cost and/or transformation (processing) cost functions are important, it may be too simplistic to linearize these functions and apply the Takayama–Judge type of quadratic programming model.[2]

In this section the functional form restrictions of the previous sections are relaxed. It is assumed that the supply and demand functions and the transformation cost functions could be of a general form such as second- or higher-order polynomial functions, exponential functions of the Cobb–Douglas form, the constant elasticity of substitution form or a combination thereof. Under this generalized environment we now postulate our generalized spatial equilibrium model (GM2) as follows.

Problem GM2 Find $(\bar{y}'\bar{x}'\bar{X}'\bar{\rho}_y'\bar{\rho}_y'\rho_x') \geqslant 0'$ that maximizes

$$Z_2 = (y'x'X'\rho_y'\rho_x')\left\{\begin{bmatrix} f(y) \\ -g(x) \\ -T(X) \\ 0 \\ 0 \end{bmatrix} - \begin{bmatrix} 0 & 0 & 0 & I & 0 \\ 0 & 0 & 0 & 0 & -I \\ 0 & 0 & 0 & -G_y' & -G_x' \\ -I & 0 & G_y & 0 & 0 \\ 0 & I & G_x & 0 & 0 \end{bmatrix}\begin{bmatrix} y \\ x \\ X \\ \rho_y \\ \rho_x \end{bmatrix}\right\} \tag{3.13}$$

subject to

[2]How important such non-linearities are in practice has not yet been extensively investigated. However, in the near future we hope to be able to report on this issue in more concrete terms.

$$\begin{bmatrix} -f(y) \\ g(x) \\ T(X) \\ 0 \\ 0 \end{bmatrix} + \begin{bmatrix} 0 & 0 & 0 & I & 0 \\ 0 & 0 & 0 & 0 & -I \\ 0 & 0 & 0 & -G_y' & -G_x' \\ -I & 0 & G_y & 0 & 0 \\ 0 & I & G_x & 0 & 0 \end{bmatrix} \begin{bmatrix} y \\ x \\ X \\ \rho_y \\ \rho_x \end{bmatrix} \geqslant \begin{bmatrix} 0 \\ 0 \\ 0 \\ 0 \\ 0 \end{bmatrix} \quad (3.14)$$

and

$$(y'x'X'\rho_y'\rho_x') \geqslant 0'$$

where $f(y)$ is a Marshallian (indirect) market demand function vector, $g(x)$ is a Marshallian market supply function vector and $T(X)$ is a transport cost function vector.

By using the same reasoning as in the previous section, and after representing the Lagrangian function for GM2 as ϕ_2, we can derive the Kuhn–Tucker necessary conditions as follows:

$$\frac{\partial\bar{\phi}_2}{\partial P} = \begin{bmatrix} f(\bar{y}) + \left(\dfrac{\partial f(\bar{y})}{\partial y}\right)'\bar{y} \\ -g(\bar{x}) - \left(\dfrac{\partial g(\bar{X})}{\partial x}\right)'\bar{x} \\ -T(\bar{X})\left(\dfrac{\partial T(\bar{X})}{\partial X}\right)'\bar{X} \\ 0 \\ 0 \end{bmatrix} + \begin{bmatrix} -\left(\dfrac{\partial f(\bar{y})}{\partial y}\right)'\bar{R}_y \\ \left(\dfrac{\partial g(\bar{x})}{\partial x}\right)'\bar{R}_x \\ \left(\dfrac{\partial T(\bar{X})'}{\partial X}\right.\bar{R}_X \\ 0 \\ 0 \end{bmatrix}$$

$$+ \begin{bmatrix} 0 & 0 & 0 & -I & 0 \\ 0 & 0 & 0 & 0 & I \\ 0 & 0 & 0 & G_y' & G_x' \\ I & 0 & -G_y & 0 & 0 \\ 0 & -I & -G_x & 0 & 0 \end{bmatrix} \begin{bmatrix} \bar{R}_y \\ \bar{R}_x \\ \bar{R}_X \\ \bar{S}_y \\ \bar{S}_x \end{bmatrix} \leqslant \begin{bmatrix} 0 \\ 0 \\ 0 \\ 0 \\ 0 \end{bmatrix}. \quad (3.15)$$

and $(\partial\bar{\phi}_2/\partial P)'\bar{P} = 0$;

$$\frac{\partial\bar{\phi}_2}{\partial D} = \begin{bmatrix} -f(\bar{y}) \\ g(\bar{x}) \\ T(\bar{X}) \\ 0 \\ 0 \end{bmatrix} + \begin{bmatrix} 0 & 0 & 0 & I & 0 \\ 0 & 0 & 0 & 0 & -I \\ 0 & 0 & 0 & -G_y' & -G_x' \\ -I & 0 & G_y & 0 & 0 \\ 0 & I & G_x & 0 & 0 \end{bmatrix} \begin{bmatrix} \bar{y} \\ \bar{x} \\ \bar{X} \\ \bar{\rho}_y \\ \bar{\rho}_x \end{bmatrix} \geqslant \begin{bmatrix} 0 \\ 0 \\ 0 \\ 0 \\ 0 \end{bmatrix} \quad (3.16)$$

and $(\partial\bar{\phi}_2/\partial D)'\bar{D} = 0$. By inspection it is clear that

$$P = D = \begin{bmatrix} \bar{y} \\ \bar{x} \\ \bar{X} \\ \bar{\rho}_y \\ \bar{\rho}_x \end{bmatrix} = \begin{bmatrix} \bar{R}_y \\ \bar{R}_x \\ \bar{R}_X \\ \bar{S}_y \\ \bar{S}_x \end{bmatrix} \geqslant 0 \tag{3.17}$$

which, if attainable will satisfy Equations 3.15 and 3.16 (and of course the non-negativity requirement in 3.17) and are a solution for Problem GM2. The question of attainability will be investigated in the next section.

The reason for the two peculiar solution characteristics mentioned above rests on the facts that (a) the quadratic matrix in 3.13 is skew-symmetric and (b) exactly the same (except for sign) expression appears in both the second term of Equation 3.15 (the per unit demand and supply price vectors and transport cost vector) and the left-hand side of inequality 3.14.

If $f(y)'y$, $-g(x)'x$ and $-T(X)'X$ are all strictly concave at y, x and X then their solution vectors must be unique. Because of the Kuhn–Tucker necessary conditions and 3.17 it can be concluded that all primal solution vectors $\bar{P}$ are unique.

Example GM2 in the Appendix deals with a single-commodity international trade case with constant elasticity demand functions with elasticity 0.7 and 0.5 in two regions, linear upward-sloping supply functions and linear transport cost functions. The efficiency of the MINOS software in solving the problem is recorded for those using different software packages.

3.5 EXISTENCE AND UNIQUENESS OF A SPATIAL EQUILIBRIUM SOLUTION

Existence of a solution for Problem SM1 can be restated (Takayama and Judge, 1971) as there exists y, x and X such that

$$0 < \hat{y}_i < \hat{x}_{ii} < \hat{x}_i \text{ for all } i \text{ and } \hat{x}_{ij} \text{ for } i \neq j \tag{3.18}$$

If $x_i = 0$ then this condition is too strict. If $x_i = 0$ for some i, then immediately 3.18 does not hold and this causes a problem for the existence condition. The argument given below will rectify this problem. A conclusion about the uniqueness of $(\bar{y}, \bar{x}, \bar{X})$ cannot be made, however, because of the possibility of linearity (non-strict concavity) of TX.

For proof of the existence and uniqueness of a solution of Problems SM2, GM1 and GM2, we start by investigating the most general case in this class of modelling, i.e. GM2. The existence depends firstly on a non-empty feasibility set (3.14). This can be checked, among other ways, by investigating the following economic logic. First, confirm that there is a price, say $\tilde{p}$, common to all i, say all n nations, i.e.

$$f_i(\tilde{y}_i) = g_1(\tilde{x}_i) = \tilde{p}_i \text{ for all } i \tag{3.19}$$

for which

$$\sum_{i=1}^{n} \tilde{y}_i < \sum_{i=1}^{n} \tilde{x}_i, \ \tilde{y}_i \geqslant 0, \ \tilde{x}_i \geqslant 0 \text{ for all } i \tag{3.20}$$

For the one-commodity two-country trade case, $\tilde{\tilde{p}}$ is the price at which

$$\sum_{i=1}^{n} \tilde{y}_i = \sum_{i=1}^{n} \tilde{x}_i \tag{3.21}$$

holds, i.e. the trade balances at zero transport cost between the two countries. Any price higher than $\tilde{\tilde{p}}$, say $\tilde{p}$, will satisfy 3.19 and 3.20 as long as the demand functions are monotone decreasing and the supply functions are monotonically increasing. In the jargon of economics, there must be an international price (conceptually a common price) at which the trading world as a whole generates excess supply of the commodity. That is, in the case of two regions, inequality 3.20 holds as

$$\sum_{i=1}^{2} \tilde{y}_i - \sum_{i=1}^{2} \tilde{x}_i < 0 \tag{3.22}$$

Next, assume that

$$T(X) \geqslant 0 \text{ for } X \geqslant 0 \tag{3.23}$$

This assumption, in accepted practical terms, is that the transport cost per unit of any internationally traded commodity is non-negative. Since $\tilde{p}_y = \tilde{p}_x$ it is easy to confirm that

$$T(\tilde{X}) - G_y'\tilde{p}_y - G_x'\tilde{p}_x = T(\tilde{X}) \geqslant 0 \tag{3.24}$$

by inequality 3.23. The final checking must be made with the following inequality:

$$\begin{bmatrix} -I & 0 & G_y \\ 0 & I & G_x \end{bmatrix} \begin{bmatrix} \tilde{y} \\ \tilde{x} \\ \tilde{X} \end{bmatrix} \geqslant \begin{bmatrix} 0 \\ 0 \end{bmatrix} \tag{3.25}$$

for which the two rows reflect quantity balance inequalities and $\tilde{y}$ and $\tilde{x}$ are already known from 3.19 and 3.20. By employing the linear programming transport cost minimization (LPTCM) problem formulation which is known to have at most $2n - 1$ non-negative solutions for any $T \geqslant 0$, existence of a solution can be demonstrated.

Problem LPTCM Find $\tilde{X}$ that minimizes

$$Z_T = T'X \text{ (for some } T \geqslant 0) \tag{3.26}$$

subject to

$$\begin{bmatrix} \tilde{y} \\ \tilde{x} \end{bmatrix} + \begin{bmatrix} G_y \\ G_x \end{bmatrix} X \geqslant \begin{bmatrix} 0 \\ 0 \end{bmatrix} \tag{3.27}$$

$$X \geqslant 0$$

We can therefore state that the feasibility set 3.14 is non-empty and, as a result, GM2 has a solution.

By strengthening functional assumptions on $f(y)$, $g(x)$ and $T(X)$, it is possible to conclude that both Problem SM2 and Problem GM1 have solutions. *A fortiori*, the solution for Problem SM2 is a maximizing solution.

Uniqueness of solutions for Problem SM2, GM1 and GM2 can be investigated in the following way. First, Problem SM2 consists of strictly concave functions $f(y)'y$ and $-g(x)'x$. Therefore $\bar{y}$ and $\bar{x}$, and consequently $\bar{\rho}_y$ and $\bar{\rho}_x$ must be unique. However, $\bar{X}$ need not necessarily be unique as the LPTCM problem above clearly points to the rather uncomfortable (or realistic) fact that a solution for any linear programming problem may not be unique. However, this general non-uniqueness should not be confused with the inherent non-uniqueness of a dual solution for the LPTCM problem.

Next, if Problem GM1 has a solution $P \equiv (\bar{y}'\bar{x}'\bar{X}'\bar{\rho}_y'\bar{\rho}_x') \geqslant 0'$, and if

$$f(y)'y \qquad -(q(x)'x) \qquad -T(X)'X \tag{3.28}$$

are each strictly concave functions in the small neighbourhood of $(\bar{y}'\bar{x}'\bar{X}) \geqslant 0'$, then $(\bar{y}'\bar{x}'\bar{X}')$ and consequently $(\bar{\rho}_y'\bar{\rho}_x')$ are both unique, as in the case of Problem SM2, and $(\bar{y}'\bar{x}'\bar{X}'\bar{\rho}_y'\bar{\rho}_x') \geqslant 0'$ is a maximizing solution.

3.6 CONCLUSION

In this paper, further generalization of the standard Samuelson–Takayama–Judge spatial equilibrium model is carried out. Restrictions on the functional form of the demand, supply and transformation cost functions are relaxed in such a way that any functional forms consistent with accepted economic theory and econometric estimation methods can be used in the new generalized modelling framework (Problems GM1 and GM2).

Existence of non-empty feasibility sets for Problems SM2, GM1 and GM2 is established under rather mild restrictions. Uniqueness of a solution, if it exists, is then stated for the small neighbourhood of a solution. It should be understood clearly that global existence and uniqueness of a maximizing solution for a general non-linear programming problem is almost impossible to establish for problems such as GM1 and GM2.

A non-linear programming software package known as MINOS was used to solve some example problems. One outstanding aspect of the solutions is that the primal solution must be exactly the same as the dual solution.[3] Consequently, because of the structure of the models, the optimal net social revenue (social revenue less social cost less social transport cost) turns out to be zero—a characteristic of a perfectly competitive market.

The above models can be further modified to accommodate more realistic economic environments surrounding present-day international trade situations. The production transformation or processing cost functions implemented by MacAulay *et al.* (1988) are one example, and the use of non-linear processing costs in oil refinery and allocation models, as presented by Takayama and Judge (1971) and Takayama and Labys (1986), is another interesting future direction of research.

In this paper the price formulation has been intentionally avoided. However, by following the traditional procedure employed in Takayama and Judge (1971) the price formulation equivalent to Problems SM1, SM2, GM1 and GM2 can be developed without undue difficulty.

ACKNOWLEDGEMENTS

T. Takayama would like gratefully to acknowledge partial support from the CTEC Special Research Fund and the Japan Foundation Research Fund. T. G. MacAulay wishes to acknowledge continuing support and encouragement from the Australia Meat and Livestock Corporation and the Australian Wool Research Trust Fund.

REFERENCES

Dafermos, S. and Nagurney, A. (1984) Sensitivity analysis for the general spatial economic equilibrium problem. *Operations Research,* **32**(5), 1069–86.

Fox, G. (1986) *Mathematical Programming with MINOS: A Handbook for Economists,* AEB/86/9, Department of Agricultural Economics and Business, University of Guelph, Guelph.

Harker, P. T. (1983) A generalized spatial price equilibrium model. *Papers of the Regional Science Association,* **54**, 25–42.

Hashimoto, H. (1977) World food projection models, Ph.D. Thesis, University of Illinois.

King, R. A. and Ho, F-S. (1974) Reactive programming: a market simulating spatial equilibrium algorithm, Economics Research Report 21, Department of Economics, North Carolina State University at Raleigh, Raleigh.

[3]From the point of view of computational efficiency it is suggested that a mechanism be installed in the algorithm used for the solution of primal–dual problems to check that $P = D$ so as to obtain a more accurate solution. However, a significant deviation of the objective function value from zero can be used to indicate problems with the formulation so that such a condition should not be imposed until the final computational step.

MacAulay, T. G., Batterham, R. L. and Fisher, B. S. (1988) Cubic programming and the solution of spatial trading systems, Paper presented to the Annual Conference of the Australian Agricultural Economics Society, La Trobe University, Melbourne, February.

MacKinnon, J. G. (1976) A technique for the solution of spatial equilibrium models. *Journal of Regional Science,* **16**(3), 293–307.

Mangasarian, O. L. (1969) *Nonlinear Programming*, McGraw-Hill, New York.

Murtagh, B. A. and Saunders, M. A. (1986) MINOS 5.1 user's guide, Technical Report SOL 83-20R, Systems Optimization Laboratory, Department of Operations Research, Stanford University, Stanford, CA.

Nagurney, A. (1987) Computational comparisons of spatial price equilibrium methods. *Journal of Regional Science,* 27(1), 55–76.

Nguyen, H. D. (1977) World food projection models and short-run world trade and reserve policy evaluation, Ph. D. Thesis, University of Illinois.

Plessner, Y. and Heady, E. O. (1965) Competitive equilibrium solutions with quadratic programming. *Metroeconomica,* **17**, 117–30.

Rathburn, J. P. and Zwart, A. C. (1985) The sandwich algorithm for spatial equilibrium analysis, Discussion Paper 92, Agricultural Economics Research Unit, Lincoln College, Canterbury.

Rowse, J. (1981) Solving the generalized transportation problem. *Regional Science and Urban Economics,* **11**, 57–68.

Samuelson, P. A. (1952) Spatial price equilibrium and linear programming. *American Economic Review,* **42**(3), 283–303.

Takayama, T. and Hashimoto, H. (1988) A comparative study of linear complementary programming models and linear programming models in multi-region investment analysis: aluminium and bauxite, in *Quantitative Methods for Market-Oriented Economic Analysis over Space and Time* (eds T. Takayama *et al.*), Gower, London.

Takayama, T., Hashimoto, H., Nguyen, D. H. and Whitacre, R. C. (1988) Application of the spatial and temporal price allocation models to the world food economy, in *Quantitative Methods for Market-Oriented Economic Analysis over Space and Time* (eds T. Takayama *et al.*) Gower, London.

Takayama, T. and Judge, G. G. (1964) Equilibrium among spatially separated markets. *Econometrica*, **32**(4), 510–24.

Takayama, T. and Judge, G. G. (1971) *Spatial and Temporal Price Allocation Models,* North-Holland, Amsterdam.

Takayama, T. and Labys, W. (1986) Spatial equilibrium analysis, in *Handbook of Regional and Urban Economics* (ed. P. Nijkamp), North-Holland, Amsterdam, Chapter 5.

Whitacre, R. C. (1979) An evaluation of Japanese agricultural trade policies with a multiregion–multicommodity model, Ph. D. Thesis, University of Illinois.

Wolfe, P. (1959) The simplex method for quadratic programming. *Econometrica,* 27(3), 382–98.

APPENDIX

In this appendix, solutions for sample problems based on Problems GM1 (similar to that reported by MacAulay *et al.* (1988)) and GM2 are presented using MINOS as the solution algorithm.

Example GM1 (linear supply and demand functions and quadratic transport cost functions).
The demand and supply functions were assumed to be

$$\rho_1 = 20.0 - 0.2y_1 \quad \rho_2 = 20.0 - 0.2y_2 \quad \rho_3 = 20.0 - 0.125y_3$$

$$\rho^1 = 5.0 + 0.1x_1 \quad \rho^2 = 2.5 + 0.05x_2 \quad \rho^3 = 5.0 + 0.1x_3$$

where y_1, y_2 and y_3 are the quantities demanded in regions 1, 2 and 3 at prices ρ_1, ρ_2 and ρ_3 and x_1, x_2 and x_3 are the quantities supplied in regions 1, 2 and 3 at prices ρ^1, ρ^2 and ρ^3. For the transport costs a quadratic cost function was assumed to apply to the trade flows between each of the regions as follows:

$$t_{ij} = 3 - 0.07x_{ij} + 0.0014x_{ij}^2$$

where t_{ij} is the transport cost from region i to region j and x_{ij} is the shipment of the good concerned between the two regions. The programming tableau used is given in Table 3.A1.

The solution to this problem was obtained using MINOS set with initial bounds of 40 on the supply and demand quantities and 20 on the trade flows. The objective function value was very close to zero at -9.09×10^{-13}. The solution values obtained were as follows.

$x_{11} = 50.0$	$x_{12} = 0.0$	$x_{13} = 0.0$
$x_{21} = 0.0$	$x_{22} = 62.90786$	$x_{23} = 35.46072$
$x_{31} = 0.0$	$x_{32} = 0.0$	$x_{33} = 46.96626$
$\rho_1 = 10.0$	$\rho_2 = 7.41843$	$\rho_3 = 9.69663$
$\rho^1 = 10.0$	$\rho^2 = 7.41843$	$\rho^3 = 9.69663$
$y_1 = 50.0$	$y_2 = 62.90786$	$y_3 = 82.42699$
$x_1 = 50.0$	$x_2 = 98.36858$	$x_3 = 46.96626$

The per unit transport cost for the shipment between regions 2 and 3 was 2.2782.

The solution to this problem was obtained on a Macintosh SE with a 20 mbyte hard disk with MINOS compiled under MS FORTRAN 77 and was given in 43 s ignoring the compile and link times for the FORTRAN subroutines involved. It should be noted that there is considerable overhead in the setting up of such a small problem.

Example GM2 (exponential supply and demand functions, fixed per unit transport costs)

For this sample problem constant elasticity demand and supply functions were combined with fixed per unit transport costs. The demand and supply

Table 3A.1 Tableau for sample Problem GM1

Minimize	Y 1	Y 2	Y 3	X 1	X 2	X 3	X 1 1	X 1 2	X 1 3	X 2 1	X 2 2	X 2 3	X 3 1	X 3 2	X 3 3	D P 1	D P 2	D P 3	S P 1	S P 2	S P 3	R H S
$RY1$	0.2																					
$RY2$		0.2																				
$RY3$			0.125																			
$RX1$				0.1																		
$RX2$					0.05																	
$RX3$						0.1																
$RX11$							0															
$RX12$								$T(X12)$														
$RX13$									$T(X13)$													
$RX21$										$T(X21)$												
$RX22$											0											
$RX23$												$T(X23)$										
$RX31$													$T(X31)$									
$RX32$														$T(X32)$								
$RX33$															0							
−20	5			0			3			3												
OBJ		−20			2.5			3			0			3								
			−20			5			3			3			0							

*RY*1	1						−1			−1			−1									≤ 0
*RY*2		1						−1			−1			−1								≤ 0
*RY*3			1						−1			−1			−1							≤ 0
*RX*1				−1			1	1	1													≤ 0
*RX*2					−1					1	1	1										≤ 0
*RX*3						−1							1	1	1							≤ 0
*RX*11							0									1			−1			≤ 0
*RX*12								$-T$									1		−1			≤ 3
*RX*13									$-T$									1	−1			≤ 3
*RX*21										$-T$						1				−1		≤ 3
*RX*22											0						1			−1		≤ 0
*RX*23												$-T$						1		−1		≤ 3
*RX*31													$-T$			1					−1	≤ 3
*RX*32														$-T$			1				−1	≤ 3
*RX*33															0			1			−1	≤ 0
*RDP*1	−0.2															−1						≤ −20
*RDP*2		−0.2															−1					≤ −20
*RDP*3			−0.125															−1				≤ −20
*RSP*1				−0.1															1			≤ 5
*RSP*2					−0.05															1		≤ 2.5
*RSP*3						−0.1															1	≤ 5

$T(X_{ij}) = -0.07X_{ij} + 0.0014X_{ij}^2$ and similarly for T in the lower part of the table.

functions used were

$$\rho_1 = 120y_1^{-0.7} \qquad \rho_2 = 100y_2^{-0.5} \qquad \rho_3 = 110y_3^{-0.6}$$

$$\rho^1 = 6.01425x_1^{0.6} \qquad \rho^2 = 10.91089x_1^{0.6} \qquad \rho^3 = 2.66083x_1^{0.6}$$

The supply functions were derived so that for a given set of demand quantities (y_i) a specific and known set of supply quantities (x_i) would result. More digits than those shown were used in the model so as to obtain an objective value as close to zero as possible.

The transportation costs assumed were as follows.

$$\begin{array}{lll} t_{11} = 0 & t_{12} = 8.0 & t_{13} = 9.0 \\ t_{21} = 8.0 & t_{22} = 0 & t_{23} = 7.20354925 \\ t_{31} = 9.0 & t_{32} = 7.20354925 & t_{33} = 0 \end{array}$$

To use MINOS gradient functions had to be derived for the objective function and for each of the non-linear constraints. The tableau for the problem is provided in Table 3.A2.

The numerical solution values obtained were as follows.

$$\begin{array}{lll} x_{11} = 10.0 & x_{12} = 0.0 & x_{13} = 0.0 \\ x_{21} = 0.0 & x_{22} = 7.0 & x_{23} = 0.0 \\ x_{31} = 0.0 & x_{32} = 5.0 & x_{33} = 15.0 \\ \rho_1 = 23.94315 & \rho_2 = 28.86751 & \rho_3 = 21.66396 \\ \rho^1 = 23.94315 & \rho^2 = 28.86751 & \rho^3 = 21.66396 \\ y_1 = 10.0 & y_2 = 12.0 & y_3 = 15.0 \\ x_1 = 10.0 & x_2 = 7.0 & x_3 = 20.0 \end{array}$$

The objective function value for this problem was close to zero at 6.04×10^{-13}.

To solve the above problem a set of rather arbitrary lower bounds of 1.0 were provided for the demand and supply quantities. If these were not provided then the algorithm gave a non-optimal solution with zero prices and quantities. This is a typical result for non-linear programming models since uniqueness of solution can only be assumed in the neighbourhood of the optimum. It is therefore important in solving realistic problems to apply a reasonableness test to the results.

Table 5A.2 Tableau for sample Problem GM2

Minimize	Y 1	Y 2	Y 3	X 1	X 2	X 3	X 1 1	X 1 2	X 1 3	X 2 1	X 2 2	X 2 3	X 3 1	X 3 2	X 3 3	D P 1	D P 2	D P 3	S P 1	S P 2	S P 3	R H S
$RY1$	$-f(y1)$																					
$RY2$		$-f(y2)$																				
$RY3$			$-f(y3)$																			
$RX1$				$g(x1)$																		
$RX2$					$g(x2)$																	
$RX3$						$g(x3)$																
OBJ								8	9	8		7.20354	9	7.20354								
$RY1$	1						−1			−1			−1									≤ 0
$RY2$		1						−1			−1			−1								≤ 0
$RY3$			1						−1			−1			−1							≤ 0
$RX1$				−1			1	1	1													≤ 0
$RX2$					−1					1	1	1										≤ 0
$RX3$						−1							1	1	1							≤ 0
$RX11$																1			−1			≤ 0
$RX12$																	1		−1			≤ 8
$RX13$																		1	−1			≤ 9
$RX21$																1				−1		≤ 8
$RX22$																	1			−1		≤ 0
$RX23$																		1		−1		≤7.20354
$RX31$																1					−1	≤ 9
$RX32$																	1				−1	≤7.20354
$RX33$																		1			−1	≤ 0
$RDP1$	$f(y1)$															−1						≤ 0
$RDP2$		$f(y2)$															−1					≤ 0
$RDP3$			$f(y3)$															−1				≤ 0
$RSP1$				$g(x1)$															1			≤ 0
$RSP2$					$g(x2)$															1		≤ 0
$RSP3$						$-g(x3)$															1	≤ 0

$f(y_i)$ and $g(x_i)$ represent the demand and supply functions.

4

Shadow pricing for natural resource goods and services, using the emergy method

GONZAGUE PILLET

4.1 INTRODUCTION

This chapter deals with non-market commodity modelling—which sounds paradoxical. In reality, this leads to shadow pricing for natural resource goods and services, which is of paramount importance for ecological–economic analysis as well as commodity, financial and monetary markets and long-run economic performance. Indeed, economic theory knows how to integrate some natural resources, those that can be considered not so differently from usual market parameters and when related opportunity costs are present–future trade-offs. These natural resources can be exchanged on organized markets.

Yet, there are other natural resource goods and services that are never exchanged and have no market at all, although contributing to economic production of all kinds. They are environmental goods and services, e.g. sun, rain, topsoil and other environmental functions, which appear as indirect and even unrecognized energy inputs to economic processes and have been known as energy externalities. In other words, this statement recognizes that an important part of energy resources which contribute to the economic output are not considered by the price mechanism (and are thus external to the monetary circle). It follows that a misuse of these resources generally occurs which, in turn, may in a way imperil market equilibria.

The task in this chapter is not primarily to discuss the possible interactions between these market and non-market models, but to focus on a possible model for establishing a shadow price for the unpaid goods and services.

This special non-market pricing model is introduced in three steps. First, prerequisites are given in some detail with respect to the method used for this purpose, the 'emergy method'. Second, the accounting of environmental non-priced goods and services is presented. Finally, the exposé is illustrated by means of three short case studies taken from Switzerland, Louisiana, USA, and Japan, adding light and diversity to the theory.

4.2 METHODOLOGICAL PREREQUISITES

The analytical idea behind the emergy method, Howard T. Odum's conceptualization of the embodied-energy theory, is that at each step of an energy chain much of the energy is used in the transformation; only a small amount is converted into a higher quality of energy, i.e. into a more concentrated form which is capable of catalytic action when fed back. The ratio of one form of energy that is required to generate another form of energy by transformation is a measure of efficiency according to the first and second laws of thermodynamics under the maximum power principle (or maximum energy flux per unit time according to the Lotka–Odum autocatalytic characterization of living as well as man-made non-living processes). In other words, this ratio is a measure of energy quality in real systems when the latter tend to operate at that efficiency which produces a maximum power output. This ratio of one form (source) of energy required to develop another (high quality) form of energy by transformation has been called transformity (symbol Tr). The term transformity thus names a ratio describing the quality of a form of energy and its measurable ability to amplify as feedback relative to the source energy consumed in its formation, and under the maximum power principle.

Embodied energy (now, emergy; symbol C) is defined by Howard T. Odum as a way to measure the cumulative action of energies in chains or webs. It is the source energy required to produce other forms of energy. As a result, if different forms of energies are to be compared with respect to the energies required for their formation (or their effect), they can be converted into the same source emergy (i.e. into equivalents of the same form) by multiplying their actual energy content by their (source) transformity.

If the joule is the current unit for actual transformation work, it is not qualified for dimensioning emergy, the unit of which is the source-equivalent joule. This new unit has been called emjoule (symbol emJ). Thus, the analysis of ecological–economic subsystems requires that every energy form participating in the product be first evaluated in actual joules and then converted into emergy (emJ) by means of the appropriate (*in situ*) transformity.

Finally, a special ratio is used in ecological-economic systems which characterizes the period-to-period relationships between the emergy used by a country and its gross national product (GNP). This emergy per dollar ratio has been called monergy (symbol mE). It is used for calculating macro-prices as well as for calibrating human services by means of emergy units. It is defined by the global emergy used within the country (in emJ) divided by the GNP of the country (in \$); it is thus expressed as an emJ/\$ ratio.

4.3 ACCOUNTING FOR UNPAID PRICES

Macro prices or shadow prices (symbol P) for environmental goods and services (symbol I; otherwise called energy externalities) can be taken into account using the concepts of the emergy method given above.

Hypothesis Based on Odum, our general hypothesis is that the shadow price of environmental goods and services, per surface and period of time, is proportional in value to the GNP in this period, and the global emergy of these goods and services per surface and period of time is proportional to the global emergy used by the country within the same period. Thus

$$\frac{P_{\mathrm{I}}\,\mathrm{ha}^{-1}\,\mathrm{yr}^{-1}}{\$\mathrm{GNP\ of\ the\ country}} = \frac{C_{\mathrm{I}}^{*}\,\mathrm{ha}^{-1}\,\mathrm{yr}^{-1}}{C^{*}\ \mathrm{used\ by\ the\ country}} \tag{4.1}$$

where P is the shadow price, I is the environmental goods and services and C^* is the global emergy of resources used by the country. Also

$$P_{\mathrm{I}}\,\mathrm{ha}^{-1}\,\mathrm{yr}^{-1} = C_{\mathrm{I}}^{*}\,\mathrm{ha}^{-1}\,\mathrm{yr}^{-1} \times \frac{\$\mathrm{GNP}}{C^{*}\ \mathrm{used\ by\ the\ country}} \tag{4.2}$$

The GNP/C^* ratio is the reciprocal of the emergy/\$ ratio (or monergy), giving

$$P_{\mathrm{I}}\,\mathrm{ha}^{-1}\,\mathrm{yr}^{-1} = C_{\mathrm{I}}^{*}\,\mathrm{ha}^{-1}\,\mathrm{yr}^{-1} \times \mathrm{mE}^{-1} \tag{4.3}$$

That is, the shadow price of environmental goods per hectare per year is obtained by dividing their emergy by the monergy of the country. Note that this can be correct up to a 10% contribution from output Y (Fig. 4.1) to the national economy. If the dollar value of Y is more than 10% of the GNP, Y must be subtracted from mE in order to avoid double counting:

$$\mathrm{mE}' = \mathrm{mE} - (Y/\mathrm{GNP}) \tag{4.4}$$

(A 10% contribution from Y to the economy means a 1% correction of mE and is negligible.)

This model of emergy calculation can be used to calibrate human labour within environmental–economic subsystems with the necessary modifications.

In this respect (also cf. Fig. 4.1), the emergy of human labour is obtained

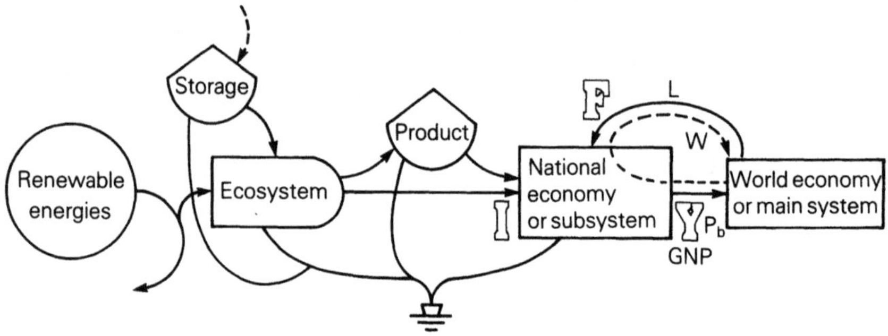

Fig. 4.1 Shadow-pricing environmental goods and services using the concept of emergy. The processing of an environmental value *I* by an economic subsystem leads to the emergy calibration of this value as an energy externality. In turn, this externality may be priced as the proportion that is environmental work in the nation's total work, both being given in energy units of the same quality, or emjoules.

by multiplying wages by the monergy of the subsystem:

$$\$W/\$P_b = C_L^*/C^*P_b \tag{4.5}$$

where W is wages, P_b is gross product, C^* is global emergy and L is human labour. Thus we obtain, for any subsystem,

$$C_L^* = \$W(C^*P_b/\$P_b) \tag{4.6}$$

i.e.

$$C_L^* = \$W \times \mathrm{mE_{subsystem}} \tag{4.6'}$$

and generally

$$C_L^*\mathrm{ha^{-1}\,yr^{-1}} = \$W\mathrm{ha^{-1}\,yr^{-1}} \times \mathrm{mE_{subsystem}} \tag{4.7}$$

Finally, the dollar value of I in any subsystem can be obtained as above.

4.4 CASE STUDIES

Using this evaluation procedure, let us consider the emergy analysis of the role of the environment and the measurement of energy externalities by means of the following three case studies:

1. vineyard cultivation and wine production in Switzerland;
2. sugar cane and sugar processing into ethanol in Louisiana, USA;
3. rice growing and saké making in Japan.

4.4.1 Emergy calibration

The Swiss vineyard totals 13 885 ha (less than 1% of the farming area in Switzerland). The must (or unfermented wine) yield is over 1 million hl per year; 75% of the area and 85% of the yield are located in Western French-speaking Switzerland. The data used are for 1986, mainly from the Lake of Geneva region.

The emergy analysis procedure is used to evaluate all the flows of the system initially in emergy units and then in dollar flows using the monergy of Switzerland.

Energy flows of the Swiss vineyard production and the processing of wine are portrayed in Fig. 4.2. Each pathway is calibrated in emjoules 1×10^{13} emJ $ha^{-1} yr^{-1}$. Calculations appear in Table 4.1.

First, the actual energy of each type of input is measured in joules (fourth column). Second, the transformity (emJ/J) of each category of flow e.g. grapes and wine in this study, is reported (or calculated) in the fifth column. Finally, the emergy (emJ per surface unit per unit time) of each energy flow is obtained as the product of the actual energy multiplied by the transformity attributed to the category under consideration (sixth column; the first three columns deal with energy forms, units and data). Emergy values are ultimately reported on the vineyard and wine energy diagram (Fig. 4.2). An aggregated three-armed diagram, calculated from data in Table 4.1 is used to help interpretation (Fig. 4.3).

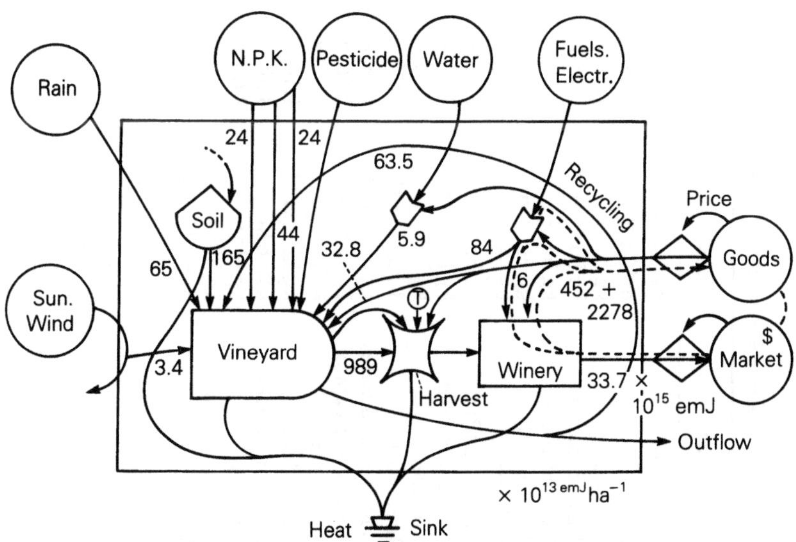

Fig. 4.2 Vineyard and wine energy diagram, Switzerland.

Table 4.1 Emergy flows in vineyard and wine production, Switzerland, Lake of Geneva region, 1986

	Energy form	*Units*	*Data*	*J ha^{-1} yr^{-1}*	*emJ J^{-1}*	*emJ($\times 10^{13}$)*
1	Direct sun	J m^{-2} yr^{-1}	3.44×10^{9}	3.44×10^{13}	1.00	3.4
2	Rain	m^{3} m^{-2} yr^{-1}	8.89×10^{-1}	4.36×10^{10}	1.50×10^{4}	65.3
3	Soil used up	kg ha^{-1}	1.17×10^{3}	2.64×10^{10}	6.24×10^{4}	165.0
4	Organic matter	kg ha^{-1} yr^{-1}	7.13×10^{2}	3.22×10^{9}	6.24×10^{4}	20.1
5	Recycled bines	kg ha^{-1}	1.50×10^{3}	1.02×10^{10}	6.24×10^{4}	63.5
6	Nitrogen (N)	kg ha^{-1} yr^{-1}	6.60×10	1.43×10^{8}	1.69×10^{6}	24.2
7	Potassium (K)	kg ha^{-1} yr^{-1}	1.32×10^{2}	9.27×10^{7}	2.62×10^{6}	24.3
8	Phosphate (P)	kg ha^{-1} yr^{-1}	6.70×10	1.07×10^{7}	4.14×10^{7}	44.4
9	Magnesium (Mg)	kg ha^{-1} yr^{-1}	3.00×10	3.78×10^{7}	2.00×10^{5}	0.8
10	Pesticide	kg ha^{-1} yr^{-1}	5.30×10	8.05×10^{8}	6.60×10^{4}	5.3
11	Direct fuels	1 ha^{-1} yr^{-1}	3.97×10^{2}	1.27×10^{10}	6.60×10^{4}	83.9
12	Machines and wires	kg ha^{-1} yr^{-1}	4.38×10^{2}	3.96×10^{7}	1.01×10^{7}	40.0
13	Services and capital	Fr ha^{-1} yr^{-1}	2.66×10^{4}		mE1.70×10^{11}	452.6
14	Must yield	hl ha^{-1} yr^{-1}	8.68×10	2.93×10^{10}	3.37×10^{5}	989.4
15	Sugar added	kg hl^{-1} yr^{-1}	1.50	2.19×10^{9}	8.39×10^{4}	18.4
16	Energy in winery	kW h hl^{-1}	8.50	2.66×10^{9}	1.59×10^{5}	42.2
17	Materials	kg hl^{-1}	8.00×10^{-1}	6.28×10^{3}	1.01×10^{7}	0.0
18	Cellulose filters	kg hl^{-1}	3.00×10^{-1}	3.24×10^{8}	1.57×10^{5}	5.1
19	Water	hl hl^{-1} must	2.00	8.50×10^{7}	1.50×10^{5}	1.3
20	Services	Fr ha^{-1} yr^{-1}	1.02×10^{3}		mE3.10×10^{11}	31.5
21	Capital	Fr ha^{-1} yr^{-1}	7.35×10^{4}		mE3.10×10^{11}	2278.1
22	Wine yield	hl ha^{-1} yr^{-1}	8.68×10	2.38×10^{10}	1.41×10^{6}	3366.0

mE, monergy.

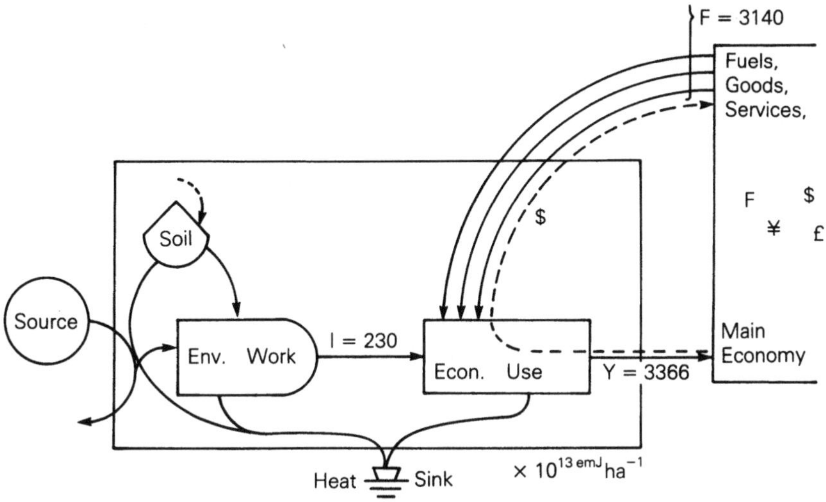

Fig. 4.3 Swiss vineyard and wine: aggregate energy diagram used for defining the externality ratio and other ratios.

The same procedure has been applied in case studies of the production of ethanol in Louisiana, USA, and saké in Japan (Odum and Odum, 1983; Pillet and Murota, in preparation). In particular, energy externalities clearly appear on the energy diagram in Fig. 4.3 (*I*). They are expressed in units of $\times 10^{13}$ emJ $\text{ha}^{-1}\,\text{yr}^{-1}$. The soil used is evaluated as 165×10^{13} emJ ha^{-1} for 1986. Together with water and wind, its work is indispensable for the continuity of the economic product.

4.4.2 Ratios and indices

The aggregated diagram in Fig. 4.3 allows us to define various ratios.

(a) *Net emergy yield ratio Y/F*

The net emergy yield ratio is the emergy *Y* of the output divided by the emergy *F* of the inputs to the process which is fed back from the economy. The results are as follows:

Swiss wine	1.04
Louisiana ethanol	1.05
Japanese saké	1.06

This measures the percentage of purchased inputs with respect to the product yield.

(b) *Emergy investment ratio F/I*

The emergy investment ratio is the ratio of the emergy F fed back from the economy to the emergy inputs I from the natural life-support system. The results are as follows:

Swiss wine	25
Louisiana ethanol	21
Japanese saké	17
US agriculture	25

This confirms the intensive use of high quality energy inputs in comparison with the use of low quality environmental inputs.

(c) *Energy externality ratio xE*

The energy externality ratio is the ratio of use that is free. It is given by $xE = I/(I + F)$. The results are as follows:

Swiss wine	0.04
Louisiana ethanol	0.04
Japanese saké	0.06

This is much more than economists usually attribute to environmental services, arguing that economic macroproducts are composed of as much as 99% of embodied human labour.

(d) *Macro prices P*

Environmental macro prices (or shadow prices) are calculated for typical energy externalities according to the monergy of the country.

With respect to the production of Swiss wine, energy externalities (i.e. indirect contributions from life-support systems) can be evaluated at SFr10 000 ha^{-1} in 1986 using the monergy of the free environmental contribution to the Swiss economy (0.24×10^{12} emJ/SFr) (Pillet and Odum, 1987). This is less than the annual capital flow per hectare (about SFr73 500), but very close to the flow of services spent per hectare in 1986 in vineyard and harvest (SFr12 800). In addition, water and soil in Switzerland, considered as energy externalities, contributed approximately 10% of the sum total of economic expenses in vineyard cultivation and wine production per hectare in 1986. With respect to the whole Swiss economy, energy externalities have been evaluated at 16% of the energy resources used by the country (Pillet and Odum, 1987). This calculation is currently applied temporally (1972–86) to vineyard cultivation and wine production in Geneva (the energy externality ratio shrank in-between). Another study is devoted to shadow pricing Swiss alpine waters in order to obtain accurate prices for these natural resource goods and services.

4.5 SUMMARY AND CONCLUSION

Ecological–economic analyses are concerned with the question of the environment-oriented implicit keys to the continuity of human societies, work and affairs. In particular, they point out the major importance of natural resource goods and services (water, wind and soil) in this respect, exploring the way macroeconomic processes use environmental resources. They tentatively evaluate the implicit contributions within interfaced environmental and economic systems by shadow pricing these energy externalities. An examination of the role of indirect environmental services conjointly with the inputs of human labour and economic goods and services is of immediate importance in order to avoid a misuse of these resources and to obtain a better understanding of the real basis of economic production processes.

ACKNOWLEDGEMENTS

The author thanks Howard T. Odum, Elisabeth C. Odum, Andrea Baranzini, Gaye Bristow, Steven G. M. Schilizzi, Maurice Villet, Gaston Gaudard, André Vifian, Takeshi Murota, Charles A. S. Hall and Fabrizio Carlevaro for critical review of earlier drafts of this chapter and/or much encouragement. Support by the Swiss National Science Foundation under Grant 1.378–0.86 and the University of Geneva Center for Human Ecology and Environmental Sciences is gratefully acknowledged.

REFERENCES AND FURTHER READING

Cleveland, C. J., Costanza, R., Hall, C. A. S. and Kaufmann, R. K. (1984) Energy and the U.S. economy: A biophysical perspective. *Science,* **225**, 890–7.

Lavine, M. J. and Butler, T. J. (1982) Use of Embodied Energy Values to Price Environmental Factors: Examining the Embodied Energy/Dollar Relationship, Final Report on NSF Award PRA-8003845, Center for Environmental Research and Department of Environmental Engineering, Cornell University, Ithaca, NY.

Murota, T. (1987) Environmental economics of the Water Planet Earth, in *Environmental Economics—The Analysis of a Major Interface* (eds G. Pillet and T. Murota), Leimgruber, Geneva, 185–99.

Odum, H. T. (1986) Emergy in ecosystems, in *Ecosystem Theory and Application* (ed. N. Polunin), Wiley, Chichester, 337–69.

Odum, H. T. and Odum, E. C. (1983) *Energy Analysis Overview of Nations,* WP-83-82, IIASA, Laxenburg, 479 pp.

Pillet, G. (1986) From external effects to energy externality: new proposals in environmental economics. *Hitotsubashi Journal of Economics,* **27**(1), 77–97

Pillet, G. (1987) Case-study of the role of environment as an energy externality in Geneva vineyard cultivation and wine production. *Environmental Conservation,* **14**(1), 53–8.

Pillet, G. (1988) Water, wind and soil: hidden keys to the Water Planet Earth and to economic macroprocesses. *The Energy Journal*, 9(1), 43–52.
Pillet, G. and Murota, T. (1987) *Environmental economics—The Analysis of a major interface*, Leimgruber, Geneva, Switzerland.
Pillet, G. and Odum, H. T. (1987) *E^S Energie, écologie, économie*, Georg, Geneva.
Pimentel, D. (1980) *Handbook of Energy Utilization in Agriculture*, CRC Press, Boca Raton, FL.

PART TWO

Application of New Methodologies to Particular Commodity Markets (Agricultural, Mineral and Energy Commodities)

5

The effectiveness of the world coffee agreement: a simulation study using a quarterly model of the world coffee market

FRANZ C. PALM AND BEN VOGELVANG

5.1 INTRODUCTION

In this chapter we present the results of a simulation analysis of a quarterly model for the world coffee market. The model has been developed by Palm and Vogelvang (1986). More details of this study can be found in Vogelvang (1988). In the model, producing and importing countries are assumed to maximize the expected utility of the present value of profits over a two-period time horizon by buying or selling on the spot market and by holding inventories, and by hedging or speculating on the futures market. Expectations are assumed to be rational, i.e. they are equal to the conditional expectation given the model and information up to the current period. The spot and futures markets clear at each time period. The model has been estimated for the period 1971–82, a period in which the quota system of the International Coffee Agreement (ICA) has almost never been effective.

The aim of the chapter is twofold. First, we give some insight into the behaviour of the model over the period of estimation. Second, and more importantly, we analyse the impact of a substantial increase in production on prices, disappearance and inventory formation and of several policy measures aimed at reducing an imbalance between demand and supply on the coffee market by decreasing production. These measures are analysed under the assumption (a) that there is no ICA and (b) that an international quota system has been agreed upon which becomes effective as soon as the spot market price drops below a certain level. Attention is also paid to the impact of the distribution of initial inventories over exporting and importing countries.

We show that it is possible to solve a medium-size model for an international commodity market assuming rational behaviour of the agents (countries) under uncertainty. The simulation results are of importance for the discussions about price stabilization through international agreements aimed at restricting trade by a quota system. Our findings lead to the conclusion that the current situation on the international coffee market, which is characterized by an excess of production compared with total world consumption, requires a substantial reduction in production in the coming years to bring supply and demand more into balance.

The chapter is organized as follows. The structure of the model is briefly outlined in section 5.2. The solution of the model under rational expectations is described in section 5.3. In section 5.4 the export quota system is discussed together with its consequences for the solution of the model. Section 5.5 contains some simulation results for the sample period. In section 5.6 the model implications of several policy measures aimed at reducing production are presented. Finally, concluding remarks are made in section 5.7.

5.2 THE MAIN FEATURES OF THE MODEL

The model is a short-term quarterly model in which the production is assumed to be predetermined. A quarterly model allows account to be taken of developments which take place within the year, e.g. the quarterly quota distribution. The model of the world coffee market has been elaborated along the lines of recent developments in the theoretical literature (e.g. Newbery and Stiglitz, 1981). A schematic summary of the theoretical model is given in Table 5.1. More details can be found in Palm and Vogelvang (1986). The list of variables is as follows:

i	country i	z_{it}	wholesale inventories
t	quarter t	z^r_{it}	retail inventories
q_{it}	production	cons_{it}	consumption
exp_{it}	exports	p_t	spot price
dis_{it}	disappearance	p^f_t	futures price
imp_{it}	imports	p^r_{it}	retail price
k_{it}	unit processing costs	cp_{it}	consumer price index
n_{it}	population	y_{it}	real disposable income
δ_i	discount factor	γ_i	$\gamma^*_i\delta^2_i$

γ^*_i	coefficient of constant absolute risk aversion
f_{it}	position on the futures market
b_i, c_i	parameters of the cost function for inventories

α_{ji}, β_i	constant parameters
var_i	variance conditional on information available at period t

At the microeconomic level we assume that market participants (individual countries in the empirical model) have access to the spot and futures markets and that they have a utility function with constant absolute risk aversion. They are assumed to take a position on these markets in such a way that the expected utility of the present value of profits for the present and next period is maximized. In this way we derive the optimal position for inventory holders at the wholesale and retail level. For price-taking inventory holders the price of storage equation 5.1 (e.g. Working, 1949) results from the two-period optimization model relating the size of inventories to the spread, i.e. to the difference between the futures and the spot price. Large producers are assumed to be price setters facing a downward-sloping demand curve. $P_{it} = \phi_{io} - \phi_{i1}\,\text{exp}_{it}(\ )$ where the coefficient ϕ_{io} and ϕ_{i1}, possibly depend on the situation on the coffee market at time t, but they are not controlled by the producer. They are able to influence the price level by varying the inventory level. For a price-setting producer, the maximization of the expected utility of the present value of profits over two periods leads to a relationship between inventories and the difference between the expected next-period spot price and the futures price, while the inventories of the other inventory holders also have an influence on his position. At the retail level, too, a price of storage equation results from arbitrage between the present and the next period. The retail inventories are related to the difference between the expected and the current retail price (5.2).

Disappearance (5.3) is equal to consumption and the change in retail inventories. Consumption per capita (5.4) is assumed to depend on the relative price of coffee with respect to the consumer price index and on per capita income. A semilogarithmic specification was chosen in order to force the income elasticity to decrease when consumption increases.

Exports by producers (5.5) are by definition equal to production minus disappearance minus the variation in wholesale inventories. For non-producing importing countries the same definition applies (5.6) with production zero.

The retail price is related to the spot market price through a cost function (5.7) where the unit costs k_{it} of roasting coffee are assumed to be proportional to the general price level, η_i denotes the profit margin (constant) and $\beta_i(L)$ is a polynomial in the lag operator L such that $\beta_i(1) = 1$.

Spot and futures prices adjust to clear the spot and futures markets at each period (see Equations 5.8 and 5.9), and expectations are assumed to be rational.

The following countries have been included in the model. On the production side, it concerns Brazil and Colombia (who are price setters) and

Table 5.1 A summary of the theoretical model

Country i	*Production* q_{it}	*Consumption* cons_{it} *Disappearance* dis_{it}	*Inventory* z_{it}	*Export* $\exp_{it}$	*Retail price* p^{r}_{it}
Producer (exporting country)	Predetermined	*Consumption* Predetermined *Disappearance*: $\mathrm{dis}_{it} = \mathrm{cons}_{it} + \nabla x^{r}_{it}$ (5.3)	*Wholesale* $z_{it} = \max\left(\dfrac{p^{f}_{t} - p_t - b_i}{2c_i}; \bar{z}_{it}\right)$ (5.1) (see below[a] for a price setter) *Retail* z^{r}_{it} predetermined	$\exp_{it} = q_{it} - \mathrm{dis}_{it} - \nabla z_{it}$ (5.5)	—
Importing country	$q_{it} = 0$	*Consumption* $\mathrm{cons}_{it} = \alpha_{0i}h_{it} + \alpha_{1i}\left(\dfrac{n_{it}p^{r}_{it}}{\mathrm{cp}_{it}}\right) + \alpha_{2i}n_{it}\ln\left(\dfrac{y_{it}}{n_{it}}\right)$ (5.4) *Disappearance* $\mathrm{dis}_{it} = \mathrm{cons}_{it} + \nabla z^{r}_{it}$	*Wholesale* $z_{it} = \max\left(\dfrac{p^{f}_{t} - p_t - b_i}{2c_i}; \bar{z}_{it}\right)$ (5.1) *Retail* $z^{r}_{it} = \max\left(\dfrac{\delta_i E_i p^{r}_{it+1} - p^{r}_{it} - b_i}{2c_i + \gamma_i \mathrm{var}_i(p^{r}_{it+1})}; \bar{z}^{r}_{it}\right)$ (5.2)	$\exp_{it} = -\mathrm{imp}_{it} = -\mathrm{dis}_{it} - \nabla z_{it}$ (5.6)	$p^{r}_{it} = (1 + \eta_i)[k_{it} + \beta_i(L)p_t]$ (5.7)

Market clearing	*Spot market*
	$$\sum_i \exp_{it} = 0 \leftrightarrow \sum_i (q_{it} - \text{cons}_{it} - \nabla z_{it} - \nabla z^{\mathrm{r}}_{it}) = 0 \qquad (5.8)$$
	Futures market
	$$\sum_i f_{it} = 0 \leftrightarrow \sum_i \left\{ \left[\frac{p^{\mathrm{f}}_t - \delta_i E_i p_{t+1}}{\gamma_i \, \text{var}_i (p_{t+1})} \right] + z_{it} + \frac{\text{cov}_i (p_{t+1} q_{it+1}, p_{t+1})}{\text{var}_i (p_{t+1})} \right\} = 0 \qquad (5.9)$$

[a]For individual countries:

$$z_{it} = \max \left\{ \frac{1}{2(\phi_{i1} + c_i)} \left[2\phi_{i1}(q_{it} + z_{it-1}) - \phi_{i1} f_{it-1} + p^{\mathrm{f}}_t - b_i - \phi_{i0} + \delta \frac{\partial E p_{t+1}}{\partial z_{it}} \frac{\delta E p_{t+1} - p^{\mathrm{f}}_t - \gamma_i \, \text{cov}_i (p_{t+1} q_{it+1}, p_{t+1})}{\gamma_i \, \text{var}_i (p_{t+1})} \right]; \bar{z}_{it} \right\}$$

groups of countries producing respectively Unwashed Arabicas, Other Milds, Colombian Milds and Robustas. On the consumption side, the USA and the European member countries of the International Coffee Organization (ICO) have been modelled individually. The specification for the wholesale inventories had to be extended by assuming a partial adjustment scheme in which the desired level of the inventories is modelled by Equation 5.1. Also, as the market for the various sorts of coffee has been modelled, the market-clearing conditions 5.8 and 5.9 are solved for the Composite Indicator Price 1968, (CIP68) (a weighted average of the spot prices) and the futures price of the New York market. Prices of the other sorts are assumed to be related to the price of Unwashed Arabicas and the New York futures price through an error correction model with a constant term reflecting the difference in quality between the various sorts of coffee. The general conclusion was that the estimation results are in fair agreement with the theoretical model.

5.3 SOLUTION OF THE MODEL

5.3.1 Solution when the export quotas are not effective

The entire model of the coffee market as specified and estimated in Palm and Vogelvang (1986) will be solved for the coffee price on the spot and futures markets and for the price expectations. The remaining endogenous variables will first be expressed in terms of prices and predetermined variables and will then be substituted in the market-clearing equations of the spot and futures markets. Equations 5.8 and 5.9 form the starting point from which the prices will be derived. Some assumptions have to be made to obtain operational equations to determine the price variable. These assumptions are as follows.

1. Price expectations are rational.
2. The conditional second moments of p_{t+1} are constant over time.
3. Given all available information at time t, q_{it+1} and p_{t+1} are independent.
4. Each country holds inventories at the wholesale level and has access to the futures market.

The rationale of assumption 2 is simplicity. On *a priori* grounds this assumption is not necessarily in agreement with the rational expectations hypothesis, but without the assumption the equations would be non-linear in the variables which would complicate the solution very much; for an example of a non-linear rational expectations model see Broze *et al.* (1986). As the simulation results appear to be plausible, there is little reason to abandon this assumption at the present stage of the research.

With assumptions 2–4, Equation 5.9 can be written as

$$v_1 p_t^f - v_2 E p_{t+1} + Z_t + EQ_{t+1} = 0 \tag{5.10}$$

with $v_1 = \Sigma_i \{1/[\gamma_i \operatorname{var}_i (p_{t+1})]\}$ and $v_2 = \Sigma_i \{\delta_i/[\gamma_i \operatorname{var}_i (p_{t+1})]\}$. Capital letters denote aggregated quantities, e.g. $Z_t = \Sigma_i z_{it}$. Equation 5.8 can be expressed as

$$Q_t + Z_{t-1} - \mathrm{DIS}_t - Z_t = 0 \qquad (5.11)$$

The variables Q_t and Z_{t-1} are predetermined in these equations. For the endogenous variables Z_t and DIS_t their components will be substituted. The world inventory Z_t is the aggregate of all the inventories of individual countries and is therefore related to futures and spot prices and predetermined variables.

The inventories can be eliminated now because Z_t is expressed as a linear function of p_t, p_t^f and Ep_{t+1} and an aggregate of predetermined variables originating from the inventory equations. Define S_t to be the sum of the above-mentioned aggregated predetermined variables and the expected production. Then Equation 5.10 can be written as

$$\alpha_0 + \alpha_1 p_t^f + \alpha_2 Ep_{t+1} + \alpha_3 p_t + \alpha_4 S_t = u_{t1} \qquad (5.12)$$

The disturbance term u_{t1} represents the aggregate of disturbances of the original model. Equation 5.1 expressed in terms of the CIP68 for simplicity rather than in terms of prices of the various types of coffee has been substituted for Z_t in 5.10 to obtain 5.12. The disturbance term is assumed to be normally independently distributed. After substitution of 5.7 for p_t^r and the expected value of 5.7 for Ep_{t+1}^r in cons_{it} (5.4) and z_{it}^r (5.2), the total disappearance DIS_t is expressed as a linear function of p_t, Ep_t and Ep_{t+1} and a number of aggregate predetermined variables from the disappearance and retail price equations, in addition to lagged values of the price variable. Let H_t be the sum of the aggregate predetermined variables in the disappearance, retail and inventory equations, plus $Q_t + Z_{t-1}$. Then Equation 5.11 can be written as

$$\beta_0 + \beta_1 p_t^f + \beta_2 Ep_{t+1} + \beta_3 Ep_t + \beta_4 p_t + \beta_5 p_{t-1} + \beta_6 H_t = u_{t2} \qquad (5.13)$$

The disturbance term u_{t2} has been introduced for the same reason as u_{t1} and is also assumed to be normally independently distributed.

The price expectations are assumed to be rational, i.e.

$$Ep_{t+1} = E(p_{t+1}|\Phi_{t-1}, \text{model}) \qquad Ep_t = E(p_t|\Phi_{t-1}, \text{model}) \qquad (5.14)$$

where Φ_{t-1} denotes the information available at time $t-1$, and the second moments are assumed to be constant.

The solution equations are now complete. The system in 5.12, 5.13 and 5.14 consists of four equations in the endogenous variables p_t, p_t^f, Ep_t and Ep_{t+1}.

When the parameters of Equations 5.12 and 5.13 have been estimated and expected prices have been computed using a solution method for rational expectation models, Equations 5.12 and 5.13 can be solved for p_t

and p^f_t. The system will be solved for CIP68 and the futures price in New York. Data on expected production are available. The US Department of Agriculture obtains and publishes production estimates in its *Foreign Agriculture Circular: World Coffee Situation*.

There are various methods to determine rational price expectations (see, for example, Blanchard and Kahn, 1980). A useful treatment for linear rational expectation models can be found in Chow (1983, pp. 356–61). His solution method has been applied here. More specifically, we estimate the parameters of 5.12 and 5.13 by the method of instrumental variables using p^f_{t-1} and p_{t+1} as proxies for Ep_t and Ep_{t+1} respectively and lagged prices as instruments. Then by eliminating p^f_t from 5.12 and 5.13, we obtain the reduced form equation for p_t, which expresses p_t in terms of p_{t-1}, Ep_t, Ep_{t+1} and the predetermined variable $x_t = \alpha_4\alpha_1^{-1}S_t - \beta_6\beta_1^{-1}H_t$. The associated final form equation for p_t is a dynamic regression model in which p_t is explained by its own lagged values and by the truly exogenous part of x_t, denoted by $x^*_t = \alpha_4\alpha_1^{-1}S^*_t - \beta_6\beta_1^{-1}H^*_t$, where S^*_t and H^*_t are the exogenous parts of S_t and H_t respectively. Notice that x^*_t is the only exogenous variable that varies in the simulations.

The final form equation for p_t is approximated by an ARMAX (Auto-Regressive Moving Average with Exogenous Variables) model and is estimated by non-linear least squares after replacing the unknown coefficients in x^*_t by consistent estimates to get $\hat{x}^*_t$. The final form equation is then used to generate the values for Ep_t and Ep_{t+1}. To compute Ep_{t+1} we need the one-step-ahead forecast errors of $\hat{x}^*_t$, which are obtained from a univariate time series model fitted to the series $\hat{x}^*_t$. Finally, the values of Ep_t and Ep_{t+1} are substituted into 5.12 and 5.13 which are solved for p_t and p^f_t (the CIP68 and the futures price in New York). Prices for the various coffee types and the futures price in London are then obtained from the error correction models mentioned above and the definition of the CIP68. These prices appear in the equations for the various coffee types and countries.

5.3.2 Solution when the quotas are effective

When the exports of coffee by producing countries are restricted by means of export quotas, it follows that $\text{exp}_{it} = \text{quota}_{it}$ and the inventories of the producing countries are $z^p_{it} = z^p_{it-1} + q_{it} - \text{quota}_{it}$. For the aggregate inventories of importing countries, we have $Z^I_t = Z^I_{t-1} + \text{QUOTA}_t - \text{DIS}_t$, where QUOTA_t denotes aggregate imports. The clearing condition 5.11 for the spot market remains unchanged.

It is also straightforward to show that the expression for the optimal position on the futures market is not changed if the quota system is introduced. The clearing condition for the futures market is therefore given in 5.10. When we solve for the rational expectations, we have to split Z_t in 5.10 and 5.11 into Z^I_t and Z^p_t respectively. For Z^I_t, we substitute expression

5.1; Z_t^p is carried along as a predetermined variable. The rational expectations solution when quotas are effective is then obtained along the same lines as explained above. Only the variables S_t and H_t have to be redefined to include Z_t^p which is now predetermined.

Before the simulation results are presented, we give a brief description of the way in which the quota system has been incorporated in the simulation study.

5.4 THE QUOTA SYSTEM MODELLED

The quota system of the ICA 1976 is based on daily developments on the coffee market (ICO, 1976). As the model is a quarterly one, the quota system has to be formulated in terms of restrictions on quarterly variables. Obviously such an approximation will be more inert than the real quota system, because only four decisions a year can be taken. Although quotas come into effect after the CIP has been below the ceiling of the price range for a period of 20 market days, the quota distribution is a quarterly matter.

The quota system has been introduced in the simulation study in the following way. Each exporting member is entitled to a basic quota, according to the provisions of the agreement. The quotas become effective in the quarter after the quarter in which the CIP76 is at or below 135 \$ct per pound, with

$$\mathrm{CIP76}_t = \tfrac{1}{2}(p_t^{\mathrm{ROB}} + p_t^{\mathrm{OM}})$$

The quarterly quotas are divided into two parts, a fixed and a variable part. The initial allocation is 97.6% of the annual quota that has been agreed in ICO meetings. The fixed part is 70% of this allocation. The variable part is 30% of the total initial allocation. It is allocated to a country in the ratio of its own inventory level to the total inventory level of the relevant coffee type (ICO, 1976).

Quotas are adjusted downwards if the CIP76 falls below 120 \$ct and again when this price average drops below 115 \$ct. If prices rise, quotas are adjusted upwards if the CIP76 rises above the level of 135, 150 and 155 \$ct respectively. Quotas must be suspended above the last-mentioned level. The size of the adjustment is also decided in ICO meetings. In the simulation experiments the quota adjustments equal 1.4 million bags, an amount decided by the International Coffee Council in the autumn of 1980.

5.5 THE BEHAVIOUR OF THE MODEL OVER THE SAMPLE PERIOD

To give the reader some insight into the performance of the model, we solved the model for the period 1977.II through 1980.I and computed standardized root mean square forecast errors (SRMSEs)[1] of the variables in the model. The largest sample period for which all variables of the model are observed is the period 1977.II to 1980.I.

The predetermined variables take the observed values. For each quarter the values of the endogenous variables are determined: first the prices by using the solution equations, then inventories, disappearance and retail prices of the individual countries. Quotas were not effective in the sample period.

Generally, the simulated series fit the observed data reasonably well. An analysis of the simulations over the sample period shows that for the price expectations and for the spot price realizations, the simulated turning point lags somewhat behind the historical turning points. In this respect, we emphasize that in the quarter 1977.II coffee prices reached their peak after the crop in Brazil had been destroyed during the harvest year 1976–7. The years 1976–7 were exceptional for the coffee market, with extreme world market price increases, which are not fully explained by the model. The SRMSE for world market prices is approximately 0.20. In contrast, the behaviour of the simulated retail coffee prices in importing countries fairly accurately describes the observed pattern. The SRMSE for retail prices varies in the range from 0.02 to 0.10.

The disappearance variable performs very well too. Most SRMSEs are in the range from 0.10 to 0.15. This is due to the fact that SRMSE depends mainly on the retail price. The simulated level of the inventories in some importing countries is rather high compared with the observed values (e.g. 0.25 for Norway, 0.27 for the UK and 0.30 for the USA), but in most cases the simulated inventory levels are rather close to the observations (SRMSEs in the range from 0.09 to 0.20).

The simulation results for the inventories in producing countries vary in quality per country. They are quite good for Colombia, reasonable for the countries producing Robusta and Other Milds and least satisfactory for Brazil. The reason for these different outcomes is not clear. The producers of Other Milds and Robusta are aggregates of many countries, and the result confirms our assumption that these countries act in a similar way with an SRMSE of about 0.20. The model for a price-setting producer performs very well for Colombia, with an SRMSE equal to 0.07. In spite of the rather good estimation result for the inventory equation of Brazil, the simulation outcome (SRMSE = 0.31) suggests that the present specification probably needs more refinement because of specific features of the coffee trade of Brazil.

[1] The SRMSE is defined as $\text{SRMSE} = [\Sigma_{j=1}^{n}(A_j - F_j)^2/\Sigma_{j=1}^{n} A_j^2]^{1/2}$ with A and F the actual value and simulated value respectively of a given variable.

5.6 THE IMPACT OF POLICY MEASURES AFFECTING THE VOLUME OF PRODUCTION

5.6.1 Introduction

Several hypothetical situations for the world coffee market will be analysed in this section. Simulation of these situations will give insight into the quantity and price effects of the quota system and will be informative about the market behaviour with respect to various levels of coffee production, which is exogenous in the model. For one-step-ahead simulations it does not matter whether the volume of production is exogenous or only predetermined (for instance, it could depend on lagged prices). For dynamic simulations with a conditional model, the exogeneity of production is required. In this section, production is assumed to be a policy variable that will be varied in the simulations. The distribution of inventories over producing and importing countries will also be varied. It will be interesting to analyse differences of the effects in the short run and in the long run. To examine long-term effects the quarterly simulations will have to be run for a longer period.

As a starting point we take the situation in 1979.IV. The end of the year 1979 was chosen because it is the beginning of a period with more or less normal market circumstances, although the price level is still rather high. The market had recovered from a period of heavy price changes. From 1979.IV on, the model will be simulated over a period of 16 years (64 quarters), a period long enough to investigate the long-term properties of the model. For this simulation period the exogenous variables such as deflators, population and exchange rates will be kept constant at the level of the base quarter 1979.IV. Relative coffee prices can therefore be compared with observed nominal prices in 1979.IV.

In the 1970s world production varied around 71 million bags (of 60 kg), and imports of coffee by member countries were around 56 million bags. Further, non-member countries imported about 6.5 million bags of coffee in these years. Therefore officially about 62.5 million bags of coffee were imported.

We present the results in terms of world production which includes imports by member and non-member countries and domestic use in coffee-producing countries. World production is set equal to $\pm 4/3$ of total production on behalf of the member countries, which is determined in the model.

5.6.2 High production level

In a first simulation, world production will be held constant at a level of approximately 100 million bags. The distribution among producing countries is as follows (in million bags): Colombia, 13.7; Ethiopia, 4; Brazil, 27;

Kenya and Tanzania, 2.9; Other Milds, 29.3; Robustas, 21.3. Production has been spread over the quarters according to the observed seasonal pattern in the harvest of the various producing countries. If there is no coffee agreement, the market collapses as a result of this high sustained production. Prices go to zero and inventories in producing countries increase tremendously so that interventions in the market cannot be avoided.

When there is an ICA, most variables in the model, in particular prices, fluctuate in the first simulation periods. The lower price bound is immediately reached and the quota system becomes effective. As a result of the reduction in the quantity brought to the market, prices cross the upper bound and the quota system becomes ineffective.

The alternations result from the simplified translation of the rules of the ICA into formulae in the model. Exports on a yearly basis vary between about 50 and 70 million bags and total export revenue equals on average about $9000 million. Disappearance equals approximately 10–11 million bags per quarter. Inventories in producing countries double to become 80 to 90 million bags whereas importing countries only marginally increase their inventories. In fact the simulation results are very similar to the actual situation on the coffee market in the 1960s when a high world production level had to be reduced via an international agreement. We simulated the effect of a high production level for various initial values for the coffee inventories and for the production in Brazil. In particular when we analyse an exceptional situation, i.e. when we set the production in Brazil equal to zero in the first year and 50% and 100% of its size in the preceding simulation in the second and later years respectively, we only find short-run effects. The price level is higher during the first and second year. In the long run the results are similar to those of the preceding simulation. Similarly, when we set the inventories in the importing countries equal to zero with inventories in producing countries unchanged or increased such that total inventories are left unchanged, the results only deviate in the short run from those in the first simulation.

When initial inventories in producing countries are set equal to zero, the deviation is stronger and lasts for about three years.

5.6.3 Decreasing the production level to stabilize the market

It is now natural to investigate to what extent world production has to be lowered to reach a situation in which the ICA becomes ineffective and coffee prices and trade become stable. When world production is decreased by 10% from 100 million bags to a constant level of 70 million bags after five years, the quota system becomes ineffective and prices become almost stable but slightly decreasing. A reduction by an additional 10% of production in the sixth year leads to slowly increasing prices. These findings are illustrated in Figs 5.1–5.6 where the outcome for aggregate inventories

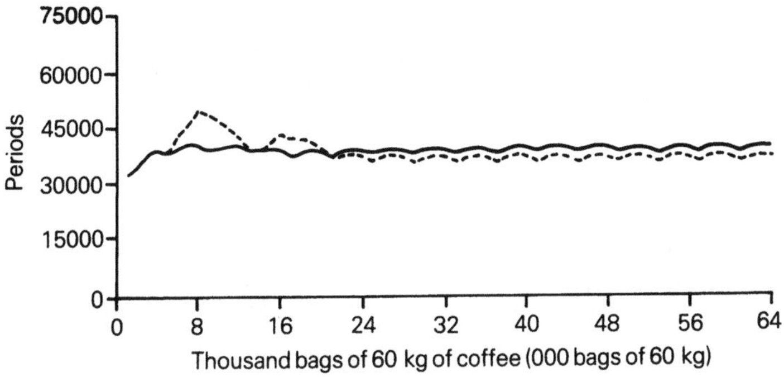

Fig. 5.1 Inventories, producing countries.

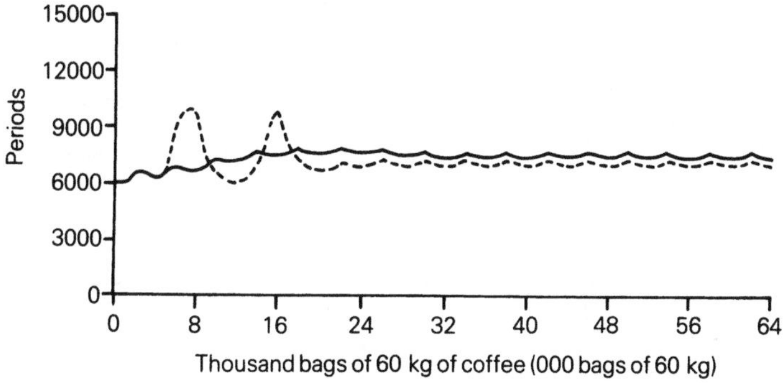

Fig. 5.2 Inventories, importing countries.

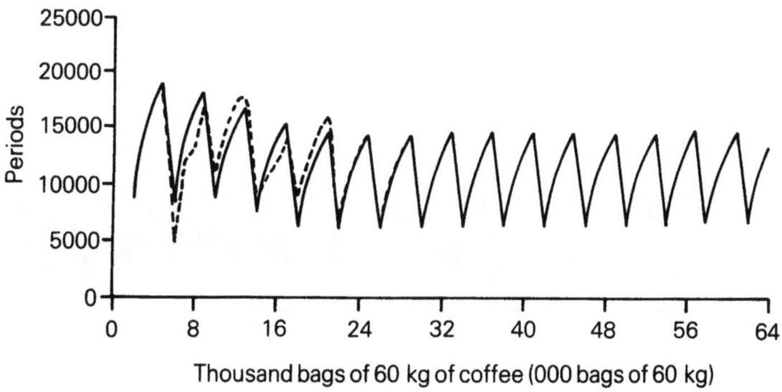

Fig. 5.3 Exports.

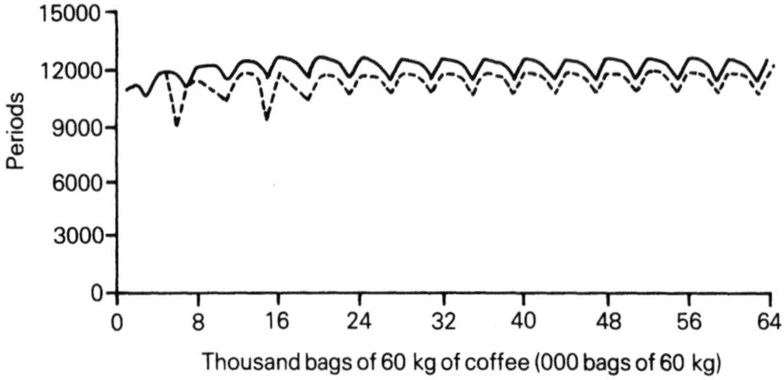

Fig. 5.4 Disappearance.

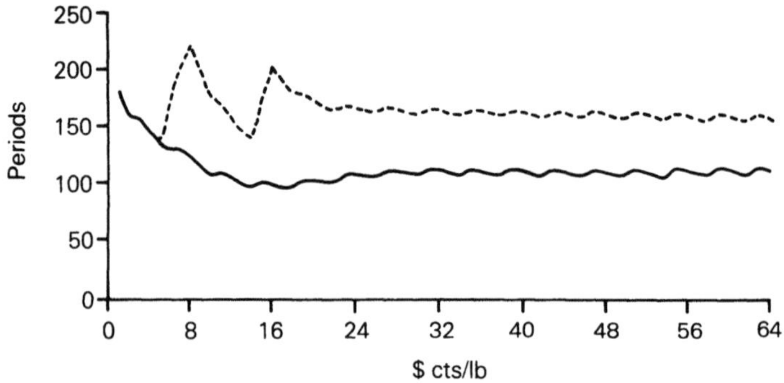

Fig. 5.5 CIP 68.

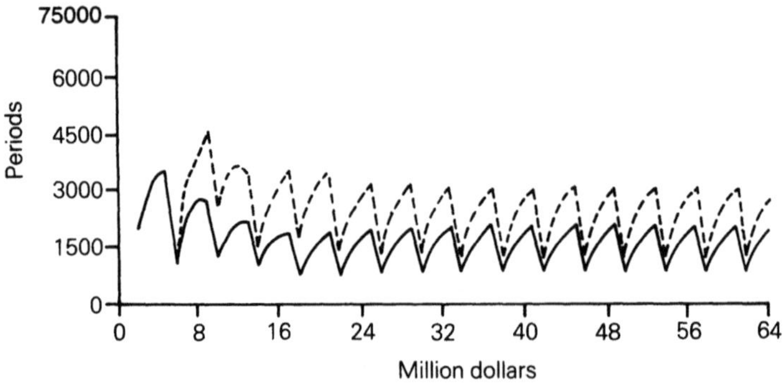

Fig. 5.6 Export earnings.

in producing and importing countries, total exports, total disappearance, the CIP68 and total export earnings are plotted.

A few interesting conclusions emerge from these findings. When there is an ICA (the broken line), the price level and export earnings are higher than in the situation without an ICA (the full line), whereas exported quantities, disappearance and inventories are lower. All series quickly reach a roughly constant level when world production remains at a level of 70 million bags. In the long run, with the ICA, exports become 43.7 million bags which are sold at a price of $1.58 per pound, yielding earnings of about $9145 million. In the case where there is no ICA, the level of exports is about 43.8 million bags in the long run, the world market price is about $1.11 per pound, yielding total export revenues of $6439 million.

These findings are similar to those of Herrmann (1988), who reports a price increase of about 30% resulting from the conclusion of an ICA (to be compared with a price increase of about 42% in our case). The difference in export earnings in his study fluctuates in the range of $1700–$1900 million, which is less than the $2706 million reported above.

Notice that the differences might be explained by the fact that he does not model the impact of uncertainty on decisions in his model and that he disaggregates his model according to countries and distinguishes between ICO member and non-member countries. In his analysis of the coffee market, the ICA roughly leads to a loss in revenue of about $2200 million for importing member countries and a gain of $140 and $1700–$1900 million for importing non-member countries and producing countries respectively.

It should be realized that there are costs involved for producing countries to earn the extra revenue from exports. As illustrated in Fig. 5.1 producing countries will temporarily increase the inventory level when there is an ICA. In the long run, however, inventories will be at a lower level compared with a non-quota regime and the extra income will more than compensate the initial costs.

Finally, we summarize the medium- and long-run impact of a reduction in coffee production on the main variables in the model. In Table 5.2 we give the elasticities for various variables when the quota system does not become effective. It shows that a reduction in coffee production is favourable for the producing countries in the long run, but certainly not for the importing countries. The former countries realize much higher export earnings in the long run. The increase in coffee price amply compensates for the decrease in disappearance.

5.6.4 Low production level

When production is held constant at a level of 70 million bags, the price level is almost stable. It very slowly decreases from ±$1.65 to ±$1.50 in 50

Table 5.2 Medium- and long-run elasticities with respect to production under stable market conditions

	After 3 years	*After 10 years*
CIP68	−1.35	−4.59
Retail price, Netherlands	−0.77	−2.77
Retail price, USA	−0.88	−3.39
Inventories, producing countries	0.48	1.10
Inventories, importing countries	−0.06	0.52
Exports	0.85	0.82
Export earnings	−0.39	−3.39
Disappearance	0.22	0.83

periods and the ICA does not become effective.

After 12 years, the CIP68 becomes $1.46 and total annual exports to member countries become 44.650 million bags. Export earnings become $8730 million. For disappearance, inventories in importing countries and inventories in exporting countries, we find a yearly average of 45.950, 7 and 37 million bags respectively. The pattern of the simulations is fairly robust with respect to the size of total inventories and its distribution among producing and importing countries in the initial period. As expected, a reduction in the size of the initial inventories leads to a higher price level in the long run, whereas a redistribution of inventories from importing countries to exporting countries or vice versa holding total inventories unchanged also has a positive effect on the price level in the short and medium run. But the order of magnitude of this effect is much smaller than when the size of the total inventories is reduced. Similar conclusions are reached for disappearance (or total demand) but, as expected, with opposite sign.

5.7 CONCLUDING REMARKS

In this chapter we simulated an econometric model for the international coffee market under various circumstances. The rational price expectations were calculated by using the solution method as presented by Chow (1983) assuming that the second moments of future values of the endogenous variables are constant. We compared the simulated values for the sample perid 1977.II–1980.I with the observations. With the exception of the price simulations in the years 1977–9, in which heavy fluctuations occurred in the coffee market, the performance of the model is quite satisfactory. Overall, the results appear to be plausible. The determination of turning points of the spot price and of the inventories in Brazil need further investigation. The simulations for other variables are good. The simulations for retail prices in general and for inventories in Colombia perform very well.

The simulation experiments for various hypothetical circumstances are very instructive and allow us to draw some policy-relevant conclusions. These experiments are concerned with the effects of the ICA when coffee production is at a high level. When production remains at a high level, the coffee market will collapse whether an ICA is concluded or not. When there is an ICA, the price will settle at the lower bound agreed upon in the ICA but the increase in inventories will necessitate an intervention leading to a reduction in total production. The ICA will only be workable if it is accompanied by restraint on the supply side.

When production is decreased to a level of 70 million bags, the market becomes stable. The ICA appears to be favourable to producing countries who earn more revenues from their exports because they are smaller in quantity but sold at a higher price. The increase in revenue (we abstract from possible negative effects of the increased price variability) of exporting countries leads to a welfare transfer from importing member countries to exporting countries. Our finding is very much in line with the conclusions reached by Herrmann (1988) who used a more disaggregated but theoretically less sophisticated model to study the effects of the ICA on the coffee market. It also becomes clear that the ICA does not lead to higher coffee consumption by importing member countries.

Recently, in October 1988, after two weeks of negotiations, the ICO agreed on a maximum export level of 56 million bags for the coffee year 1988–9 to bring the world market price to between $1.20 and $1.40, as the average price had been decreased to $1.14 because of the over-supply of coffee. In 1987, it was already agreed that production should be substantially reduced in the following years. Our simulation results show that such an export quantity, together with a reduction in total coffee production to this level, may be successful in reaching a stable situation in the long run.

ACKNOWLEDGEMENTS

The authors thank R. D. M. Molenaar and G. A. van Pruissen for their help in carrying out the computations and preparing the plots, and T. E. Petzel for his useful comments on an earlier version of the chapter.

REFERENCES

Blanchard, O. J. and Kahn, C. M. (1980) The solution of linear difference models under rational expectations with an application to German hyperinflation. *Econometrica,* **48**, 1305–11.

Broze, L., Gouriéroux, C. and Szafarz, A. (1986) Bulles spéculatives et transmission d'information sur le marché d'un bien stockable. *Revue d'Analyse Economique*, **62**, 166–83.

Chow, G. C. (1983) *Econometrics*, McGraw-Hill, Tokyo.

Herrmann, R. (1988) *Internationale Agrarmarktabkommen, Analyse ihrer Wirkungen auf den Märkten für Kaffee und Kakao*, Kieler Studien 215, J. C. B. Mohr (Paul Siebeck), Tübingen.

International Coffee Organization (1976) *International Coffee Agreement 1976*, London.

Newbery, D. M. G. and Stiglitz, J. E. (1981) *The Theory of Commodity Price Stabilization*, Clarendon Press, Oxford.

Palm, F. C. and Vogelvang, E. (1986) A short-run econometric analysis of the international coffee market. *European Review of Agricultural Economics*, **13**, 451–76.

US Department of Agriculture (1957–87) *Foreign Agricultural Circular, Coffee, World Coffee Situation*, Department of Agriculture, Washington, DC, several issues.

Vogelvang, E. (1988) *A Quarterly Econometric Model of the World Coffee Economy*, Free University Press, Amsterdam.

Working, H. (1949) The theory of the price of storage. *American Economic Review*, **31**, 1254–62.

6

Modelling the world fibre market

M. ELTON THIGPEN

AND DONALD O. MITCHELL

6.1 INTRODUCTION

The model described herein represents the first attempt to model the world textile industry. Previous models have focused on a specific country or fibre. For example, Ward and King (1973) modelled inter-fibre competition in the USA with an emphasis on cotton.

The four basic fibre types, cotton, wool, cellulosic (acetate, rayon and triacetate) and non-cellulosic (acrylics, nylon and polyester), comprise the basic raw materials for the textile industry.[1] These fibres are used to produce textile apparel, home furnishings and industrial fabrics. The demand for textiles and for the fibres as a whole is mainly driven by population, prices and income. The derived demand for individual fibres, such as cotton, is determined in turn by their relative competitiveness in the numerous end-uses. This competition for fibre market share involves both relative prices and technical performance in the end-use. Examples of fibres that have been replaced in specific end-uses on technical performance grounds include extra-long staple cotton's loss of the tyre cord and parachute markets to the man-made fibres which are stronger, lighter and more durable then cotton. Likewise, the popularity of the nylon double-knit dress shirt in Western Europe was short-lived because it was uncomfortable to the wearer.

The share of individual fibres in each product group is an indication of the composite competitiveness on all relevant grounds—quality, cost and

[1]Many other fibres that compete only tangentially with those listed above are excluded from this study.

performance—in meeting consumers' demands. As indicated by its market shares in the US, cotton has met the requirements of the apparel market much better than those of industrial uses. Cotton's moisture absorbency and dyeability are very important to the comfort and appearance of garments but are of little value in many industrial fabrics. Similarly, the non-cellulosics' strength, low specific gravity and inability to absorb moisture make them favoured in many industrial uses, but they are usually blended with cellulosic fibres in order to be acceptable in many apparel uses. In the home furnishings segment, the fibre characteristics are also end-use specific. Cotton does not meet the resilience requirement of face yarns for tufted carpets, but it is ideally suited for sheets and towels on comfort and performance grounds owing to its moisture absorbency.

Very often end-use technical requirements are best met by a blend of fibres which combine the desirable properties of the two fibres and tend to mask the undesirable ones. When consumers are indifferent to a fairly wide range of blend ratios, relative prices of the substitute fibres tend to influence the amount of individual fibre in the blend. In extended periods of high cotton prices relative to man-made fibre prices two substitute actions are commonly taken. Rayon, a regenerated cellulosic fibre with absorbency and dyeing characteristics similar to cotton, is more widely substituted for cotton in cotton/polyester blends. Also, the proportion of polyester fibre in cotton/polyester blended fabrics is increased.

Cotton prices were quite stable up to 1969. During that time, prices of cotton's primary substitute, polyester staple, were declining very sharply as rapid capacity expansion in large units utilizing more efficient processes reduced unit costs. By the mid-1970s polyester's penetration into cotton products had fully exploited its opportunities based on technical advantage, and the focus of competition shifted to the ratio of fibres in blends, which is sensitive to relative prices.

The supply of fibres differs greatly between the natural fibres (cotton and wool) and the man-made fibres (cellulosic and non-cellulosic). The natural fibres are produced in competition with other agricultural products and thus are directly influenced by relative agricultural commodity prices and policies. Production is also governed by the biological constraints of the annual production cycle and the variability of climatic conditions. The response of production to market conditions is therefore neither quick nor assured. Non-cellulosic fibres are produced by industrial processes from long-chain synthetic polymers, usually of petroleum origin. The levels of production can be adjusted quickly to market conditions within the limits of plant capacity. The cellulosic man-made fibres are industrially regenerated cellulose, usually of wood fibre origin.

The market structure of the natural fibres industry also contrasts sharply with that of the man-made fibre industry. The natural fibre industry comprises many small producers, each producing too small a quantity to affect the market price, i.e. the wool and cotton markets are perfectly competitive at the producer level. At the national and international market level, governments often intervene by limiting imports, providing price supports to producers or organizing monopoly marketing agencies to perform marketing services such as ginning, warehousing, transportation and merchandising.

Man-made fibres are produced by industrial processes which give producers a considerable degree of control of output over a relatively short period of time and within the limits of total capacity. The synthetic fibre industry comprises a relatively small number of firms with individual firms large enough to influence the level of production and price. Thus an oligopolistic structure exists. The oligopolistic structure is supported by (a) the patent protection given to fibre discoveries, (b) substantial economies of scale in fibre production, (c) large capital requirements per economic production unit, (d) high research and development costs to maintain technical advantage and (e) large expenditure for market development. The industry entry requirements have been met most readily by large multinational companies, often major chemical or petrochemical companies. In the market restructuring that has occurred since man-made fibre consumption growth slowed decidedly after 1973, producers have tended to withdraw from those products where they were not a significant market factor, further increasing the concentration in specific products.

International trade in textiles operates under special provisions outside regular General Agreement on Tariffs and Trade (GATT) rules.[2] Special trading arrangements were agreed for cotton textile trade in 1961 and were expanded to include trade in textiles of other fibres in 1974. Initially the special arrangements were adopted to give the US and European industries time to reorganize in preparation for competition from the large quantities of textiles becoming available in other countries. Under these arrangements, bilateral trade agreements are negotiated to set quantitative limitations on trade in various textile categories between the two countries.

6.2 DATA

Textile manufacturing involves several intermediate stages and products. The products include carded yarns, combed yarns, filament yarns, grey cloth, bleached cloth, printed cloth, knitted goods and garments made from fibres of a single fibre or from blended fabrics containing proportions of two or more fibres. Many of these products are heavily traded domestically and internationally.

Data on the textile industry are generally difficult to obtain because of the

[2]The Arrangement Regarding International Trade in Textiles, GATT.

large number of intermediate products and stages of production. The most readily available data are for the production of the agricultural fibres cotton and wool. Production, consumption and trade data are available for these commodities from the Food and Agriculture Organization of the United Nations (FAO) and the US Department of Agriculture (USDA). Supply data for the man-made fibres are more difficult to obtain but are available for most countries from industry sources such as the Textile Economics Bureau and The International Committee for Rayon and Synthetic Fibres. Data on consumer demand, trade, prices and stock levels for fibres are often unavailable and modelling of the world textile industry is constrained by this lack of data. For example, textile product trade is identified by value and type, such as value of stocks, but the fibre proportions or weights are not included. The most readily available and perhaps most reliable data of fibre use are at the mill level where raw fibre is weighed before processing begins. At this stage, however, fibre use relates to industrial demand, which may or may not be closely related to consumer demand in the country where the manufacturing takes place. For example, in 1980 the cotton available for use by domestic consumers in the Republic of Korea was estimated at 38% of the cotton utilized by Korean spinning mills, taking into account the net trade in cotton textiles. In order to model the demand for fibres, an acceptable procedure is required for adjusting mill consumption levels by net textile trade to estimate final consumer demand.

We have addressed this consumption adjustment problem in the following way. The FAO publishes world fibre consumption survey results and the data are fairly consistent for the periods 1964–75 and 1979–84. The country/regional coverage is comprehensive for cotton, flax, wool and cellulosic and non-cellulosic fibres. Fibre consumption in individual countries was estimated by adding to mill consumption (defined as the volume of each fibre utilized by industry and village craftsmen) the balance of net trade (imports minus exports) in processed products made from the fibre (converted to raw fibre equivalent). This quantity is described as the fibre stocks of intermediate and end products to be taken into account. In order to convert the mill consumption data into a series of availability for consumption, adjustment factors were calculated for each country/region from the FAO reports for the years 1964–75 and 1979–84. Adjustment factors for the 'gap' years were estimated from the trends in the series or from other available mill consumption and trade data. The adjustment factor is defined as the ratio of 'fibre available for the home use' and mill consumption.

6.3 MODEL

The complexity of the textile industry as well as limitations on data precluded modelling every stage of the industry. The model includes three of the basic fibre types: cotton, cellulosics and non-cellulosics. Wool has not been included at this stage because it has the smallest share of the four basic fibres. A total of 19 countries and six regions are modelled individually and these are shown in Table 6.1. Each fibre type is modelled at two levels, the raw fibre and the texile level. Figures 6.1 and 6.2 show the structure of the cotton model and the man-made fibre model (which is similar for both the cellulosic and non-cellulosic models).

The fibre model is an econometric model estimated primarily by ordinary least squares. The data is annual time series extending from 1965 to 1984 for cotton and from 1970 to 1984 for the man-made fibres. The three fibre types are modelled separately with cross-linkages in demand and trade.

Table 6.1 World fibres model, countries and regions

Region	*Countries*
Developed countries	
1 Australia	Australia
2 Canada	Canada
3 EC-12	Belgium, France, Italy, Luxembourg, Netherlands, the Federal Republic of Germany, United Kingdom, Ireland, Denmark, Greece, Portugal, Spain
4 Japan	Japan
5 Other developed countries	Austria, Finland, Iceland, Malta, Norway, Sweden, Switzerland, New Zealand, Yugoslavia
6 United States	United States
Centrally planned economies	
7 Eastern Europe	Albania, Bulgaria, Czechoslovakia, East Germany, Hungary, Poland, Romania
8 USSR	Union of Soviet Socialist Republics
Developing countries	
9 Argentina	Argentina
10 Brazil	Brazil
11 Central Africa	Botswana, Lesotho, Namibia, Swaziland, Kenya, Malagasy Republic, Malawi, Mozambique, Tanzania, Uganda, Zambia, Angola, Burundi, Cameroon, Central African Republic, Chad, Congo, Côte d'Ivoire, Ethiopia, Djibouti, Benin, Gabon, Gambia, Ghana, Guinea, Equatorial Guinea, Guinea-Bissau, Liberia, Mali, Mauritania, Mauritius, Niger, Reunion, Rwanda, Senegal, Sierra Leone, Somalia, Sudan, Togo, Upper Volta, Zaire, Zimbabwe
12 China	China

Table 6.1 (*Continued*)

13 East and South Asia	Burma, Kampuchea, Laos, Vietnam, Hong Kong, Singapore, Brunei, Malaysia, Philippines, North Korea, Mongolia, Pacific Islands, Afghanistan, Bangladesh, Bhutan, Nepal, Sri Lanka, Papua New Guinea, Fiji Islands
14 Egypt	Egypt
15 India	India
16 Indonesia	Indonesia
17 Republic of Korea	Republic of Korea
18 Latin America and Caribbean	Bahamas, Barbados, Bermuda, Belize, other Caribbean Islands, Cuba, Dominica, Dominican Republic, Jamaica, Trinidad and Tobago, Honduras, Nicaragua, Panama, Costa Rica, El Salvador, Guatemala, Haiti, Bolivia, Chile, Colombia, Ecuador, French Guiana, Guyana, Paraguay, Peru, Surinam, Uruguay, Venezuela
19 Mexico	Mexico
20 Nigeria	Nigeria
21 North Africa and Middle East	Algeria, Bahrain, Cyprus, Iran, Iraq, Israel, Kuwait, Libya, Oman, Qatar, Saudi Arabia, United Arab Emirates, Jordan, Lebanon, Morocco, Syria, Tunisia, Yemen Arab Republic, Yemen D.M.
22 Pakistan	Pakistan
23 Republic of South Africa	Republic of South Africa
24 Thailand	Thailand
25 Turkey	Turkey

The cotton model consists of a cotton fibre component and a cotton textile component. The two components are separately solved but are linked through cotton manufactured into textiles. Quantities are measured in raw fibre equivalent in both models. Only a single quality of cotton and cotton textiles is considered, even though many grades of cotton are produced and traded. The flowchart for the single country model is shown in Fig. 6.1.

Cotton fibre production is the product of area and yield. Area is determined by lagged cotton area, lagged cotton revenue and the lagged revenue of one or more competing crops such as coarse grains or soybeans. Yields are determined by linear trend, lagged cotton price and the current fertilizer price. The fertilizer price and the competing crop revenue are exogenous to the cotton model.

Stocks from the previous period are added to production to determine current period supply. The cotton supply is then disposed of through net trade, milled into cotton textiles or retained as ending stocks of the current period. The disposition of cotton supply and the determination of cotton

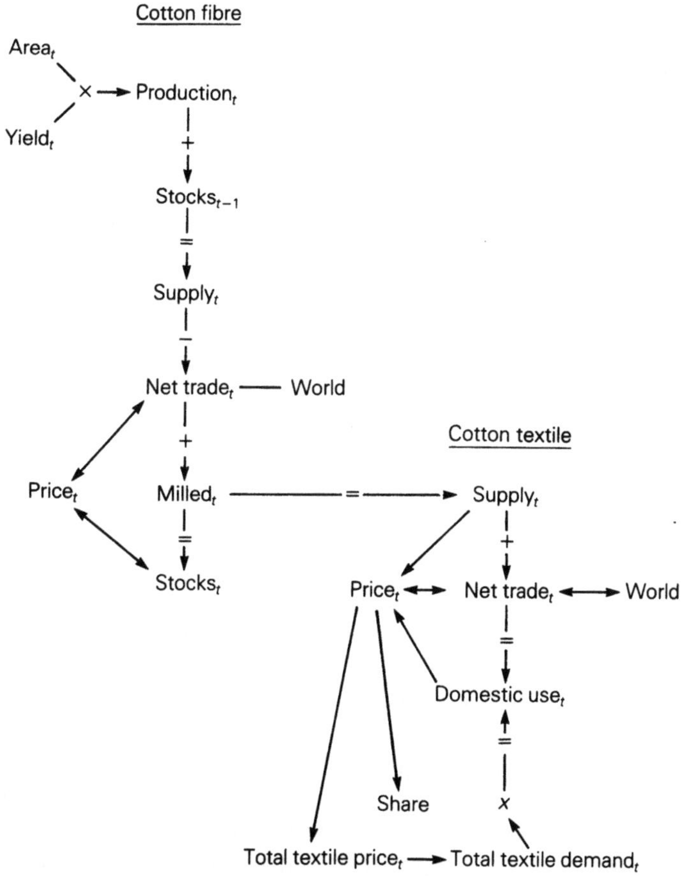

Fig. 6.1 Flowchart.

prices are simultaneous.

Mill demand is the quantity of cotton processed by the mills. Since no information is available on stock holdings by cotton traders and the mills, the stocks of cotton held by the mills are assumed to remain constant at the level needed to assure a continuous supply of cotton to the mill for processing. Stocks of cotton are the residual of supply minus net trade and mill demand.

Cotton prices are simultaneously determined with net trade, mill demand and the level of ending stocks. Prices are inversely related to the level of ending stocks in any period. The ending stocks of each period become available for use in the next period.

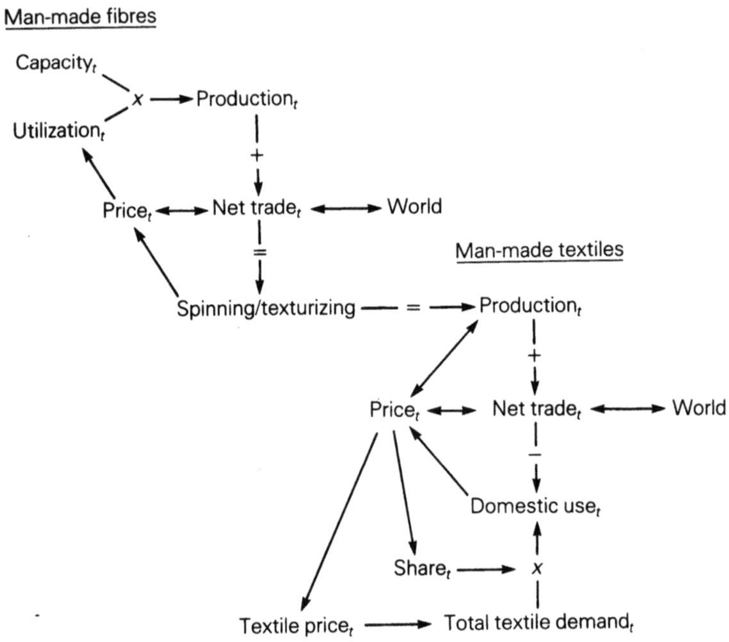

Fig. 6.2 Flowchart of synthetics model.

The allocation of cotton supplies to mill demand, net trade or stocks is government controlled in many countries, through cotton marketing boards which purchase the crop from producers and allocate supplies between domestic cotton mills and exports. In other cases, export policies such as quantity restrictions and taxes influence the flow of cotton into the domestic and export market. Many countries import cotton in years of low production in order to assure the supply of raw material for their cotton mills. Some countries both import and export cotton within the same year in order to achieve the desired blend of the various qualities of cotton for their mills.

While cotton supplies are assumed to be perfectly inelastic in a single year, the cotton mills are assumed to be price responsive within the year. This results in additional cotton being made available to the domestic mills through a reduction in exports or ending period stocks and also through an increase in net imports. In a year of abundant supplies, mills may obtain their requirements at lower prices and larger quantities may be exported or accumulated as ending stocks.

Domestic demand for cotton textiles can be satisfied by output from domestic mills or from imports, and mill production in excess of local requirements may be exported. Many countries also simultaneously import and export textiles in order to achieve the consumers' desired balance of products. As stated previously, the domestic demand for cotton is derived from cotton's share of total textile demand. Total textile demand is determined by population, income and textile prices. The share of cotton in this total textile demand is determined by cotton prices relative to other textile prices and by certain rigidities relating to the particular final use, production technologies and consumer preferences.

The model adjusts to changes in economic conditions such as income or the price of a substitute crop by adjusting cotton fibre and textile prices, production, trade, consumption and stocks. Consider an example of the response in the cotton sector to an increase in the level of income in a cotton producing and manufacturing country. The enhanced income would first be reflected in the demand for all textiles, which would increase the demand for cotton and other textiles in proportion to their relative price elasticities of demand. In response, prices of cotton and other textiles would increase and, in turn, stimulate increased domestic production and net imports. The higher cotton and total textile prices would increase the amount of fibres milled and the supply of cotton and other textiles, while at the same time tending to dampen the increase in demand. The inelastic nature of short-run supply of natural fibres would reduce the cotton available for ending stocks and raise cotton prices which would cause imports to increase. The short-run static equilibrium effects are (a) higher total textile demand, (b) higher cotton textile demand, cotton textile supply and net imports, (c) higher cotton fibre price, cotton milled and net imports, and (d) lower cotton fibre stocks. In the second period, the higher cotton price would increase cotton production. The share of increased cotton fibre demand and cotton milled demand which is satisfied by domestic supplies or trade would depend also on government policies as well as relative price elasticities.

Man-made fibres are either cellulosic (rayon, triacetate, acetate) or non-cellulosic (acrylics, nylon, polyester). Both fibre groups have similar production and marketing patterns. The two fibre types are substitutes in use, but not in production. A production plant cannot easily be shifted from producing one fibre to the other. The flowchart for a country model for either fibre is shown in Figure 6.2. The model is separated into a fibre model and a textiles model with separate prices and market-clearing conditions. The fibre and textile components are linked by the production of textiles from man-made fibres. Any country may have either a fibre or a textile industry or both.

The production of man-made fibres depends on plant capacity and utilization rates. Plant capacity is fixed in the short run while the utilization rate is variable even in the short run. When capacity is not fully utilized, production can respond to price increases. Man-made fibre production is technology and capital intensive and economies of scale generally require large plants to attain international competitiveness. Although man-made fibres are often produced primarily for domestic textile industries, large volumes of these fibres are traded internationally.

Synthetic fibres are produced with properties such as denier, tensile strength and other factors that make them readily adaptable for a wide range of end-uses. The cost of producing synthetic fibres is dependent on the cost of the raw material, labour and other operating costs and the long-run capacity charges. In the short run, the variable costs determine the minimum cost of production and in the long run the fixed costs associated with production capacity also become a determinant of costs.

Once the man-made fibres are processed into textiles, the textiles are either traded on the world market or sold in the domestic market. Countries may also import man-made textiles or textile blends for sale in their domestic markets. The domestic demand for man-made textiles is derived from the total textile demand since consumer demand is partially indifferent to fibre blends. For example, a shirt which is 35% cotton and 65% polyester would be a nearly perfect substitute for a shirt which is 40% cotton and 60% polyester. Many other final uses also have a range of fibre blends over which the consumer is partially indifferent. The fibre demand is therefore dependent upon the total textile demand and the share of each fibre in total demand. The share depends on relative fibre prices and technical specifications for textiles. These shares are very sensitive to relative prices, however.

The total textile price is composed of the individual textile fibre prices and other demand and production cost factors. In the model, each textile fibre price enters the total textile price in proportion to its share in textile production. For example, change in the price of non-cellulosic textile such as nylon would have two effects on nylon use. It would increase the price of total textiles because nylon price is a component of this price and it would also reduce the share of nylon used in textile production.

Textiles are durable goods which typically have a useful life exceeding one period. Therefore purchases in a particular year will increase the level of stocks of the goods and influence purchases in future years. Further, even though a good is durable, it depreciates each period, which leads to a replacement demand. These attributes of durability have been included in the estimation of demand for a number of consumer goods. Chow's (1960) work on the demand for automobiles specified the following simple model. Consumers are assumed to have a desired ending period stock of automobiles, S_t^*, which is a function of price and income:

$$S_t^* = a_0 + a_1 p_t + a_2 y_t$$

If the demand for annual purchases adjusted instantly to the desired level of stocks, then demand in each period would be given by

$$X_t = S_t^* - (1 - \delta) S_{t-1}$$

where δ is the rate of depreciation per year. Purchases would then be the difference between desired stock at the end of the year and the depreciated old stock from the previous year. This could also be written as

$$X_t = (S_t^* - S_{t-1}) + \delta S_{t-1}$$

where the first part $(S_t^* - S_{t-1})$ is the demand for the desired change in stocks during the year and the second part (δS_{t-1}) is the demand for replacement of old stock.

The estimating equation, obtained by substituting for S_t^*, becomes

$$X_t = a_0 + a_1 p_t + a_2 y_t - (1 - \delta) S_{t-1}$$

Income and price have the expected signs and stocks held at the beginning of the year (i.e. S_{t-1}) have a negative influence on purchases during the year. This basic model presents some estimation problems because the level of stocks is often not available—as is the case for textiles. To overcome this, the stock variable is eliminated by using the Koyck transformation to restate the equation in terms of price, income and lagged variables (Deaton and Muellbauer, 1980, p. 352). A number of refinements have been made to this basic model including the use of multiple depreciation rates for stocks to allow current year purchases to depreciate more slowly than previous year stocks and the introduction of partial adjustment schemes. However, according to Deaton and Muellbauer the basic model has been widely used to estimate the demand for consumer durables. Houthakker and Taylor (1966, 1970) estimated the demand for clothing in the USA using the equation

$$X_t = f(X_{t-1}\ \Delta Y_t, Y_{t-1}, P_t, P_{t-1})$$

As a first attempt at estimating the total demand for textiles, the following model was fitted:

$$D_t = f(Y_t, Y_{t-1}, P_t, P_{t-1}, D_{t-1})$$

where D_t is per capita textile demand, P_t is a real weighted textile price and Y_t is per capita real gross domestic product.

The share of each fibre in total textile production is assumed to depend on (a) relative fibre prices, (b) individual fibre characteristics and (c) historical market relationships. Since many textiles are blends of several fibres, textile manufacturers can adjust the proportions of each fibre in response to relative prices. However, limits on the shares of each fibre exist for various end-uses and it is possible that limits on the share of each fibre will be observed in the overall textile market. This could result from fibre characteristics which make a certain fibre undesirable for a particular use. For example, cotton has very desirable properties for use in some clothing items while synthetics are much less desirable. This may effectively place a lower limit on the share of cotton in the textile market of a particular country regardless of the relative price between cotton and synthetic fibres.

Historical trends in the textile industry may also influence the share of a particular fibre in total textile production. This could occur because of less than perfect substitutability of machinery for processing the various fibres. It could also occur because of the reluctance of textile manufacturers or distributors to vary fibre blends due to concerns about consumer acceptance. It could also result from restrictions on trade of fibres or textiles.

The quantity of a particular fibre in total textile production will be assumed initially to be a function of relative fibre prices and past fibre use. This leads to the model

$$F_i = f(Q_T, P_i/P_T\ F_{i,\,t-1})$$

where F_i is the fibre i used in production of total textiles Q_T, and P_i/P_T is the ratio of the price of fibre i to the price of textiles.

6.4 PRELIMINARY MODEL RESULTS

The model is not fully operational at this stage; however, we have completed the production and consumption components. Results are presented for consumption for the three major regions: industrial countries, centrally planned economies and developing countries. The solution was derived with fibre prices exogenous and growing at historical trend rates. The historical data and preliminary forecast give an interesting perspective of the world fibre markets. We present it as a preliminary result from the model.

The demand for fibres, being heavily influenced by per capita income growth and relative price developments, changes in an irregular pattern over time. This is particularly noticeable for cotton and non-cellulosic fibres in the industrial countries (Fig. 6.3). The non-cellulosic fibres were well established in the USA, Western Europe and Japan by the mid-1960s and quickly penetrated cotton's high value and end-uses as large-scale increasingly efficient production capacity came on-stream, permitting a steadily declining price trend for these fibres.[3] The consumption gains for the non-cellulosic fibres came both from substitution for other fibres in present end-uses or from new levels of consumer acceptance of products formerly either too expensive or of unacceptable quality made from other fibres for broad consumer use. An example of the former is the widespread displacement of men's cotton hosiery by nylon and acrylic products, and the latter is characterized by the near-total adoption of tufted floor covering made with non-cellulosic face yarns for residential and commercial buildings. The growth rate of non-cellulosic fibre consumption in the industrial countries averaged 10.2% per annum during the period 1964–79 while the demand

[3] The non-cellulosic fibre most competitive with cotton is polyester.

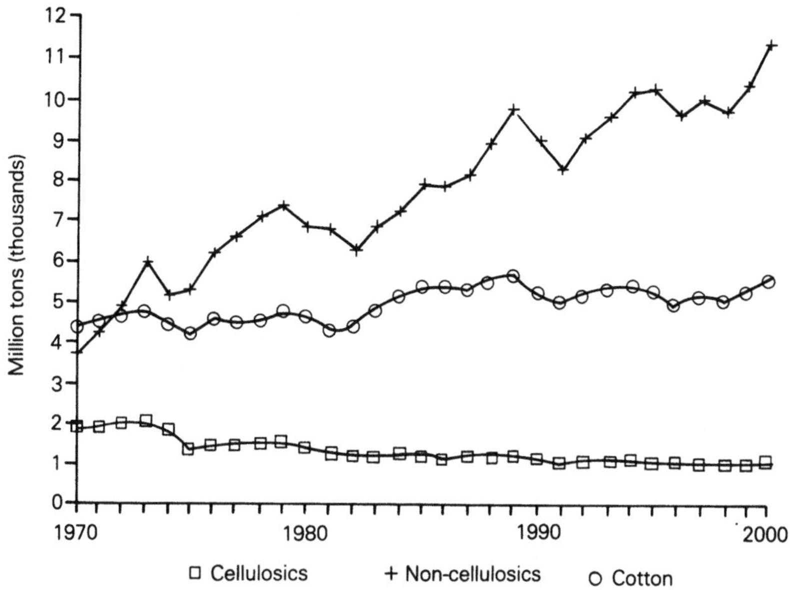

Fig. 6.3 Availability—industrial countries.

for cotton increased at a rate of less than 0.1% per annum. The impact of non-cellulosic growth during this period was greatest in the USA where cotton demand declined at a rate of 2.6% per annum, in spite of a favourable income growth trend. Cotton prices became more competitive relative to non-cellulosic fibre prices during the 1980s, especially during 1982, 1985 and 1986. As a result of this change in relative prices, a shift in consumer preferences towards fibre products and a period of sustained income growth after 1982, cotton demand in the industrial countries increased by 3.8% per annum during the 1980–5 period and demand for cotton textiles is estimated to have continued to increase at a high rate through 1987. By comparison, the demand for non-cellulosic fibres in the industrial countries increased at only 1.8% per annum during the 1980–5 period.

It is believed that cotton demand in the industrial countries peaked in 1987. Evidence of the slowing of demand for cotton products, particularly those made from heavyweight fabrics, is seen in lower daily rates of US mill consumption of cotton, reduced operating hours per week in some European cotton mills and announcements of extended holiday mill closings in early July 1988. Weakness in cotton yarn prices is also indicated by the 11% decline in the Cotlook index over the last year to early September.

The projections for cotton, cellulosic and non-cellulosic fibre consumption in the industrial countries are shown in Fig. 6.3. The projected cotton demand growth rates for the 1979–81 to 2000 period is 1.1% per annum. Cotton's market share is lowest in the industrial countries where the man-made fibres have experienced the greatest increase in use. Its share in these countries declined from around 40% during the 1969–71 period to about 36% in the mid-1980s and is projected to decline to below 30% in the late 1990s. The cellulosic fibres lost price competitiveness with the non-cellulosic fibres during the 1970s and with cotton during the 1980s. As a result, cellulosic fibre demand in the industrial countries declined by 3.1% per annum during the 1964–85 period. Unless a substantial reduction in the cost of producing cellulosic fibres relative to the cost of producing cotton and non-cellulosic fibres can be achieved, the consumption of cellulosics will continue to be limited to specialty items in the industrial countries. The non-cellulosics have become established as the predominant fibre group in the industrial and home furnishing markets in the industrial countries. These fibres had captured one-third of the market in the industrial countries by the 1969–71 period and one half by the early 1980s. With continued favourable growth in their major end-uses in the industrial countries, the non-cellulosic fibres market share in 2000 is projected at around 60%.

The demand for cotton in the non-market region is currently distributed as follows: 80% in the USSR and 20% in the East European countries. The USSR is a major cotton producer and a substantial net exporter of the fibre, while only a small volume is produced in Eastern Europe. Accordingly, the textile industry in the USSR is a major consumer of cotton and domestic demand increased by 3% per annum during the 1964–74 period. In Eastern Europe cotton demand growth was 2.1% per annum over the same period. The consumption of non-cellulosic fibres increased faster than cotton use in recent years and during the 1975–85 period cotton demand growth declined to 1.3% per annum in the USSR and to only 0.3% per annum in Eastern Europe. With plans to continue expanding man-made fibre capacities to supply an increasing share of domestic fibre requirements, cotton demand growth in the non-market region is projected to increase by only 0.9% per annum from 1987 to 2000 (Fig. 6.4). Consequently, cotton's share of fibre use in the non-market region is projected to decline from 46% in the mid-1980s to around 43% in 2000.

Cotton consumption in the developing countries amounts to over one-half of the world total, and its growth rate during the period 1964–84, at 2.8% per annum, was greater than for either of the other two regions. However, cotton consumption in the developing countries lost considerable momentum during the last decade. Consumer incomes and expenditures in many developing countries were adversely affected by the consequences flowing from the increase in energy costs during the 1970s and by the economic consequences of low commodity prices and high interest rates in the 1980s. These two periods stand out distinctly in the graph of cotton and non-cellulosic fibre consumption in Fig. 6.5. The use of non-cellulosic man-made

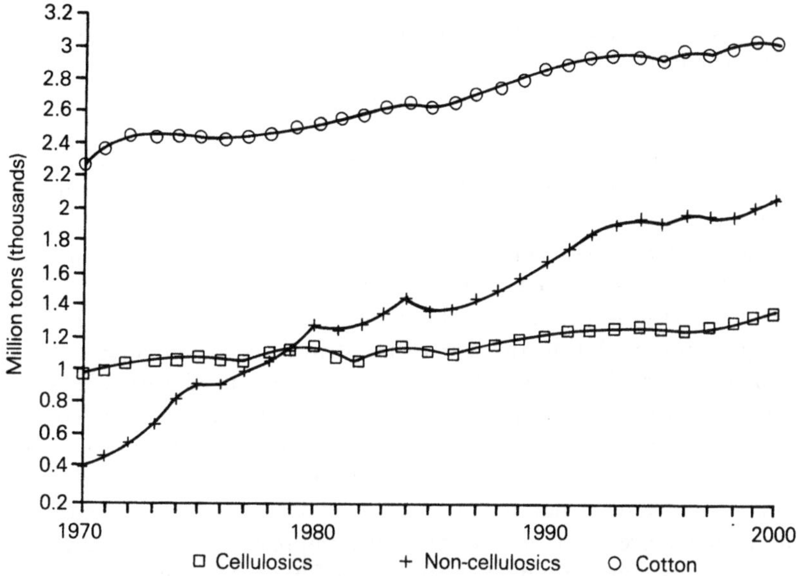

Fig. 6.4 Availability—CPE countries (Europe).

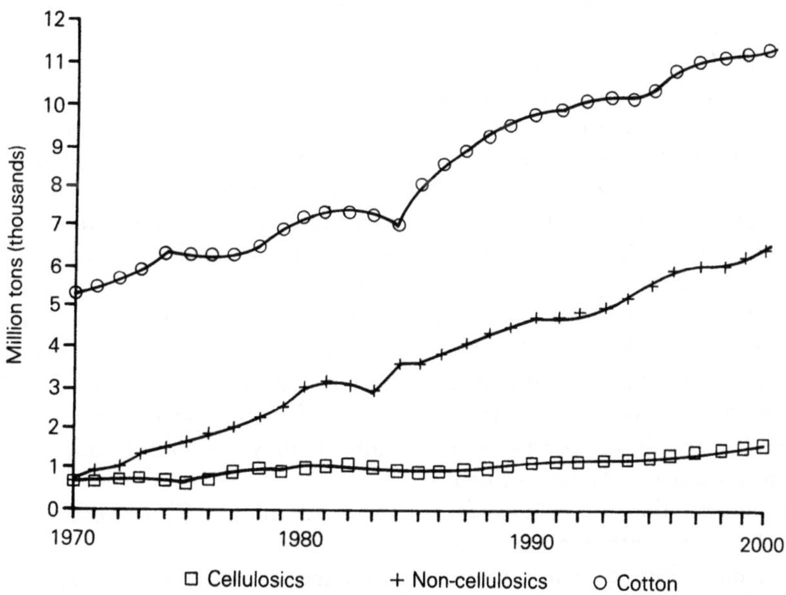

Fig. 6.5 Availability—developing countries.

fibres has experienced rapid growth in the developing countries from a low base in the early 1970s, and its share of total fibre consumption is projected to increase to more than 30% in 2000 as increased domestic supply permits fuller utilization of the technical advantages of non-cellulosic fibres—high strength and abrasion resistance—in blends with cotton and cellulosic fibres. Rayon production and use is still expanding in some developing countries as a domestic-source substitute for imported cotton. However, the cellulosic fibres' share of total consumption is declining from the early 1980s level of around 8% and is projected to be around 6% in 2000.

Cotton demand has been quite buoyant in the Asian developing countries (3.2% per annum during the 1964–84 period) as a consequence of their continued favourable income growth. China experienced particularly strong growth in cotton consumption—5% per annum—but expanded cotton production enough to change from a major cotton importer in the 1970s to a major cotton exporter in the mid-1980s. A sustained high rate of investment in China's man-made fibres production and textile manufacturing industries since 1970 has dramatically improved the indigenous supply and diversity of textile products available in China. Moreover, the export of Chinese textiles has increased to the point where it accounts for a quarter of the country's foreign exchange earnings. With the pent-up demand for textiles now being satisfied and more limited expansion possibilities for textile exports, the rate of investment in the Chinese textile industry is slowing. Even so, favourable income growth is expected to sustain indigenous demand growth by around 2.6% per annum for cotton and by around 4% per annum for the non-cellulosic fibres during the period 1987–2000.

Cotton accounted for about 70% of total fibre consumption in Africa during the early 1980s and is projected to hold around 65% of the market in 2000. Cotton consumption in Africa increased at a rate of about 3% per annum from the mid-1960s to the late 1970s but the distressed income situation in many countries during the 1980s resulted in a decline in cotton use of over 5% per annum. Strenuous efforts are being made to improve economic performance in African countries and a modest growth in the demand for fibres is projected during the 1990s.

The demand for cotton in Southern Europe increased by an average of 2.4% per annum between 1964 and 1984, in spite of a sharp drop during the 1981–2 period. Consumption recovered during the mid-1980s and is expected to increase in proportion to projected income growth through the 1990s. The textile industries in Greece and Turkey are important suppliers of cotton textiles to Western Europe.

Cotton and the non-cellulosics are the predominant fibres consumed in Latin America, with market shares in the early 1980s of 53% and 40% respectively. Fibre consumption in Latin America increased by 1.9% per annum from the mid-1960s through the 1970s but declined during the first half of the 1980s owing to the difficult economic circumstances resulting from the slowing of world trade growth, the decline in commodity prices and the increase in external debt. Cotton manufacturing in Argentina and Brazil has made an encouraging recovery since 1985. Based on prospects of reasonably good income growth, fibre demand is projected to increase moderately during the 1990s.

Prospects for fibre demand growth in the various countries and regions indicate a level of world fibre consumption in 2000 of about 48 million tons. This projection represents an annual average growth rate of 2.2% from the 1979–81 period. With the non-cellulosic fibre consumption expected to continue growing faster than the use of other fibres, the market share of the non-cellulosics and cotton are projected to be equal at around 43% each in 2000 compared with their 1979–81 market shares of 47% for cotton and 36% for the non-cellulosics. During the period the market share of the cellulosics is expected to decline by 3 percentage points.

ACKNOWLEDGEMENT

The World Bank does not accept responsibility for the views expressed herein which are those of the authors and should not be attributed to the World Bank or to its affiliated organizations. The findings, interpretations and conclusions are the result of research supported by the Bank; they do not necessarily represent official policy of the Bank. The designations employed and the presentation of material used in this chapter are solely for the convenience of the reader and do not imply the expression of any opinion.

REFERENCES AND FURTHER READING

Agbadi, I. (1988) An econometric analysis of the U.S. cotton, man-made fibers and textile sectors, unpublished Ph.D. dissertation, Texas Tech University, May.

Chow, G. (1960) Statistical demand functions for automobiles and their use for forecasting, in *The Demand for Durable Goods* (ed. A. C. Heberger). University of Chicago Press, Chicago, IL.

Collins, K. J. and Glade, E. H., Jr (1983) Cotton in the U.S. economy: an inter-industry analysis of international trade in cotton and textiles, Staff Report, National Economics Division, Economic Research Service, US Department of Agriculture, Washington, DC.

Deaton, A. and Muellbauer, J. (1980) *Economics and Consumer Behavior,* Cambridge University Press, New York.

Houthakker, H. S. and Taylor, L. D. (1966) *Consumer Demand in the United States, 1929–1970, Analyses and Projections*, Harvard University Press, Cambridge, MA.

Houthakker, H. S. and Taylor, L. D. (1970) *Consumer Demand in the United*

States, 1929–1970, Analyses and Projections, second and enlarged edition, Harvard University Press, Cambridge, MA.

Naylor, T. H., Wallace, W. H. and Sasser, W. F. (1967) A computer simulation model of the textile industry, *Journal of the American Statistical Association*. **62**, 1338–64.

Ward, L. F. and King, G. A. (1973) Interfibre competition with emphasis on cotton: trends and projections to 1980, Technical Bulletin 1487, Economic Research Service, US Department of Agriculture, Washington, DC, December.

7

Technical change, relative prices and intermaterial substitution

TIMOTHY J. CONSIDINE

7.1 INTRODUCTION AND SUMMARY

Durable goods, containers and structures are redesigned continually to accommodate changing preferences and new technology. This process seems to have accelerated over the past decade, however. For instance, automobiles are now lighter and more fuel efficient than cars produced during the 1960s. As a result, the average car now contains more plastic and aluminium and lighter higher strength steel parts. These shifts in the material composition of products are not unique to the automobile industry. Household appliances, industrial equipment and buildings have been redesigned along similar lines.

Flexibility, strength and other technical characteristics are carefully considered in the selection of materials to meet certain criteria. Relative material costs are another critical factor. If markets are competitive, material prices reflect these technical characteristics. Changes in material-using technology, however, are not reflected completely in relative prices. Design changes induced by government policy, (e.g. fuel economy standards) or by market forces unrelated to materials are examples of exogenous or disembodied technical change. Furthermore, these design changes did not occur at an even pace, as a simple time trend used to represent technological change in many neoclassical models would imply. Consequently, in this study we use average miles per gallon for the US automobile fleet as a proxy for design changes that affect the material composition of products.

A crucial first step in analysing material substitution is the careful measurement of prices and apparent consumption. The materials considered in this study include sheet, strip, pipe and wire products made from steel, copper and aluminium and fabricated plastic products, including moulded parts, sheets, pipes and containers. Transactions prices rather than producer prices are used because material users make decisions based upon prices including discounts. Prices and apparent consumption for steel, copper, aluminium and plastics over the period 1960–86 are presented in section 7.2. Two major trends stand out. First, steel prices rose significantly relative to prices for aluminium and plastics from 1960 to 1978. This fact lends some support to the view that 20 years of trade protection may have contributed to the long-term erosion of steel markets. The second major trend is the remarkable growth in the consumption of plastics.

Given the limitations of end-use consumption data for steel and plastics, the analysis is conducted at the aggregate level. In this way, measurement error introduced from any arbitrary scheme to allocate aggregate materials to end-use sectors can be avoided. The specification of the steel, copper, aluminium and plastic substitution model assumes that material users minimize cost. In addition, these materials are assumed to be weakly separable from other inputs, such as labour, capital and energy inputs. The role of dynamic adjustments and other relevant theoretical considerations are presented in section 7.3.

Maintaining downward-sloping demand relationships is particularly important in forecasting and policy analysis. Several studies have shown that many flexible functional forms used to approximate cost functions are 'well behaved' for only a limited range of relative prices (Diewert and Wales, 1987; and Christensen and Caves 1980). The translog functional form often yields positive own-price elasticities for applied demand problems with wide swings in prices and quantities (Considine, 1989a).

The basic reason for these irregularities is that, even though flexible functional forms yield linear share equations, the elasticities are non-linear in shares. A log-linear share system that has linear share elasticities and therefore more desirable global properties is the logit model of cost shares (Considine and Mount, 1984). Bewley (1986) examines a very similar formulation for consumer demand, and also finds that the elasticities are more stable for extreme values. Although symmetry can be imposed in the logit model for only one set of cost shares, if the concavity conditions hold at the mean, then the own-price elasticities remain negative for all shares (Considine, 1989a). Another advantage of the logit specification is that dynamic adjustments can be easily introduced (Considine and Mount, 1984). Since the data used here embody large swings in relative prices and material users quite probably make decisions based upon expected prices, the logit specification is adopted here and is described in section 7.4.

The empirical results are presented in section 7.5. A comparison is made between translog and logit estimates for selected own-price elasticities. The translog form yields positive own-price elasticities while the linear logit specification does not. This result has been found for other empirical problems (Considine, 1989a, b). Accordingly, the logit specification is used for subsequent analysis. Three hypotheses are tested. The first test is for a dynamic model. This hypothesis could not be rejected. The absence of scale effects on cost shares, or homotheticity, also could not be rejected. This supports the specification of material substitution models that assume homothetic weak separability. Third, the presence of neutral technical change is tested and strongly rejected, which suggests that technological changes unrelated to shifts in relative material prices are important in explaining shifts in material use. In addition to biased technical change, relative prices are found to be very significant factors explaining shifts in material market shares. Plastics are found to be relatively strong substitutes for steel and copper products. Steel and aluminium and copper and aluminium are estimated to be substitutes. Complementarity is estimated between steel and copper and aluminium and plastics.

7.2 MEASURING MATERIAL CONSUMPTION AND PRICES

Material substitution possibilities can be analysed at several stages of the production process. For instance, steel firms make material choices between scrap iron, iron ore, and metal alloys in the production of various steel products. Similarly, automobile companies choose between several grades of steel and other materials in the manufacture of their products. Hence the term 'materials' can be defined differently depending upon the stage in the production process. In this chapter the focus is on consumption of finished metal and plastic products.

The trade-offs between theory and measurement are particularly vexing in the analysis of material consumption. Data on material consumption by stage of processing are needed. Unfortunately, these data are available infrequently and with a considerable time lag. Some researchers have developed procedures to generate missing observations. Rather than introduce possible measurement error through arbitrary data generation methods, the basic philosophy followed here is simply to use the reported data.

There is a great diversity of product attributes that are included in a material aggregate. For instance, total steel consumption includes sheet, strip wire, tube, bars and many other product classes. In addition, there are three different grades of steel—carbon, alloy and stainless—for most of the product shapes. Each of these steel products has a different unit price that largely reflects its physical properties. Simple addition of the tonnages for each shipment category would ignore these quality differentials. It is well known that this method of aggregation implies that the components are perfect substitutes. Consequently, the Divisia index, which weights each product by its value share of total expenditures, is used here because it does not make such a restrictive assumption.

Another important measurement issue concerns price. Many metal demand studies use producers' posted prices. The problem with this approach is that material producers generally subtract discounts from these posted prices. In this study, transactions prices are estimated by dividing the value of shipments by the quantity of shipments net of interplant transfers. Once consistent price indices are computed for the material aggregates, apparent consumption can be obtained by dividing net domestic expenditures by the appropriate divisia price index. Trade data collected by the US Bureau of the Census are used to estimate import and export quantities and prices. Exports are valued at domestic prices. Import prices include customs value plus insurance, freight and tariffs.

Certain metal products are specifically excluded from this analysis. For instance, semifinished material products, such as ingots, wire rods and slabs, are not included. Metal forgings and castings are also not considered. In addition, plate and heavy structural steel are not included in the steel aggregate because these products probably are not substitutes for the other three materials under consideration. For instance, steel plates are used in the construction of ships, barges and oil storage facilities, which generally cannot be built with alternative materials. Structural steel and some plate steel may be substitutes with cement and concrete products, which can best be addressed in a separate study.

The data for apparent consumption of steel, aluminium, copper and plastics are displayed in Fig. 7.1 and reveal some interesting patterns. First is the 11.5% compound annual rate of increase in plastic consumption over the sample period. Aluminium consumption also increased at a strong 5.2% annual rate. In contrast, apparent consumption of steel and copper grew at 0.7% and 1.5% respectively. In addition, each of the materials displays substantial cyclicality, which is reflected in the plot of total material consumption presented in Fig. 7.2. Note that total material consumption in

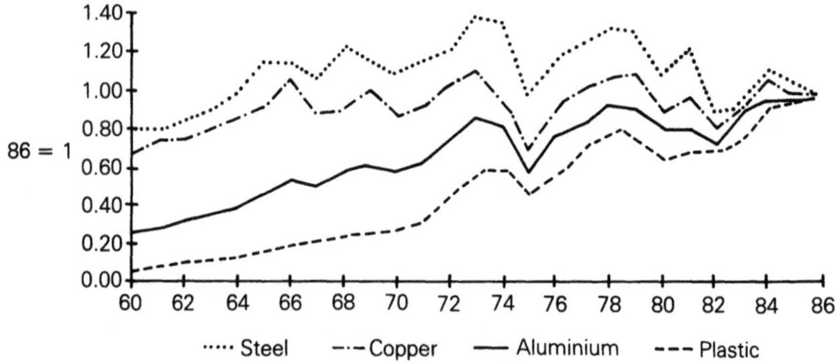

Fig. 7.1 Apparent consumption of steel, copper, aluminium and plastic products in the USA, 1960–86.

1986 is just about at the 1978 peak. It is quite likely that this peak has been surpassed since 1987 with the export-led boom in the US manufacturing sector.

Movements in the ratio of metal prices to plastics prices are presented in Fig. 7.3. Over most of the sample period, relative metal prices increased. With the exception of copper, it was not until the early 1980s that relative metal prices began to fall substantially. Trends in relative metal prices are illustrated in Fig. 7.4. The steel to aluminium price ratio steadily increased over most of the sample period. The copper to aluminium price ratio exhibited large swings, increasing sharply during the 1960s and falling since then.

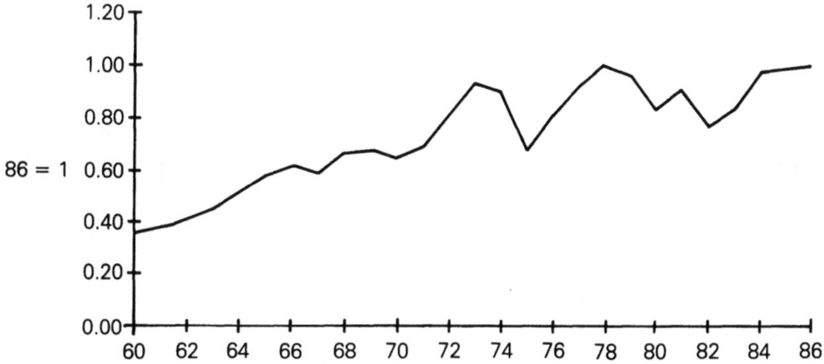

Fig. 7.2 Divisia index for apparent consumption of steel, copper, aluminium and plastic products in the USA, 1960–86.

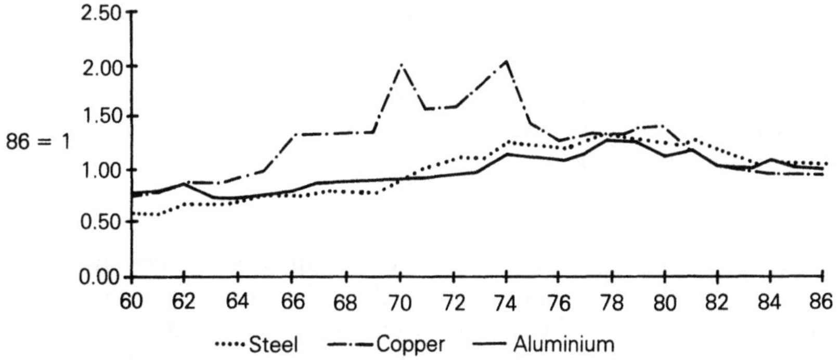

Fig. 7.3 Prices of steel, copper, and aluminium products relative to prices for plastic products.

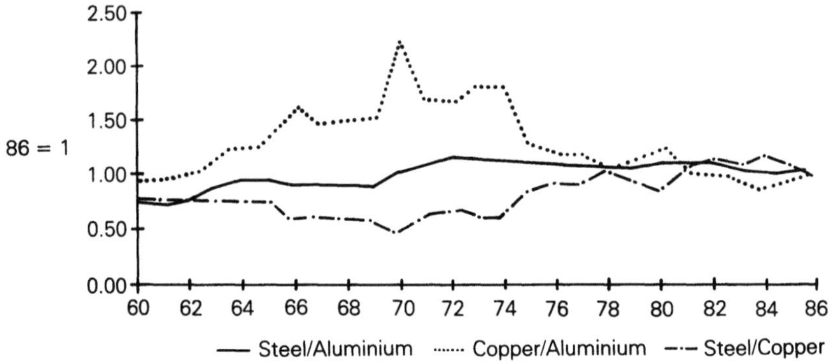

Fig. 7.4 Relative metal prices, 1960–86.

7.3 THEORETICAL FRAMEWORK

The material substitution model is specified assuming that producers minimize cost. If competitive input markets are assumed then producers select input levels to produce a certain level of output at minimum cost. These assumptions allow the first partial derivatives of cost with respect to prices to be equated with input quantities via Shephard's lemma: $Q_i = \partial C/\partial P_i$. The associated set of conditional input demand equations are zero-degree homogeneous in prices with symmetric cross price effects and non-negative input levels. To ensure concavity of the underlying cost function, the matrix of second-order partials of cost with respect to prices $(\partial^2 C/\partial P_i \partial P_j)$ should be negative semi-definite.

Certain separability assumptions must be employed in order to reduce the number of parameters that must be estimated. First consider an aggregate cost function for the major material-consuming sectors:

$$C = C[(P_{k1}, \ldots, P_{kn}), (P_{l1}, \ldots, P_{ln}), (P_{el}, \ldots, P_{en}), (P_{ml}, \ldots, P_{mn})] \quad (7.1)$$

where P_{ki}, P_{li}, P_{ei}, and P_{mi} represent prices for the ith type of capital, labour, energy and material input respectively. If we assume that materials are weakly separable from labour, capital and energy, we can specify the following material cost function:

$$P_m = C_m[P_s, P_c, P_a, P_p] \quad (7.2)$$

where P_s, P_c, P_a, P_p are the prices for steel, copper, aluminium and plastics respectively. Weak separability means that the cost-minimizing mix of materials is independent of the optimal mix and level of labour capital and energy even though the level of total material use is not (Fuss, 1977).

Another advantage of the weak separability assumption is that a general equilibrium framework that explicitly recognizes inter-industry transactions is not needed. If aggregate labour and capital are included in 7.2 then there would be a double-counting problem in which an aggregate 'prime' input, such as labour is used to make steel as well as automobiles. Anderson (1981) has shown that, for studies of aggregate manufacturing, the endogeneity of prices for intermediate goods such as steel does not permit the derivation of input demand relationship via Shepard's lemma because material prices would be endogenous.

Furthermore, this study considers apparent consumption, inclusive of net exports, which in recent years comprise more than 20% of the value of total material consumption. Hence, the assumption of exogenous material prices is not untenable.

Static neoclassical economic theory assumes that producers achieve their cost-minimizing input combination in each period. There are several reasons for this response to span several periods. Firms may not be sure whether price changes are temporary or permanent. Also, metals are used in conjunction with capital equipment and skilled labour which are both costly to adjust.

How best to represent this dynamic process is an empirical question. Two major approaches exist. The first approach assumes that firms minimize short-run variable costs subject to output and fixed factors of production (Brown and Christensen, 1981). Firms then minimize long-run total costs by adjusting their accumulation of fixed factors, such as equipment. Within this context, Considine (1989c) develops a two-stage generalized Leontief model of material substitution. The approach adopted below is theoretically less formal but more flexible. Specific *ceteris paribus* assumptions as to what input remains fixed during the adjustment process are not needed.

7.4 FUNCTIONAL FORM

Economic theory does not suggest any particular functional form but only that it satisfies the regularity conditions—positive inputs and negative own-price elasticities. Flexible functional forms circumvent the limitations associated with the constant elasticity of substitution (CES) model, which satisfies the regularity conditions for all data points. For more than two inputs the CES is very restrictive, requiring that the partial elasticities of substitution be equal for all pairs of inputs (Uzawa, 1962). A functional form often used in modelling interfuel substitution is the translog cost function.

Given the variable elasticities in the translog model, the concavity conditions vary with cost shares, or equivalently relative prices. Christensen and Caves (1980) define a range of relative prices over which the concavity conditions are satisfied. The size of this so-called regular region depends upon the estimated elasticities of substitution and the cost shares. For instance, the translog collapses to a Cobb–Douglas when the elasticities of substitution equal unity. In this limited case, the translog satisfies the regularity conditions everywhere and is said to be globally concave. As elasticities move further away from unity and as cost shares become increasingly unbalanced, the regularity conditions are satisfied for a smaller range of relative prices (Considine, 1990).

Another variable elasticity model with interesting global properties is the linear logit model of cost shares. The first point to note about the logit approach is that it does not require the specification of a cost function. This feature, however, does not make the logit approach *ad hoc*. Under constant returns to scale there is no theoretical reason requiring the formulation of a cost function because the cost shares or demand equations yield all the necessary information on the cost structure. Furthermore, the logit formulation can be used to estimate non-homothetic cost structures under variable returns to scale (Considine, 1988b).

To see that the logit model is not *ad hoc*, first consider a system of N share equations that can be written in terms of Shepard's lemma:

$$w_i = P_i \frac{\partial C}{\partial P_i} \Big/ \sum_{j=1}^{n} P_j \frac{\partial C}{\partial P_j} \tag{7.3}$$

Rather than deriving 7.3 by specifying some cost function and differentiating with respect to prices, as in the translog approach, we can specify a functional form for the price-weighted partial derivatives in 7.3. Instead of a linear formulation, we could use an exponential one:

$$w_i = \frac{\exp f_i}{\Sigma_{j=1}^{n} \exp (f_j)}$$

where

$$f_i = \eta_i + \sum_{j=1}^{n} \varphi_{ij} \ln P_j + \tau_i Z \tag{7.4}$$

and Z is an index of technical change. If we take the logarithm of the logistic function given by 7.4 we obtain the linear logit model of cost shares:

$$\ln w_i = f_i - \ln\left[\sum_{j=1}^{n} \exp (f_j)\right] \tag{7.5}$$

The dependent variable is the 'logit', a term first used by Berkson (1944) to describe a logarithm of any variable that sums to unity with values falling between zero and unity. The equivalent properties in demand theory are adding-up and non-negativity. Hence the logit form automatically satisfies two very basic conditions of neoclassical demand theory.

In assessing the applicability of logit models to applied demand analysis, the distinction between the inherent properties of this function form and external restrictions is very important. Hausman (1975) and Oum (1979) find irregularities in price elasticities in some applications of logit models. These problems largely result from the restrictions implied by the independence of irrevelant alternatives. This axiom states that the choice of A over B is independent of the presence or absence of an unchosen third alternative. Obviously, this is an implausible assumption to employ when specifying a demand system. The logit model of cost shares used here does not make this assumption.

The unconstrained share elasticities can be found by differentiating 7.5:

$$H_{ik} = \frac{\delta \ln w_i}{\delta \ln P_k} = \varphi_{ik} - \sum_{j=1}^{n} w_j \varphi_{jk} \tag{7.6}$$

The own-price and cross-price elasticities of demand are functions of these share elasticities:

$$E_{ik} = H_{ik} + w_k \qquad E_{ii} = H_{ii} + w_i - 1 \tag{7.7}$$

Allen–Uzawa elasticities of substitution are given by

$$\sigma_{ij} = E_{ij}/w_j \tag{7.8}$$

The sufficient conditions of neoclassical demand theory can be imposed by linear restrictions on the parameters in 7.4. Homogeneity in prices can be imposed with the following constraints:

$$\sum_{j=1}^{n} \varphi_{ij} = d \qquad \text{for all } i \tag{7.9}$$

where d is an unknown scalar that can be normalized to zero (Considine and Mount, 1984). The symmetry conditions can be imposed with the following constraint:

$$\varphi^*_{ik} = \varphi^*_{ki} \qquad \text{for all } i \neq k \tag{7.10}$$

in which $\varphi^*_{ik} = \varphi_{ik}/w^*_i$ where w^*_i are specific cost shares. The mean cost shares are the logical choice. Anderson and Thursby (1986) argue that valid statistical inference in variable elasticity models, such as the translog, can only be conducted based on the mean of the actual cost shares. Although this argument does not provide a justification for imposing symmetry at the mean, it suggests that elasticities derived from the logit model are subject to the same statistical qualifications. If valid statistical inference is feasible only at the mean, then testing the neoclassical theory of derived demand requires the imposition of symmetry and homogeneity restrictions at that point.

Given the homogeneity (7.9) and symmetry (7.10) constraints, the elasticities of substitution at the point of symmetry are

$$\sigma^*_{ii} = \left(- \sum_{j \neq i} w^*_j \varphi^*_{jk} + w^*_i - 1\right)\Big/ w^*_i \qquad \sigma^*_{ik} - \varphi^*_{ik} + 1 \qquad \text{for } i \neq k \tag{7.11}$$

The form of the own-price elasticities of substitution in 7.11 follow from the homogeneity constraints (7.9). The elasticities of substitution for cost shares other than the mean can be expressed as a share-weighted function of these point estimates of the elasticities of substitution, σ^*_{ik}:

$$\sigma_{ii} = \left[- \sum_{k \neq i} w_k \sigma_{ik} \right] \Big/ w_i \qquad \sigma_{ik} = \frac{w^*_k}{w_k} \left(\sigma^*_{ik} - \sum_{j=1}^{n} w_j \sigma^*_{jk} \right) \qquad \text{for } i \neq k \tag{7.12}$$

An examination of these expressions for the elasticities of substitution reveals that the logit model of cost shares has some important links to the CES model. For example, a logit cost share model for two inputs collapses to a CES model. The logit cost share model also collapses to a CES model when the elasticities of substitution are all equal. For less restrictive cases, however, the logit model is a variable elasticity model for more than two inputs.

This relationship with the CES is important for the global properties of the linear logit model of cost shares. Non-negativity of input levels is ensured by the logit model because the exponential functions are always positive. Futhermore, the concavity conditions for the linear logit model are continuous functions of the cost shares. These conditions require that the matrix of second partial derivatives of cost with respect to price is negative semidefinite. This implies that the eigenvalues of this matrix are all less than or equal to zero or equivalently that the principal minors alternate in sign starting with negative.

The second partial derivatives of cost can be expressed in terms of the share elasticities as follows:

$$\frac{\delta^2 C}{\delta P_i \delta P_j} = \frac{w_i Q_i}{P_i} + \frac{C w_i}{P_i P_j} \left(\frac{\delta \ln w_i}{\delta \ln P_j} \right) + \mu \frac{Q_i}{P_j} \tag{7.13}$$

where $\mu = 1$ when $i = j$ and $\mu = 0$ otherwise. For the logit model these derivatives are

$$\frac{P_{it} P_{jt}}{C} \left(\frac{\delta^2 C}{\delta P_{it} \delta P_{jt}} \right) = w_i \left(w_j + w^*_j \varphi^*_{ij} - \sum_{k=1}^{n} w_k \, w^*_j \varphi^*_{kj} - \mu \right) \tag{7.14}$$

The second partial derivatives can be conveniently expressed in matrix notation:

$$\frac{pp}{C} \left(\frac{\delta^2 C}{\delta p \, \delta p} \right) = \Omega = (ww' - W^D)[I - \varphi W^{D^*}] \tag{7.15}$$

where w is a $n \times 1$ vector of cost shares, W^D is an $n \times n$ diagonal matrix formed with w (the asterisk indicates the mean shares), and φ is an $n \times n$ matrix of the estimated parameters φ^*_{ik}. Given that non-negativity is assured in the logit model, the first matrix in 7.15 is negative definite. The sign of the second matrix will vary with the parameter estimates. So there is no mathematical property of the logit model that would bias the elasticities of substitution in either direction. At the mean shares, the matrix given by 7.15 reduces to

$$\Omega^* = w^* w^{*\prime} - W^{D^*} + W^{D^*} \varphi W^{D^*} \tag{7.16}$$

If Ω^* is negative semidefinite, then own-price elasticites remain negative for all shares (Considine, 1989a). Note that, away from the mean cost shares, symmetry does not hold, and so the model deviates from cost-minimizing behaviour. A non-linear globally symmetric version of the model is developed by Considine (1990). For policy analysis and forecasting, however, the linear logit model can be quite useful, as demonstrated below.

7.5 EMPIRICAL RESULTS

The base model assumes non-neutral technical change, dynamic adjustments and homotheticity. Three alternative models, which examine each of these three assumptions, are estimated. Both translog and logit formulations are estimated for each of the four models. Rather than a simple time trend, technical change is represented by the average fuel economy of the US automobile fleet. For the linear logit model, the dynamic formulation is essentially a simple Nerlovian partial adjustment process. Instead of shares, the adjustment process is specified in input levels. If the adjustment parameter falls between zero and unity, this formulation guarantees that short-run price elasticities are less than their long-run counterparts (Considine and Mount, 1984). A similar formulation with a set of translog cost shares is not possible because the adding-up constraints would force the adjustment parameter to zero. Consequently the dynamic translog model uses lagged cost shares.

The specification of the translog estimating equations is well known and therefore will not be presented here. The logit cost share model for the four-input material cost model with zero-degree homogeneity in prices and symmetry imposed at the mean cost shares is as follows:

$$\begin{aligned}
\ln\left(\frac{w_{1t}}{w_{4t}}\right) = {} & (\eta_1 - \eta_4) - [w_2^*\varphi_{12}^* + w_3^*\varphi_{13}^* + (w_1^* + w_4^*)\varphi_{14}^*]\ln\left(\frac{P_{1t}}{P_{4t}}\right) \\
& + (\varphi_{12}^* - \varphi_{24}^*)w_2^*\ln\left(\frac{P_{2t}}{P_{4t}}\right) \\
& + (\varphi_{13}^* - \varphi_{34}^*)w_3^*\ln\left(\frac{P_3}{P_{4t}}\right) \\
& + (1 - \lambda)\ln\left(\frac{Q_{1t-1}}{Q_{4t-1}}\right) + (\tau_1 - \tau_4)Z_t + (\varepsilon_{1t} - \varepsilon_{4t})
\end{aligned}$$

$$\begin{aligned}
\ln\left(\frac{w_{2t}}{w_{4t}}\right) = {} & (\eta_2 - \eta_4) - [w_1^*\varphi_{12}^* + w_3^*\varphi_{23}^* + (w_2^* + w_4^*)\varphi_{24}^*] \\
& \times \ln\left(\frac{P_{2t}}{P_{4t}}\right) + (\varphi_{12}^* - \varphi_{14}^*)w_1^* \ln\left(\frac{P_{1t}}{P_{4t}}\right) \\
& + (\varphi_{23}^* - \varphi_{34}^*)\omega_3^* \ln\left(\frac{P_{3t}}{P_{4t}}\right) + (1 - \lambda)\ln\left(\frac{Q_{2t-1}}{Q_{4t-1}}\right) \\
& + (\tau_2 - \tau_4)Z_t + (\varepsilon_{2t} - \varepsilon_{4t}) \\
\ln\left(\frac{w_{3t}}{w_{4t}}\right) = {} & (\eta_3 - \eta_4) - [w_1^*\varphi_{13}^* + w_2^*\varphi_{23}^* + (w_3^* + w_4^*)\,\varphi_{34}^*]\ln\left(\frac{P_{3t}}{P_{4t}}\right) \\
& + (\varphi_{13}^* - \varphi_{14}^*)w_1^* \ln\left(\frac{P_{1t}}{P_{4t}}\right) \\
& + (\varphi_{23}^* - \varphi_{24}^*)\omega_2^* \ln\left(\frac{P_{2t}}{P_{4t}}\right) \\
& + (1 - \lambda)\ln\left(\frac{Q_{3t-1}}{Q_{4t-1}}\right) + (\tau_3 - \tau_4)Z_t + (\varepsilon_{3t} - \varepsilon_{4t}) \qquad (7.17)
\end{aligned}$$

where the input order is steel, copper, aluminium and plastic, Z_t is the index of technical change and $(\varepsilon_{it} - \varepsilon_{nt})$ are assumed to be normally distributed random disturbances. The estimator is full information maximum likelihood. The multiplicative error specification in the logit model is consistent with the theoretical requirement that shares lie between zero and unity (Chavas and Segerson, 1986). Maximum likelihood estimates of 7.17 are invariant to the selection of the base input (Considine and Mount, 1984).

Each of the translog models violates the concavity conditions and yields positive own-price elasticities for many observations in the sample. In contrast, all own-price elasticities are negative for the linear logit model. To illustrate some of the problems associated with the translog, Fig. 7.5 presents a time series plot of the own-price elasticity for copper products estimated by translog and logit formulations of the model that is static and homothetic with non-neutral technical change. Early in the sample the translog price elasticities fall between −0.25 and −0.40. After 1975, the translog estimates become positive (Fig. 7.5). In contrast, the logit elasticities steadily decline from −0.15 in 1960 to −0.40 in 1986.

The reason for these differences lies with the mathematical form of the elasticities for both models. For instance, the translog own-price elasticities

$$E_{ii}^{\mathrm{TL}} = \frac{\beta_{ii}}{w_i} - (1 - w_i) \qquad (7.18)$$

are inversely proportional to the cost shares. Movements in the translog elasticities tend to become magnified because the denominator in 7.18 is less

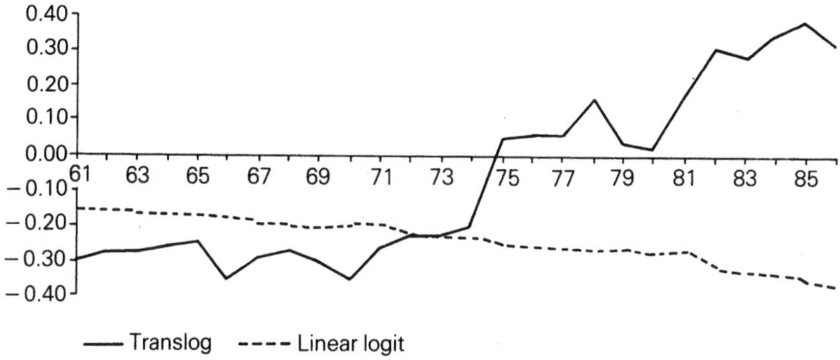

Fig. 7.5 Comparison of translog and logit estimates for the own price elasticity of copper products.

than unity. Thus the rather unusual behaviour of the translog own-price elasticity for copper in Fig. 7.5 occurs as the cost share for copper declines, which drives the elasticity upwards. Such large swings in the elasticities are counter-intuitive.

The own-price elasticities in the linear logit model of cost shares do not have a ratio form

$$E_{ii}^{\mathrm{LL}} = w_{ii}^{*}\left(\varphi_{ii}^{*} - \sum_{j=1}^{n} w_j \varphi_{ji}^{*}\right) - (1 - w_i) \tag{7.19}$$

and therefore do not exhibit exaggerated movements as shares change. The estimated own-price elasticity for the logit model is far more stable. Consequently, subsequent analysis is conducted using the linear logit model.

Hypothesis testing proceeds under the maintained assumptions of a dynamic homothetic model with non-neutral technical change. The hypothesis of neutral technical change in material use is decisively rejected. The log likelihood value is 118.04 for model 1 and 103.5 for the model assuming neutral technical change. The resulting χ^2 statistic is 29.2 and far exceeds the critical value of 7.8 at the 5% significance level. The static model has a likelihood value of 115.3. This implies a χ^2 statistic of 5.5, which also exceeds the 5% critical value of 3.8 at one degree of freedom. Hence the static model also cannot be accepted.

Many time series studies have avoided testing for homotheticity in the presence of non-neutral technical change because output tends to be highly correlated with time. The correlation coefficient between the index of technical change used here and deflated material expenditures is only 0.30. Hence, a non-homothetic cost structure is tested assuming non-neutral technical change with partial adjustments in quantities. Estimation of this model yields a log likelihood value of 119.6 and a χ^2 statistic of 3.0 when compared with the base model, which is less than the critical value. Hence, a non-homothetic formulation cannot be accepted.

The parameter estimates and summary fit statistics for the base model are presented in Table 7.1. Eight of the ten price parameters are significantly different from zero at the 5% level of significance. The price parameters are significant even for the rejected models. The technical chance coefficients and adjustment parameter are also very significant. The fit of the predicted shares is quite good and the residuals do not display significant serial correlation. Although the degree of serial correlation is reduced by the dynamic specification, the reduction is not dramatic. Similarly, the fit of the rejected neutral technical change model is fairly good. Thus adding the index of technical changes does not alone contribute to the excellent fit.

The price elasticities and Allen partial elasticities of substitution evaluated at the mean cost shares are presented in Table 7.2. The estimates confirm the stylized fact that materials have inelastic short-run demand. The short-run demand for plastics, however, is a fairly price-responsive −0.92.

Table 7.1 Parameter estimates and summary fit statistics with asymptotic *t* ratios in parentheses

$\eta_1 - \eta_4$	2.555 (7.571)	φ_{23}	−0.325 (−0.807)
$\eta_2 - \eta_4$	0.857 (3.053)	φ_{24}	0.806 (2.834)
$\eta_3 - \eta_4$	−0.082 (−0.435)	φ_{33}	6.777 (6.323)
φ_{11}	0.331 (2.512)	φ_{34}	−1.161 (−4.874)
φ_{12}	−1.906 (−9.702)	φ_{44}	−0.717 (−1.410)
φ_{13}	−0.781 (−4.346)	$\tau_1 - \tau_4$	−1.973 (−6.181)
φ_{14}	0.503 (1.863)	$\tau_2 - \tau_4$	−1.668 (−6.161)
φ_{22}	6.302 (13.924)	$\tau_3 - \tau_4$	−0.907 (−5.007)
		$1 - \lambda$	0.441 (5.894)

Fit statistics for cost shares

	R^2	Durbin–Watson
Steel	0.933	2.091
Copper	0.980	2.252
Aluminium	0.805	2.677
Plastic	0.968	1.859

Table 7.2 Price and substitution elasticities evaluated at the mean cost shares

	Steel	*Copper*	*Aluminium*	*Plastic*
Price elasticities				
Short run				
Steel	−0.355	−0.104	0.024	0.435
Copper	−0.439	−0.159	0.075	0.523
Aluminium	0.106	0.078	−0.137	−0.047
Plastic	0.728	0.208	−0.018	−0.918
Long run:				
Steel	−0.805	−0.237	0.055	0.987
Copper	−0.995	−0.360	0.170	1.186
Aluminium	0.241	0.176	−0.311	−0.106
Plastic	1.651	0.472	−0.041	−2.083
Allen–Uzawa partial elasticities of substitution				
Short run				
Steel	−0.733	−0.906	0.219	1.503
Copper		−1.380	0.675	1.806
Aluminium			−1.238	−0.161
Plastic				−3.173
Long run				
Steel	−1.662	−2.054	0.498	3.409
Copper		−3.129	1.531	4.097
Aluminium			−2.808	−0.366
Plastic				−7.196

The short-run demand relationships for copper and aluminium are very inelastic, which may help to explain price volatility in those markets. The own-price elasticity of steel demand is −0.36 in the short run and −0.80 in the long run. Copper and aluminium remain price inelastic in the long-run while the long-run demand for plastic is price elastic.

Also displayed in Table 7.2 are the Allen–Uzawa partial elasticities of substitution also evaluated at the mean cost shares. Complementarity is estimated between copper and steel and between aluminium and plastics. Steel and aluminium and copper and aluminium are estimated to be substitutes. Plastics are a strong substitute for both steel and copper products. The price elasticities (Fig. 7.6) display some interesting movements over time. First, the short-run own-price elasticity of demand for steel rises (in absolute value) from −0.20 in 1961 to −0.60 in 1986, primarily due to greater competition from plastics, which is reflected in a cross-price elasticity rising from 0.24 to 0.72 over the same period. The demand for copper also becomes more price responsive, with the own-price elasticity rising from −0.11 in 1961 to −0.23 in 1986, again due to more competition from plastics. The own-price elasticity for aluminium does not really change over the sample period. Even though plastic and aluminium are complements at

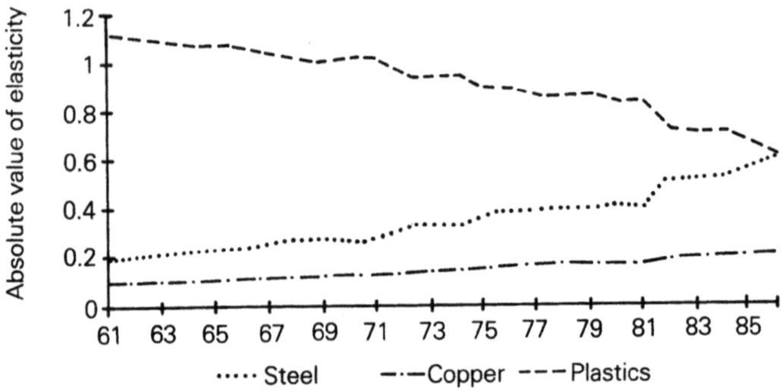

Fig. 7.6 Linear logit estimates of own-price elasticities for steel, copper and plastics, 1961–86.

the mean of the sample, after 1982 they are estimated to be substitutes. Also during the later period aluminium and steel are estimated to be very weak complements, which may reflect the growing importance of speciality steels in the steel aggregate. The short-run own-price elasticity of demand for plastics steadily decreases from −1.11 in 1960 to −0.63 in 1986. This trend reflects a classic pattern of market penetration for a new product in which supply-side innovations lead to falling relative prices and rising consumption until market saturation is reached. The concavity conditions hold at the mean and therefore for all predicted cost shares.

The bias and rates of technical change for each material also can be determined. Figure 7.7 presents the share elasticities with respect to the technical change index from 1961 to 1986. The negative share elasticities for steel and copper suggest that technical change is baised toward saving these materials. Furthermore, both share elasticities increase steadily over the sample period, which suggests a faster pace of technical change, especially during the 1980s (Fig. 7.7). The positive share elasticities for

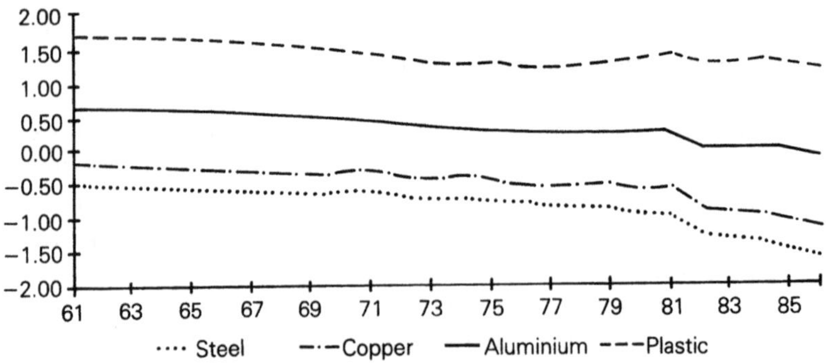

Fig. 7.7 Share elasticities with respect to technical change, 1961–86.

aluminium and plastic suggest that technological forces, in addition to relative price movements, promoted the use of these materials. Both the last two share elasticities decline over the sample period but the aluminium share elasticity declines rather sharply. Finally, the size of these share elasticities suggests that exogenous technical change is at least as important as relative prices in explaining material market shares.

REFERENCES AND FURTHER READING

Anderson, R. A. (1981) On the specification of conditional factor demand functions in recent studies of U.S. manufacturing, in *Measuring and Modeling Natural Resource Substitution* (eds E. Berndt and B. Field), MIT Press, Cambridge, MA, pp. 119–43.

Anderson, R. A. and Thursby, J. G. (1986) Confidence intervals for elasticity estimators in translog models. *Review of Economics and Statistics*, 647–56.

Berkson, J. (1944) Application of the logistic function to bioassay. *Journal of the American Statistical Association*, **39**, 357–65.

Bewley, R. (1966) *Allocation Models*, Ballinger, Cambridge, MA.

Brown, R. S. and Christensen, L. R. (1981) Estimating elasticities of substitution in a model of partial static equilibrium, in *Measuring and Modeling Natural Resource Substitution*, (eds E. Berndt and B. Field), MIT Press, Cambridge, MA, pp. 207–29.

Chavas, J. P. and Segerson, K. (1986) Singularity and autoregressive disturbances in linear logit models, *Journal of Business and Economic Statistics*, **4** (2), 161–69.

Christensen, L. R. and Caves, D. W. (1980) Global properties of flexible functional forms. *American Economic Review*, **70**, 422–32.

Considine, T. J. (1989a) Estimating the demand for energy and natural resource input: trade offs in global properties. *Applied Economics*, **21**, 931–45.

Considine, T. J. (1989b) Separability, functional form, and regulatory policy in models of interfuel substitution. *Energy Economics*, **11**, 82–94.

Considine, T. J. (1989c) Modelling material substitution with quasi-fixed inputs: a two-stage generalized Leontief formulation, Department of Mineral Economics Working Paper, Pennsylvania State University, July.

Considine, T. J. (1990) Symmetry constraints and variable returns to scale in logit models. *Journal of Business and Economic Statistics*, **8**, (3), 347–53.

Considine, T. J. and Mount, T. D. (1984) The use of linear logit models for dynamic input demand systems. *Review of Economics and Statistics*, **66**, 434–43.

Diewert, W. E. and Wales, T. J. (1987) Flexible functional forms and global curvature conditions. *Econometrica*, **55**, 43–68.

Fuss, M. A. (1977) The demand for energy in Canadian manufacturing. *Journal of Econometrics*, **3**, 89–116.

Hausman, J. A. (1975) Project independence report: An appraisal of U.S. energy needs up to 1985. *Bell Journal of Economics* **6**, Autumn, 517–51.

Hazilla, M. and Kopp, R. J. (1984) A factor demand model for strategic nonfuel minerals in the primary metals sector. *Land Economics*, **60** (4), 328–39.

Oum, T. H. (1979) A warning on the use of linear logit models in transport mode choice studies. *Bell Journal of Economics*, **10**, 374–88.

Uzawa, H. (1962) Production functions with constant elasticities of substitution. *Review of Economic Studies*, **29**, 291–9.

8

Spectral interpretation of stock adjustment processes in mineral markets

AHMAD AFRASIABI, MASSOUD MOALLEM AND WALTER C. LABYS

8.1 INTRODUCTION

In this chapter we investigate long-run market adjustment processes in the copper, lead and zinc markets. These adjustments have been particularly problematic for both producers and consumers of minerals; they embody long-term cyclical variations which cannot be explained by conventional short-term adjustment mechanisms. The possibility of the existence of these long-run processes has been hinted at in several mineral studies which suggest periodic components of 5–15 years (e.g. Slade, 1981). Particularly, in the case of copper, lead and zinc in particular much of the variance of the market time series has been found to be in the corresponding frequency bands (e.g. Labys and Afrasiabi, 1983).

Previous theoretical explanations have related these long-term cyclical variations to swings in capacity adjustment. There is a consensus among various authors that lumpy investments and the long gestation periods common to mineral projects are among the most important causes of low frequency cycles in mineral prices and quantities. While much of the discussion of this phenomenon has been limited to expository research, attempts to confirm it econometrically have been hampered primarily by the problems of dynamic specification and secondarily by the lack of accurate capacity data.

Excluding the studies based on rational expectations, there is every indication that the question of dynamic specification in econometric models of mineral commodities is generally treated on an *ad hoc* basis (e.g. Labys,

1980a). Yet it should not be necessary to mention that a model's ability to explain cyclicalities of a certain nature is primarily dependent on its dynamic specification. In particular, in mineral industries where the speeds of adjustment are low and lags are lengthy, the specification of appropriate lag structures is essential for constructing realistic models.

The purpose of this chapter is to attempt to explain this dynamic behaviour, employing an econometric model of mineral markets based on spectral regression. The spectral approach has two important advantages. First, the model can be estimated and tested only over the particular frequency range of interest—hence the noisy high frequency variations that disrupt results based on standard regression techniques are avoided. Second, the model can be estimated without imposing strong parametric restrictions on the lag structures.

8.2 MODEL FRAMEWORK

The theoretical background of a commodity model such as the one employed here can be found in detail in Labys and Pollack (1984). This model, although of a mini-variety, embodies most of the features deemed necessary to explain long-run market adjustments. Limitations are imposed on the size and the scope of the model by the lack of needed data and by the sheer size of the computational requirements. The model described below is based on a set of static equations; because these equations are estimated and interpreted in the frequency domain they already possess a dynamic interpretation. The infinite-dimensional parameter space implied by this approach ensures that a model's parameters are estimated within the most general dynamic setting that can be afforded for a given sample size. The simplest model specification that can be proposed describing mineral market adjustments is as follows.

$$D_t = D(P_t, \mathrm{PS}_t, Y_t) \tag{8.1}$$

$$K_t = K(P_t/C_t, Y_t) \tag{8.2}$$

$$\mathrm{QS}_t = K_t U_t \tag{8.3}$$

$$U_t = U(P_t, Y_t) \tag{8.4}$$

$$P_t = P(I_t/D_t, U_t) \tag{8.5}$$

$$\mathrm{NET}_t = \mathrm{NET}(\mathrm{PDF}_t, I_t/D_t) \tag{8.6}$$

$$I_t = I_{t-1} + Q_t - D_t + \mathrm{NET}_t \tag{8.7}$$

Definitions of the variables are as follows: D, demand; P, price, PS, price of substitute, Y, activity indicator of the using sectors; K, capacity; QS, quantity supplied; U, capacity utilization; NET, net foreign trade; I, inventories; I/D, coverage ratio; P/C, ratio of prices to cost of capital, PDF, domestic–international price difference. Further definitions of the variables and their supporting data can be found in the Appendix (Table 8.A1).

Demand Equation 8.1 is a simple derived demand function obtainable from the first-order conditions of cost minimization. The price of aluminium is used in the copper and zinc demand equations but not in the lead equation since aluminium is not a substitute for lead in its major use category, i.e. storage batteries. Since all three metals are used widely in many industrial sectors, the Federal Reserve index of industrial production is employed as the activity variable.

Equations 8.2, 8.3 and 8.4 together form the supply side of the model. The capacity Equation 8.2 can be derived along the lines of the standard neoclassical investment model. Briefly we can write[1]

$$K = A(P/C)^{\sigma}Q^{*} \tag{8.8}$$

where Q^* is the expected level of output and σ is the elasticity of substitution. Assuming that firms' expectations about their future rate of growth of output is a constant multiple β of their expectation about the rate of growth of industrial activity, we can write

$$K = A(P/C)^{\sigma}Y^{\beta}$$

Reserve formation R in turn, depends on lagged prices P.D.F.

Once the level of productive capacity has been determined, the decision about the short-run level of output is assumed to be essentially the same as that of determining what fraction of this total capacity the firm wants to utilize, given short-run market conditions. This is reflected in Equations 8.3 and 8.4 which represent the supply identity and capacity utilization equations respectively. In Equation 8.4 it is assumed that capacity utilization in the short run is governed by the level of industrial activity and prices.

Major advantages of this formulation, as originally tested in the context of the copper market by Labys (1980), are as follows:

1. It avoids the problems associated with the partial adjustment specification of the supply equation which represents a predetermined short-run –long-run relationship. For further details see Lonoff (1983).
2. It permits specification of two separate lag structures for short- as well as long-run expectations.
3. It allows a distinction to be made between *ex post* and *ex ante* excess capacity. This is necessary since capacity utilization is treated as a decision variable rather than as an *ex post* ratio of current output to current capacity. This is an important feature when the industry being modelled is suspected to plan for excess capacity.

[1]The capacity data available to us do not, strictly speaking, measure the productive capacity of the industry and only reflect the size of fixed capital in terms of output.

Equation 8.5 is a conventional specification of the price equation. In this equation it is assumed that prices are adjusted until desired ratios of inventories to demand and of output to capacity (capacity utilization) are reached. Finally, the effect of foreign trade on the market is summarized in the net trade Equation 8.6. The flow diagram of the model is presented in Fig. 8.1.

8.3 STATISTICAL APPROACH

Since the pioneering work of Hannan (1963), regression analysis in the frequency domain has been considered by economists such as Sims (1971), Dhrymes (1971) and Fishman (1968). The properties of the related regression estimators have been studied under a variety of conditions by Engle (1978, 1980), Espasa (1977) and Espasa and Sargan (1977). The approach adopted here is similar to these studies except that the estimated frequency response parameters are not transformed back to the time domain. Since the focus of the study is on cyclical variations, we prefer to interpret the parameters in the frequency domain.

A transfer function is estimated for each frequency band in the following manner.[2] Let W be the matrix of Fourier elements defined as

$$W_{jt} = T^{-1/2} \exp(\mathrm{i}\omega_j t)$$

where $\omega_j = 2\pi j/T$, $\mathrm{i}^2 = -1$, and $t = 1, 2, 3, \ldots, T$. The periodogram of two vector series Y and X and the cross-periodogram between them are given by

$$I_{yy}(\omega_j) = |(WY)j|^2$$

$$I_{xx}(\omega_j) = |(WX)j|^2$$

$$I_{yx}(\omega_j) = (WY)_j^{\mathrm{H}}(WX)_j$$

where superscript H denotes the Hermitian of the complex matrix. These estimates are smoothed to yield consistent estimates of the spectral and cross-spectral density matrices S_{yy}, S_{xx} and S_{yx} respectively (dependence on frequency is dropped for convenience).[3] If Y is related to X by a system of linear filters, $Y = L(X)$, then least squares estimates of the transfer function of L are obtained from

$$\hat{B} = S_{yx}S_{xx}^{-1}$$

Given the estimates of this transfer function, gain, phase and coherency

[2]For development of the theory of spectral regression see Koopmans (1974) or Brillinger (1970).

[3]In band spectrum regression the periodograms are not smoothed simply because the sum of periodograms approaches the sum of spectral values which is the total variance of the variable. But since we interpret the estimates for individual frequency bands, we need to smooth the periodograms to obtain consistent estimates.

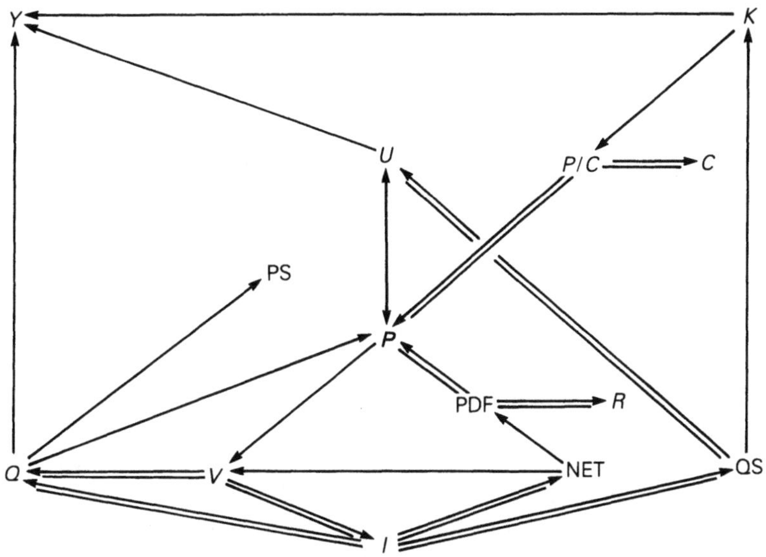

Fig. 8.1 The flow diagram of the model.

values can be estimated using their standard definitions (Koopmans, 1974; Fuller, 1976). The choice of a smoothing function is particularly difficult in this case since we have only a limited number of observations. A wide bandwidth had to be chosen to release enough degrees of freedom for the regression estimates. The resulting loss of resolution over the lower frequency bands is significant but this is not as important in a multivariate study as it would be in a univariate study, where the primary goal is the detection of cycles. A Daniel (rectangular) window of length 7 was chosen to gain 14 degrees of freedom. Experiments with other smoothing functions (some in conjunction with cosine or trapezoidal tapers) have also been conducted. However, any gains obtained in resolution are more than offset by the loss in degrees of freedom. With fewer degrees of freedom, tests of the parameters of equations such as demand (with three explanatory variables) become less meaningful.

Although the development of multivariate spectral estimators is routine by virtue of their analogy to standard regression and correlation theory, the task of determining the sampling distributions of these estimators is exceedingly difficult. The asymptotic distributions of partial and multiple coherencies have been computed by Goodman (1963) and Khatri (1964) and the distribution of regression parameter estimates by Hannan (1970) and Brillinger (1970). Despite the complexity of the asymptotic distributions of these estimators, simple statistical tests of significance and of goodness of fit have been developed.[4]

These test statistics have been calculated for every frequency band and have been employed to test formally the validity of the causality postulated by the model at the frequency bands of interest. Given the bandwidth selected above, the stated periodicity limits of 5–15 years correspond to frequency band limits of 0.34–0.85 to 0.93–1.44 radians per year. The latter two bands will be referred to as band A and band B respectively.

8.4 EMPIRICAL RESULTS

Data employed in the model estimation include the years 1906–80, a period long enough to examine long-run cycles in capacity adjustment. The copper, lead and zinc industries were selected for testing the model because of the strong cycles appearing in their capacities and because of the availability of data. A further description of the data and their characteristics can be found in the Appendix. The estimation of the above model for the three minerals for all possible frequency points has generated a very large number of observations, the presentation of which is both cumbersome and unnecessary. In order to make the results tractable we present them in Tables 8.1–8.3 for bands A and B only and only for the frequency at which peaks in multiple coherence occur. For each variable the estimates of partial gain, time shift and the F ratio for the test of zero gain are reported. Time shift, referred to as 'shift' in the tables, has been calculated from the estimates of the phase shift function.

In order to facilitate the interpretation of results, please note the parallelism between transfer function analysis in the frequency domain and regression analysis in the time domain. In this respect, gain and multiple coherence are equivalents of partial regression coefficients and the R^2 respectively. The time delays for each explanatory variable at each frequency are inferred from the time shift parameters. Since time shift values are direct as well as naive translations of the phase shift functions, it is necessary to be cautious in their interpretation.[5]

[4]Under the assumption of zero partial coherence (or zero gain) it can be shown that $(\mathrm{df} - 2P)k^2/2(1 - k^2)$ has an F distribution with 2 and $\mathrm{d}f - 2P$ degrees of freedom, where k is the partial coherence, df is the degrees of freedom of the spectral density estimates, and P is the number of explanatory variables. Similarly, under the assumption of zero multiple coherence it is shown that $(\mathrm{df} - 2P)k^2/2P(1 - k^2)$ has an F distribution with $2P$ and $\mathrm{df} - 2P$ degrees of freedom.

[5]In general, for periodic components, lags can be interpreted as leads and vice versa. Furthermore, whether the relation between two time series is direct or inverse is also recorded by phase differentials.

Table 8.1 Estimates of the parameters of the copper model[a]

Equation		*Price*	*Industrial activity*	*Aluminium price*	*Price interest ratio*	*Inventory demand ratio*	*Capacity–utilization ratio*	*Price differential*
Band A								
Demand								
	Gain	0.58	1.36	0.25				
	Shift	−0.29	0.55	6.10				
	F	3.64	14.43	0.95				
K^2 = 0.88								
F' = 9.78								
Utilization								
	Gain	0.64	0.78					
	Shift	−0.42	−0.07					
	F	4.83	3.21					
K^2 = 0.79								
F' = 9.40								
Capacity								
	Gain		0.14		0.06			
	Shift		−1.78		−2.16			
	F		0.52		0.19			
K^2 = 0.19								
F' = 0.59								
Price								
	Gain					0.69	1.66	
	Shift					6.07	8.10	
	F					8.78	5.20	
K^2 = 0.76								
F' = 7.92								

Net trade							
	Gain					71.55	5.19
	Shift					−2.15	6.77
	F					3.17	0.20
$K^2 = 0.40$							
$F' = 1.67$							
Band B							
Demand							
	Gain	0.62	2.26	0.38			
	Shift	−1.53	−0.11	−1.33			
	F	2.51	9.40	0.69			
$K^2 = 0.79$							
$F' = 5.02$							
Utilization							
	Gain	0.35	1.25				
	Shift	−0.48	−0.02				
	F	1.97	7.71				
$K^2 = 0.80$							
$F' = 10.00$							
Capacity							
	Gain		0.10		0.12		
	Shift		0.12		−2.28		
	F		0.24		2.15		
$K^2 = 0.35$							
$F' = 1.35$							

Price				
	Gain	0.21	0.59	
	Shift	1.83	1.25	
	F	11.47	6.54	
K^2 = 0.83				
F' = 12.21				
Net trade				
	Gain	35.03		12.79
	Shift	0.30		−0.76
	F	7.29		7.76
K^2 = 0.74				
F' = 7.16				

[a]See note 5 for explanation of the table entries.
Source: Based on the authors' computations.

Table 8.2 Estimates of the parameters of the zinc model

Equation		*Price*	*Industrial activity*	*Aluminium price*	*Price interest ratio*	*Inventory demand ratio*	*Capacity–utilization ratio*	*Price differential*
Band A								
Demand								
	Gain	0.39	0.95	0.11				
	Shift	1.49	−0.25	0.88				
	F	2.49	9.20	0.27				
$K^2 = 0.84$								
$F' = 7.00$								
Utilization								
	Gain	0.93	0.79					
	Shift	0.47	−0.18					
	F	10.24	3.76					
$K^2 = 0.85$								
$F^{\circ} = 14.17$								
Capacity								
	Gain		0.73		0.50			
	Shift		0.83		−3.85			
	F		10.31		11.75			
$K^2 = 0.75$								
$F' = 7.50$								
Price								
	Gain					0.05	0.74	
	Shift					2.23	0.34	
	F					1.50	21.49	
$K^2 = 0.89$								
$F' = 20.23$								

Net trade							
	Gain					164.84	16.38
	Shift					−0.05	3.55
	F					0.37	3.81
$K^2 = 0.44$							
$F' = 1.96$							
Band B							
Demand							
	Gain	0.93	2.04	0.93			
	Shift	1.52	0.32	−0.56			
	F	9.03	18.24	6.00			
$K^2 = 0.85$							
$F' = 7.50$							
Utilization							
	Gain	0.37	1.37				
	Shift	0.58	0.02				
	F	4.10	13.11				
$K^2 = 0.77$							
$F' = 8.37$							
Capacity							
	Gain		0.21		0.19		
	Shift		1.95		−0.67		
	F		3.46		17.36		
$K^2 = 0.79$							
$F' = 9.40$							

Price				
	Gain	0.22	0.60	
	Shift	−1.73	0.74	
	F	4.77	1.01	
$K^2 = 0.63$				
$F' = 4.26$				
Net trade				
	Gain	204.11		13.94
	Shift	0.80		−1.61
	F	1.90		22.55
$K^2 = 0.83$				
$F' = 12.21$				

Table 8.3 Estimates of the parameters of the lead model

Equation		*Price*	*Industrial activity*	*Aluminium price*	*Price interest ratio*	*Inventory demand ratio*	*Capacity–utilization ratio*	*Price differential*
Band A								
Demand								
	Gain	0.18	0.69					
	Shift	0.18	0.26					
	F	1.57	11.72					
$K^2 = 0.83$								
$F' = 12.21$								
Utilization								
	Gain	0.43	0.57					
	Shift	0.28	−0.21					
	F	4.18	3.65					
$K^2 = 0.76$								
$F' = 7.92$								
Capacity								
	Gain		0.11	0.19				
	Shift		3.13	−4.76				
	F		0.45	3.86				
$K^2 = 0.45$								
$F' = 2.04$								
Price								
	Gain				0.09	1.20		
	Shift				1.31	−0.21		
	F				0.56	11.68		
$K^2 = 0.73$								
$F' = 6.76$								

Net trade						
	Gain				273.43	46.40
	Shift				3.24	0.69
	F				0.63	2.52
$K^2 = 0.41$						
$F' = 1.74$						
Band B						
Demand						
	Gain	0.20	0.89			
	Shift	1.21	0.02			
	F	2.54	12.53			
$K^2 = 0.83$						
$F' = 12.21$						
Utilization						
	Gain	0.19	1.09			
	Shift	−1.53	0.19			
	F	0.90	8.99			
$K^2 = 0.69$						
$F' = 5.56$						
Capacity						
	Gain		0.42	0.25		
	Shift		2.12	0.47		
	F		3.04	3.73		
$K^2 = 0.45$						
$F' = 2.04$						

Price			
	Gain	0.06	0.93
	Shift	−1.96	−0.55
	F	0.11	−2.57
$K^2 = 0.38$			
$F' = 1.53$			
Net trade			
	Gain	1278.98	40.23
	Shift	1.56	−0.41
	F	7.97	8.18
$K^2 = 0.71$			
$F' = 6.12$			

Consider, for example, estimates of the copper demand equation for band A in Table 8.1. Judged by the high multiple coherence (0.88), the equation as a whole performs well over this frequency band. The gain in price shows that, *ceteris paribus*, a unit change in the amplitude of price will change the amplitude of the matching component of demand by 0.58. The time shift parameter, interpreted without qualification, shows a lag of demand behind prices of slightly less than four months (0.29 years). More importantly, the gain coefficient for price at this frequency is significant at the 10% level but not at the 5% level.

The gain for industrial activity is 1.36. Since the percentage change transformation of the original series has been used in the estimation, the series are unit free and a gain value larger than unity implies that this periodic component of industrial activity is amplified by the demand equation. The gain in industrial activity, as can be seen, is significant at the 5% level. The aluminium price coefficient is interpreted in the same manner.

8.5 CONCLUSION

Given this background, we can now proceed to interpret the results. The degree of support provided for the model varies between the three metals. This can be seen in Tables 8.1–8.3.[6]

In general, the demand equations perform well in both bands for all three metals. Industrial activity explains the largest portion of the cyclical variations in the consumption of these metals. This might seem intuitive for the higher frequencies but the fact that it is also valid for longer-term cycles is significant. Combined with the fact that the estimated gains are larger than unity, this result shows that long-term fluctuations in total industrial activity are amplified and reflected in the consumption of these metals.

Demand responses to their respective prices, in contrast, are limited. Own price variables are only marginally significant in the case of copper and not significant for lead. Only for zinc in band B can a cyclical interaction between consumption and prices be accepted with certainty. The same is true with the price of each substitute. That is, the estimated coefficients for the price of aluminium are not significantly different from zero except for zinc in band B.

Similar to the case of demand, the capacity utilization equations show a good fit in all cases. Capacity utilization has a significant association with industrial activity as well as with prices. While the cyclical dependence on industrial activity occurs mostly in band B, prices seem to have caused the

[6]For each equation, GAIN refers to the partial gain coefficient, SHIFT refers to the time shift (lag) and F is the statistic for testing the significance of individual coefficients. Also reported are the square of the coefficient K of multiple coherence and the F statistic F' for the joint test of significance.

longer-term fluctuations in band A. It is also important to note the magnitude of the gain coefficients. It can be seen that when capacity utilization is significantly correlated with industrial activity, the estimated gain values are larger than unity. Price gains, however, are in general smaller than unity. This implies that cyclical variations in prices are attenuated by these equations.

So far the results have closely followed the postulated hypotheses. The proposed capacity equation, however, has a satisfactory performance only for zinc. Results are marginally significant in the case of lead but extremely poor for copper. This total lack of response of copper capacity to prices and industrial activity might seem to be unwarranted. However, a close inspection reveals the underlying cause. More than 87% of the variations in copper capacity are explained by a simple regression on time, but only 49% are explained in this way for lead and 33% for zinc.

It thus appears that the results of detrending have been severe in that the residual variations in copper and lead capacities are very small. Quite contradictorily it also suggests that capacity formation in the copper industry behaves in such a way as to follow a simple linear trend. Should this implication be true, then one must conclude that producers have increased their productive capacity consistently, even while facing wide fluctuations in prices, demand and short-run supply. Could it be that this provides a confirmation of one popular belief that mineral investors prefer to neglect price signals and instead base their investment decisions on the desire to maintain or expand corporate market shares over time?

The results of the estimation of the price equations confirm to some extent the nature of forces affecting mineral prices over the adjustment cycle. This outcome is stronger for the case of copper than for lead and zinc. For copper, the estimated coefficients for both explanatory variables are significant in both bands. One can thus conclude that the lower frequency components of prices in band A are better explained by variations in capacity utilization, while higher frequency components are better explained by fluctuations in inventory coverage.

The net trade equation also performs very similarly for all three markets, but significance occurs only in band B. Variations in price differentials seem to be more significant than inventory coverage as explanatory variables. In fact the inventory coverage variable lacks significance in the zinc equation. Given the dependence of the domestic zinc market on imports of concentrates, this result seems to be counter-intuitive.

ACKNOWLEDGEMENTS

Any opinions, findings, conclusions and recommendations expressed in this publication are those of the authors and do not necessarily reflect the views of Allegheny College, Rockford College or West Virginia University.

NOTES

1. Please note that the capacity data available to us do not, strictly speaking, measure the productive capacity of the industry and only reflect the size of fixed capital in terms of output.
2. For development of the theory of spectral regression see Koopmans (1974) or Brillinger (1970).
3. In band spectrum regression the periodograms are not smoothed simply because the sum of periodograms approaches the sum of spectral values which is the total variance of the variable. But since we interpret the estimates for individual frequency bands, we need to smooth the periodograms to obtain consistent estimates.
4. Under the assumption of zero partial coherence (or zero gain) it can be shown that $(df - 2P)k^2/2(1 - k^2)$ has a F distribution with 2 and $df - 2P$ degrees of freedom, where k is the partial coherence, df is the degrees of freedom of the spectral density estimates, and P is the number of explanatory variables. Similarly, under the assumption of zero multiple coherence it is shown that $(df - 2P)K^2/2P(1 - K^2)$ has a F-distribution with 2P and $df - 2P$ degrees of freedom.
5. For each equation, GAIN refers to the partial gain coefficient, SHIFT refers to the time shift (lag) and F is the statistic for testing the significance of individual coefficients. Also reported are K^2 which is the square of the coefficient of multiple coherence and F' which is the F statistic for the joint test of significance.

REFERENCES AND FURTHER READING

Bischoff, C. W. (1971), Business investments in 1970's: a comparison of models, *Brooking Papers on Economic activity*, The Brooking Institute, Washington, D. C.

Brillinger, D. R. (1970), The frequency analysis of relations between stationary spatial series, *Proceedings of the Biannual Seminar*, Canada Math Congress, 12th R. Pyke (ed.), Canada Math Congress, Montreal.

Dhrymes, P. J. (1971), *Distributed Lags: Problems of Estimation and Formulation,* Holden Day, San Francisco.

Engle, R. F. (1978), Band Spectrum Regression. *International Economic Review,* **15**, (1), 1–11.

Engle, R. F. (1980), Exact maximum likelihood methods for dynamic regressions and band spectrum regressions. *International Economic Review,* **21**, (2), 391–407.

Espasa, A. (1977), *The Spectral Maximum Likelihoode estimation of Econometric Models with Stationary Errors,* Vandenhoeck and Ruprecht, Gottingen.

Espasa, A. and J. D. Sargan (1977), The spectral estimation of simultaneous equation systems with lagged endogenous variables, *International Economic Review,* **18**, 583–605.

Fishman, G. S. (1969), *Spectral Methods in Econometrics,* Cambridge, Harvard University Press, Massachusetts.
Fuller, W. A. (1976), *Introduction to statistical time series,* John Wiley and Sons, New York.
Goodman, N. R. (1963), Statistical analysis based upon a certain multivariate complex gaussian distribution (an introduction). *Annals of Mathematical Statistics,* **34**, 152–177.
Hannan, E. J. (1963), Regression for time series, in M. Rosenblatt (ed.), *Proceedings of a Symposium in Time series analysis*, John Wiley and Sons, Inc., New York.
Hannan, E. J. (1970), *Multiple Time Series,* John Wiley and Sons, New York.
Khatri, C. G. (1964), Distribution of the generalized multiple correlation matrix in the dual case, *Annals of Mathematical Statistics,* **35**, 1801–6.
Koopmans, L. H. (1974), *Spectral Analysis of Time Series,* Academic Press, New York.
Labys, W. C. (1980), A Model of disequilibrium adjustments in the copper market. *Materials and Society,* **4**, 153–164.
Labys, W. C. (1987), Contributions of stock-flow adjustments to long run commodity modeling. Working Paper, department of mineral and energy Resource Economics, West Virginia University, Morgantown.
Labys, W. C., and Afrasiabi, (1983), Cyclical disequilibrium in the U. S. copper market, *Applied Economics,* **4**, (2).
Labys, W. C., and Pollack, P. K. (1984), *Commodity Models for Forecasting and Policy Analysis,* Croom-Helm, London.
Lonoff, M. (1983), Economic aspects of the copper industry. Ph. D. Dissertation, Massachusetts Institute of Technology.
Sims, C. A. (1971), Are there exogenous variables in short-run production relations, Discussion Paper No. 10, Center for Economic Research, University of Minnesota, August.
Slade, M. (1981), Cycles in natural commodity price: an analysis of the frequency domain, *Journal of Environmental Economics and Management,* 122–137.
Winston, G. (1974), The theory of capacity utilization and idleness, *Journal of Economic Literature,* **12**, (4), 1301–20.

APPENDIX: THE DATA

The data used in this paper consist of annual time series covering the period 1906–80. This period was selected because it is sufficiently long to capture the long-run cycles inherent in investment and capacity adjustments. Before 1906 no reliable data on capacity could be found. In addition to the problem with capacity data which was discussed above, a number of other data problems exist. The most important ones are described here.

Appropriate data on the rental price of capital services (variable *C*) cannot be compiled for the whole century. Since we were interested only in long-term cyclical variations, we assumed that these variations are sufficiently captured by the interest rates. Even the interest rates series had to be obtained by splicing two series together (see Table 8.A1).

Another major problem was the occurrence of a number of outliers. Even though 75 observations are considered a small sample size for this purpose,

Table 8.A1 Definition of variables[a]

D	Total refined consumption
P	US producers' price
LME	London Metal Exchange price
PS	Aluminium price
Y	Federal Reserve Index of Industrial Activity[b]
K	Refining capacity[c]
QS	Total refined production
I	Refined inventories (at the end of the period)
C	Long-term interest rates[d]
NET	Imports − exports of refined metal
U	Capacity utilization ratio QS/*K*
V	Inventory coverage ratio *I*/*D*
PDF	*P* − LME

[a]Except as noted, all data for the variables are from annual publications of the *US Geological Survey* and *Minerals Yearbook*, US Bureau of Mines, Washington, DC.
[b]*Federal Reserve Bulletin*, Federal Reserve Board, Washington, DC.
[c]*ABMS Yearbook*, American Bureau of Metal Statistics, New York, various years.
[d]This series present the yield on long-term government bonds. Before 1921, however, no data existed for this and therefore the series had to be spliced to the series available for long-term railroad bonds. *Source*: *Statistical Abstracts of the US*, Department of Commerce, Washington, DC, various years.

the sample period includes wars, depressions and major technological changes which obviously influenced the estimated results. The definitions of variables and their sources are given in Table 8.A1.

9

The linkages between the markets for petroleum products and the market for crude oil: an econometric–linear programming study

F. GERARD ADAMS, EUGENE A. KROCH AND

VYTIS DIDZIULIS

Crude oil and petroleum products are by far the most important commodities in international trade, whether measured by weight or measured in monetary value. Collectively, they comprise roughly half the tonnage, two-thirds the ton miles, and a quarter of the value of all commodities traded worldwide. The petroleum markets occupy the centre stage in the world economy. Even relatively small disturbances in petroleum markets can have a substantial impact on the business climate. The major disruptions of the 1970s and early 1980s thwarted the attainment of such national goals as steady growth and stable prices worldwide. It follows that understanding the markets for crude oil and products is critical to planning and devising policies for the future. This is true for crude oil exporting countries (whether or not part of OPEC) as well as for net importers.

To understand the impact of crude oil production on consumers of petroleum products analysis must go beyond the crude oil market. A model needs to include the transformation process, petroleum refining, which connects the two sides of the oil market. The demand for crude oil is a derived demand, a demand for refinery inputs. The intrinsic value of the crude oil lies entirely in the value of products which can be produced from it; but the supplies of refined products are constrained both by the available supply of crude petroleum and by refining unit capacities. This means that the model of the petroleum market must simultaneously equilibrate the market for crude oil and the markets for petroleum products. This chapter is a progress report on an ongoing study integrating markets for petroleum products with the market for crude oil.

A simple model can capture the basic features of the oil market price determination process. The underlying theory is based on optimization. The linkages involve three principal economic agents: producers, refiners and consumers. Refiners play a pivotal role in that they interact with the crude suppliers on one side and with the consumers of refined products on the other. This basic configuration can be used in a variety of modes depending on the purpose of the analysis. One mode is to set product prices for refiners, who then select a feasible product mix to maximize revenues given the demands of consumers and the crude supplies of producers. Another mode allows product prices to be determined endogenously given consumer demand schedules and a level of crude throughput. Still another mode focuses on the optimal supply schedule for crude: refiner decisions in response to product demand give rise to refinery netbacks (the value of the product yield). Variation in product demands will shift the derived demand schedule for crude. These shifts may in turn alter the optimal crude supply level, tracing out a crude supply curve. In the overview (Fig. 9.1) we show

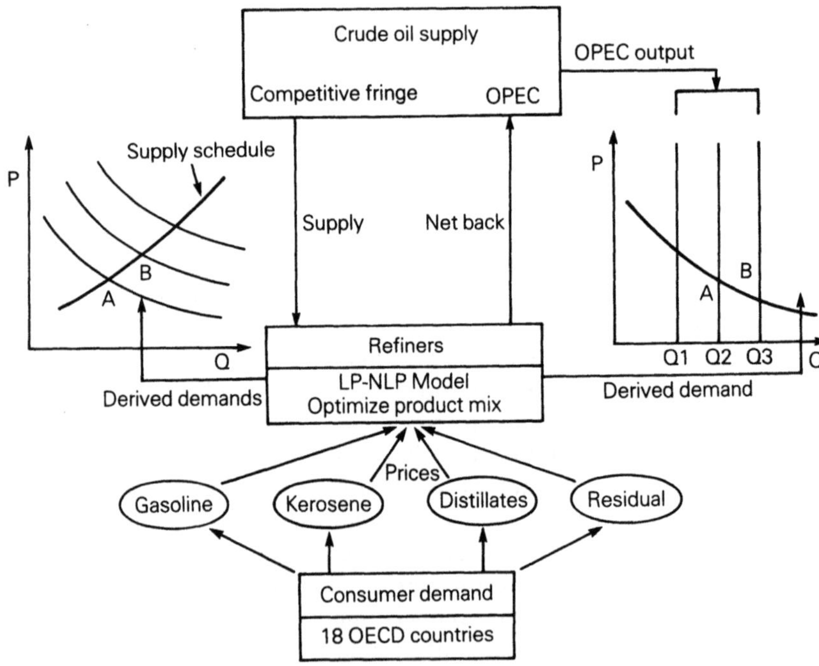

Fig. 9.1 Linkages between the market for petroleum products and the market for crude oil.

consumer demand for products at the bottom, the refinery LP in the centre and the effect of shifts in the demand and supply of crude oil in the side diagrams at the top of the graph.

Product demands play an important role in the model. They have been painstakingly specified and estimated for the OECD countries in a companion econometric study (Kroch and Didziulis, 1988) which has provided us with some valuable insights into the lag structure of these product markets. The demand forces are the underlying mechansims that determine crude oil resource value and ultimately crude supplies.

The unique aspect of our research has been the linkage of these demand-side results with a linear programming supply-side model, whose elements are illustrated in Fig. 9.2. Crude oil enters the diagram at the left, through alternative refining units, to produce products on the right. In this model, the refining capacities of the USA and alternatively of the other major OECD countries are aggregated into a six-process refinery, which includes atmospheric and vacuum distillation, reforming and thermal, catalytic and hydro cracking. This prototype refinery assumes capacities of

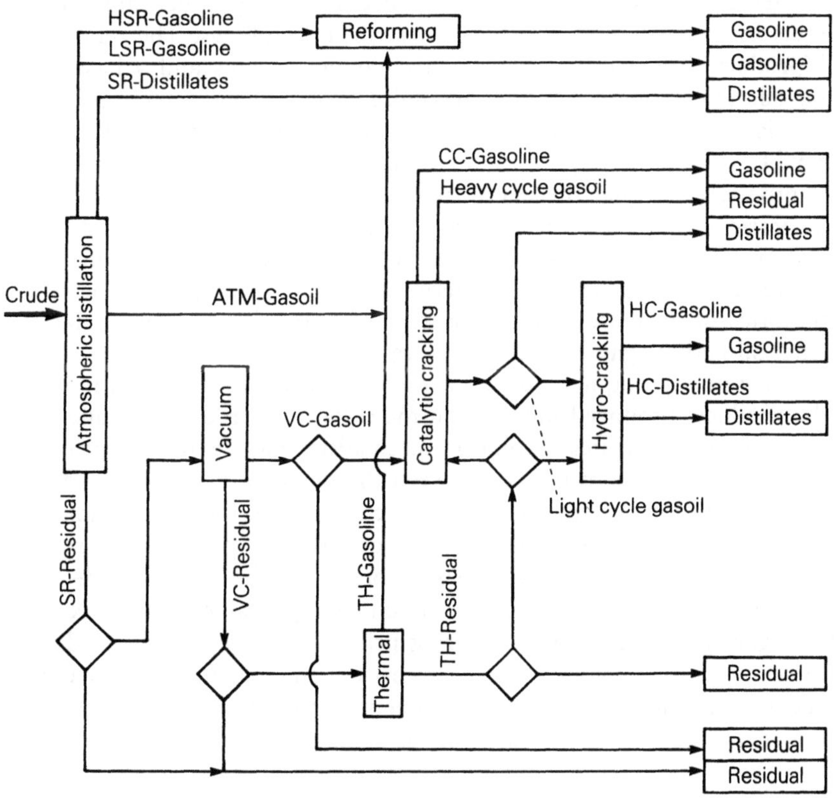

Fig. 9.2

refining units as in the base year. We can also examine the impact of additions to or contractions from available refining unit capacities. The petroleum products considered are gasoline, middle distillates, residual oil and other products. The level and type of crude input can be varied to affect the output and mix of these products.

The main purpose of such a linkage from products to crude is to generate a derived demand for various types of crude oil and to test hypotheses about crude oil output optimization. The shadow prices that result from the LP give the refinery netbacks and a measure of the extent to which petroleum exporters are able to capture the rents that result from changes in product demands. We shall be able to play a variety of counterfactual exercises and test how close the OPEC members have come to maximizing their profits.[1]

9.1 BASIC MODELLING DEVICE

The interaction between the refiner's decision on product mix and the markets for refined products can be described as follows. The refinery is a multiproduct firm with a variety of technology options and a basic input, crude oil. The refiners want to maximize profit, given market prices and production technology, subject to the supply of crude available. The optimality conditions require the marginal revenue product of crude for all joint products to equal the marginal value of crude itself. Another way to characterize these conditions is that for any pair of refinery products the rate of product transformation equals the inverse ratio of the product prices. The revenue maximizing mix is determined by the price of gasoline relative to the price of distillates. The same would be true for distillates relative to residual oil and for gasoline relative to residual oil.

But the refinery outputs must clear the market; hence the quantity of each of the products produced must lie on the respective product demand curve. This requirement dictates a solution mechanism that iterates on prices and quantities of the three products between the demand curve on the demand side and the LP on the supply side. At each iteration, the LP determines the optimal mix of products given a set of product prices. The resulting quantities give a set of market-clearing prices on the demand side. These new prices are used by the LP to obtain an optimal product mix again, which again gives new market-clearing prices from product demands. This iterative process goes on until the prices and quantities converge in successive iterations. This gives the competitive equilibrium in the market.

It is tempting to incorporate the demand curves directly into the LP

[1]This is an implicit test of the exercise of market power in the crude oil industry (Griffin, 1985).

(making it a non-linear program) to eliminate the need to iterate the model, but this would be equivalent to supposing that the typical refinery is a monopolist with market power to increase prices by limiting output. Prices would be higher and outputs much reduced compared with the competitive solution, and way below historical levels.[2]

The competitive solution assumes that each refiner faces a flat demand curve at the prevailing price but that overall demand and supply are equilibrated. This approach provides our model's benchmark. Varying the level of the throughput constraint would affect both product prices and the value of crude oil. As crude becomes more abundant, shifting out the transformation locus, the equilibrium prices of the refined products decline, as does the value of crude (i.e. the refinery netback). Hence varying crude throughput gives us a derived demand schedule for crude oil, as illustrated at the top right of Fig. 9.1. Derived demand for crude is thus linked directly to the product demands. Alternatively, a shift in the product demand schedules will generate a new derived demand curve for crude oil via this procedure. In Fig. 9.1, top left, schedule B represents an exogenous increase in product demand relative to schedule A. Moreover, for each of these demands for crude oil there is a level of supply that will maximize revenues to crude suppliers. The locus of these points traces out the optimal crude supply schedule sketched on the top left of Fig. 9.1. The confrontation between the derived demand for crude oil and the supply curve sets the equilibrium quantity and price of crude oil.

With this mechanism in place, it is possible to explore a variety of policies and model sensitivities. The effect of shifts in demand for and supply of crude can be studied. By running the model with different API ratings of crude oil, it will be possible to give values to differences in the quality of crude. We can study the effect of introducing additional (new) refining units and technologies. Finally, we can explore the consequences of tax and conservation policies in product markets, including advances in energy-efficiency technologies.

9.2 THE DEMAND MODEL

In this section we summarize the demand submodel utilized in this study.[3] The focus is on the demand for refined petroleum products: gasoline, kerosene, distillates (diesel or light oil) and residual oil. The approach is an econometric interpretation of observations drawn from 18 OECD (industrialized) countries over the 26 year period since 1960.

[2]Such a mode of operation is an interesting modification of the model and frequently leads to slack in the throughput constraint. Since the availability of crude is no longer a binding constraint, the shadow price on crude oil goes to zero.

[3]Bohi (1981) and Dahl (1982, 1986) give excellent surveys of previous investigations of petroleum products demand.

The study design allows price changes to have lasting effects on demand through time, reflecting decisions on fuel-using equipment turnover and utilization rates. The lags in behaviour have been incorporated into the model of demand for petroleum products by including polynomial distributed lags (PDL) on prices over time periods as long as 10 years.

A typical specification of such a PDL model is that for the demand for gasoline:

$$\ln\left(\frac{GAS}{N}\right)_t = \beta_0 + \beta_1 \ln\left(\frac{Y}{N}\right)_t + \sum_{i=0}^{10} \beta_{2i} \ln(P)_{t-i} + \beta_3 \ln\left(\frac{CAR}{N}\right)_t$$

The dependent variable is the log of gasoline consumption per capita, and the explanatory variables are (in the order presented) the logs of real income per capita, the real retail price of gasoline, the automobiles per capita. The income and price variables, measured in 1980 domestic currencies, were transformed into 1980 US dollars by using Kravis–Summers–Heston purchasing power parities for 1980 (for details see Summers and Heston, 1984). The lags in the price terms and the corresponding price elasticities are indexed by i from 0 (the current or short-run price elasticity) up to a lag of 10 years. Summing all these β_{2i} gives the long-run price elasticity of demand.

Table 9.1 gives the estimation results for four categories of refined product demand based on the pooled OECD data. Although not reported in the table, the models control for prices of competing fuels and for country effects, such as differences in the stock of fuel-using equipment. The table is

Table 9.1 Elasticities of demand

	Gasoline	*Kerosene*	*Distillates*	*Residual*
Income	0.50	1.79	1.27	0.73
Automobiles	0.41			
Price				
Short run (lag 0)	−0.22	−0.15	−0.14	−0.20
Lag 1	−0.19	−0.06[a]	−0.06	−0.11
Lag 2	−0.16	−0.04[a]	−0.02[a]	−0.07
Lag 3	−0.13	−0.06	−0.01[a]	−0.06
Lag 4	−0.11	−0.08	−0.02[a]	−0.06
Lag 5	−0.08	−0.08[a]	−0.04	−0.08
Lag 6	−0.06		−0.06	−0.10
Lag 7	−0.05		−0.09	−0.12
Lag 8	−0.03		−0.10	−0.13
Lag 9	−0.02		−0.10	−0.12
Lag 10	−0.01		−0.07	−0.08
Long run	−1.06	−0.46	−0.70	−1.13
R^2	0.89	0.88	0.96	0.97

[a]Not significantly different from zero.

typical of a wide ranging exploration of alternative models: both quadratic and cubic polynomials with lag lengths from 5 to 10 years.

The lags are quite long: easily 10 years in all cases but kerosene.[4] This explains why the long-run price elasticities are about five times the size of the short-run elasticities for all products but kerosene, for which the long-run effect is only three times the short-run effect. With the exception of gasoline, these lag distributions display typical cubic (third-degree polynomial) patterns. The short-run and immediate lagged response represents relatively rapid changes in utilization rates of equipment either to avoid increased costs when prices rise or to take advantage of lower cost operation when prices fall. The second 'hump' at a lag of about 7 or 8 years represents decisions about the stock of fuel-using equipment.

These results are derived from pooling time series data for 18 OECD countries and hence represent average responses over the last 25–30 years, but, as indicated by the high R^2 values, the patterns fit all the countries individually rather well over this period.[5]

An issue is whether our model system, maximizing the interests of refiners and crude suppliers, should use short- or long-run demand elasticities. There is no definitive answer. Some of these suppliers may be short-term maximizers whereas the objectives of others seem to be dominated by long-run considerations. In contrast, at the petroleum products level where short-run demand elasticities are very low, it is clear that short-term market adjustment occurs largely through quantity change. If product markets were 'flex price' systems, price variations would have to be considerably more violent in the short run that what is actually observed on the market. We have operated our system alternatively on the basis of both short- and long-run elasticities, but the long-run elasticities appear to yield more realistic results.

9.3 THE PETROLEUM REFINING MODEL

Linear programming models, used at most refineries for scheduling purposes, have been widely adapted for studies of regional or worldwide petroleum products supply.[6]

[4]Kerosene is by far the smallest component of product demand, consisting mainly of jet fuel. Moreover, the estimates for kerosene demand are confined to the USA only, because of data limitations.

[5]The study period has been dominated by rising real prices of petroleum products. Hence, the price 'changes' discussed above can be understood as price *increases*. New evidence is beginning to emerge that indicates that responses to falling prices are *not* equal and opposite to responses to rising prices and that most of the movement of capital stocks to greater fuel efficiency are irreversible. Dermot Gately and Peter Rappoport (1988) explore this asymmetry.

[6]Among studies that used econometric and input–output techniques in estimating refined oil product supply are those by Hudson and Jorgenson (1976), Manne (1958), Marshack (1966), Adams and Griffin (1972) and Kennedy (1975).

Our petroleum refining model is a simple linear programming built specifically for our purpose. We could have made use of commercially available models as did Adams and Griffin (1972) and Kennedy (1975),[7] but the size and complexity would have made it difficult to integrate with other parts of our system. In any case, it was one of the objectives of this study to demonstrate that the essential characteristics of the petroleum markets can be described with relatively simple and understandable model structures.

In our linear programming model two types of crudes are used as input in a six-process refinery (Fig. 9.2). The output of the refinery is grouped in four products: gasoline, distillates, residual oil and 'other' products that includes gases, petroleum coke, asphalts and lubricants. Figure 9.2 traces the possible flows through the refinery. To illustrate: crude oil is first sent to the atmospheric unit where it is separated into four straight-run products by the boiling range of the fractions.[8] The light straight-run gasoline is sold as it is, and the heavy straight-run gasoline is sent to the catalytic reforming unit. Distillates go directly to the market. The atmospheric gasoline is sent to the catalytic cracking unit, and residual can be sold or sent to the vacuum unit. Vacuum gasoil can be sold as residual or sent to the catalytic cracking unit. Likewise, vacuum residual can be sold or sent to the thermal unit etc.

Two types of crude oil are used in this model, Alaskan North Slope 25.7° API and Saudi Light-Berri 37° API, and each one is treated independently. Each crude has its own yields for each refinery process and its own output. The sum of the output of both crudes is the total output of the refinery, and the sums of feedstocks from each crude are constrained by the capacity of each process. The proportion of each crude is set exogenously, taking into account information regarding the average characteristics of the crudes actually used as approximated from production and trade data. There are several advantages of using two types of crude instead of one that approximates the average quality. First, it adds flexibility to the model. Since each crude has its own yields, it is easy to alter the proportion of each crude to calibrate the model. Second, it is possible to adjust the model to obtain the crude oil price differential by quality. The Alaskan North Slope and the Saudi Light-Berri crudes were chosen because they cover a wide range of crude characteristics from which an average quality can be estimated.

The objective function of the LP is profit maximization. The prices of the refined oil products used in the model correspond to wholesale prices net of taxes. The cost of crude oil is not included in the model, but the crude oil netback is computed by the model and represents the value of refined oil

[7]They used different versions of the Bonner and Moore model, a highly detailed proprietary system.
[8]The detailed refinery specifications used are available from the authors.

products. A further adjustment for transportation of the crude oil should be made to yield a netback at the source.[9]

9.4 ILLUSTRATIVE APPLICATIONS

This is an ongoing project and considerably more work is needed for realistic applications. We present here only some illustrative applications.

The model has been calibrated on information for the USA only for 1980. The structure of demand and of refinery capacity is very different in Europe and the results of the model would be substantially modified in an application to European conditions.[10]

9.4.1 The supply of products

The supply curve of a particular product can be evaluated by postulating alternative values of product prices for that product, holding crude throughput and prices of other products fixed. Given the refinery capacity constraints and other product prices, the LP computes the prices needed to overcome the various refinery constraints. The supply of gasoline at a refinery crude throughput of 13.5 million barrels per day (mbd) is depicted by curve A in Fig. 9.3(a).

At low gasoline prices, the supply of gasoline is insensitive to price since the product comes directly from the distillation unit with little or no further processing. Over a range of prices from approximately $0.60 to $1.10 per gallon[11] gasoline output is highly responsive to price, ranging from approximately 4 mbd to 6.5 mbd as the catalytic and hydrocracker units come into use. We note that the actual level of gaoline output in 1980 was near the upper end of this range. Of course, the gains in gasoline output as gasoline price rises are obtained at the expense of reductions in the output of other products. Finally, beyond 6.5 mbd gasoline, the catalytic unit reaches a capacity limit; with fixed refinery throughput there is no additional output of gasoline regardless of price.

Alternatively, we assume a lower level of crude throughput of 0.4 mbd. Curve B corresponds to the supply of gasoline with a lower level of throughput: 13.1 mbd. The shape of the curve is similar to curve A but

[9]An additional adjustment should be made at the product level to allow for product transport and distribution margins.

[10]The capacities of the processing units, the throughput and the output of refined oil products in 1980 were obtained from the Department of Energy's *Petroleum Supply Annual* (Energy Information Administration, 1987). The model was calibrated to 1980 historical outputs by making small adjustments. The prices of the 'other' products were set to be equal to 20% of the prices of residual.

[11]These prices are indicative. They are not meant to be realistic ex-refinery prices. Allowances for crude costs, taxes, transportation etc. have not been made.

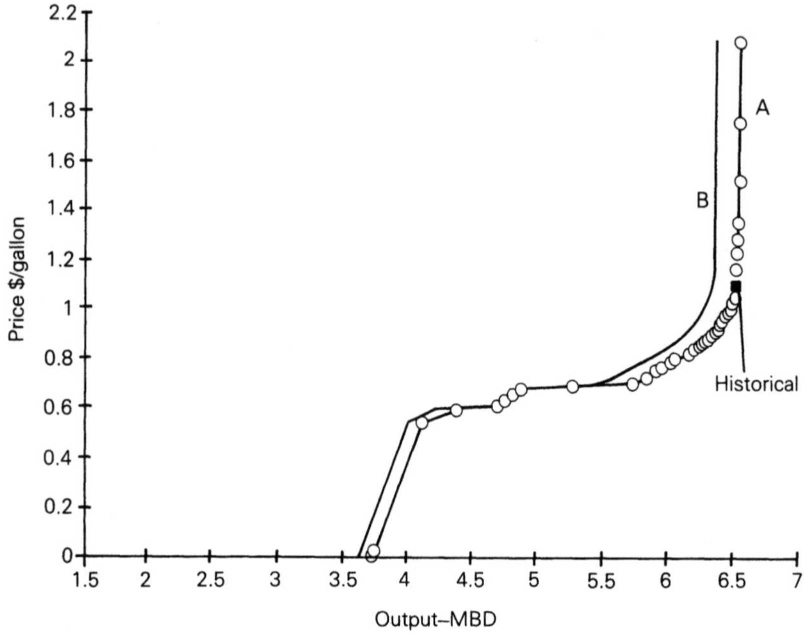

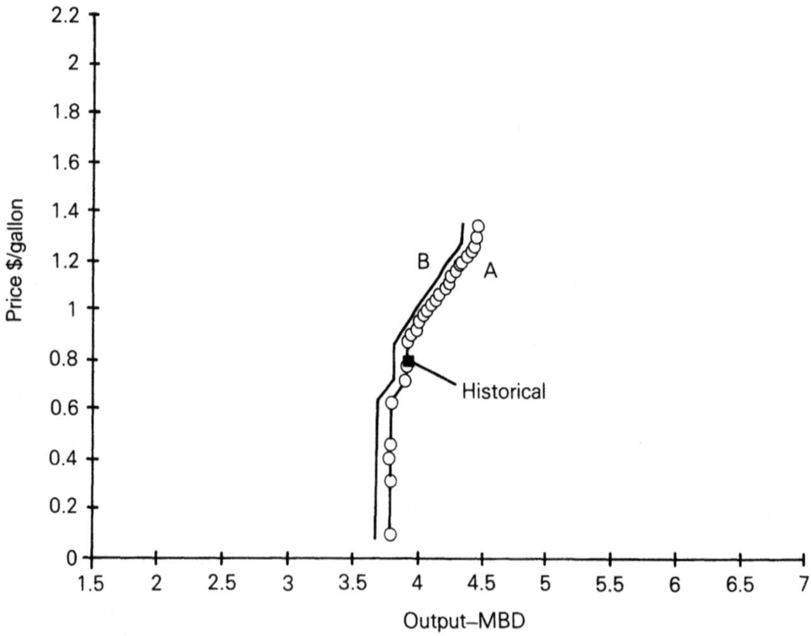

Fig. 9.3 (a) Gasoline supply, 1980; (b) distillates supply, 1980; (c) residual supply, 1980.

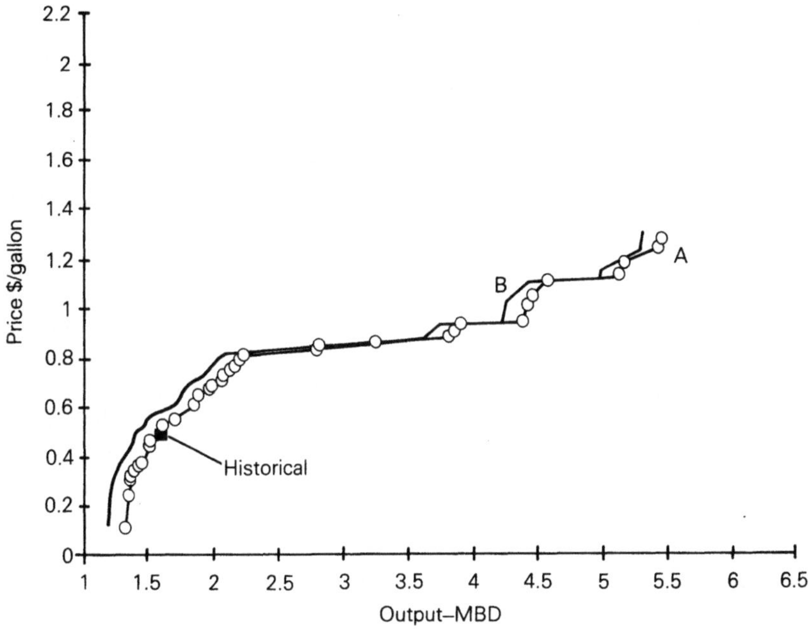

shifted to the left by 0.2 mbd at all price levels. The constraints become binding at approximately 6.3 mbd of gasoline output.

Similar computations can also be made for the other products. Figure 9.3(b) shows the supply curves for distillate fuels. As above, we assume crude inputs are fixed and we vary the price of one product. We might have anticipated that the supply curve for distillates would be flatter than the supply curve for gasoline since the high level refining units are not required and do not limit distillate fuel output. In fact, Fig. 9.3(b) shows the supply to be quite steep since the refining units in our LP refinery do not permit the product cycles which would produce more distillate. Different assumptions about the refining units, e.g. assumptions appropriate to European refiners, would yield different results.

For residual fuel, the picture shown in Fig. 9.3(c) is different again. At low prices there is a fairly steep relation between residual fuel output and price. The output curve then flattens as it becomes profitable to put the product on the market directly as residual fuel in lieu of further processing into light fractions.

These estimates of the sensitivity of product output to price depend on the nature of the refining capacity assumed, on the quality of the crude oil used and on the volume of crude throughput. Only the supply side has been taken into account here, and so there is no consideration so far whether the quantities of products produced can be accommodated in the market at the price assumed. We leave that question for the next simulations.

9.4.2 Product demands and market equilibrium

Implicit demand curves can be computed from our model by solving the system for various levels of crude oil throughput and tracing the market equilibrium relationships between product prices and quantities obtained. The price elasticities for products correspond to the estimates given in Table 9.1. The equation used is a stock adjustment version of our demand model. The computed demand curves are shown for both long- and short-run demand elasticities in Fig. 9.4(a)–9.4(c). Behind these calculated demand curves lie very different price–quantity responses among the various products when refiners seek to maximize profit given crude supplies and product demands.

The point is that the demand curves and the supply curves for the various products differ. An increase in crude throughput yields an adjustment in supply and price for each product which equilibrates product markets. The

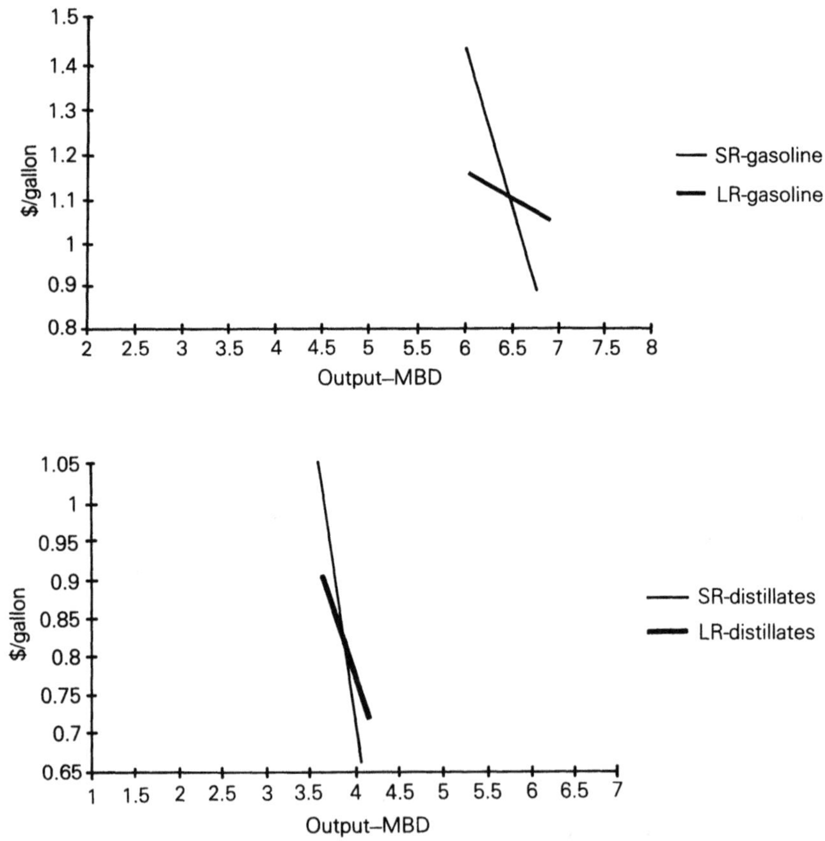

Fig. 9.4 (a) Demand for gasoline; (b) demand for distillates; (c) demand for residual.

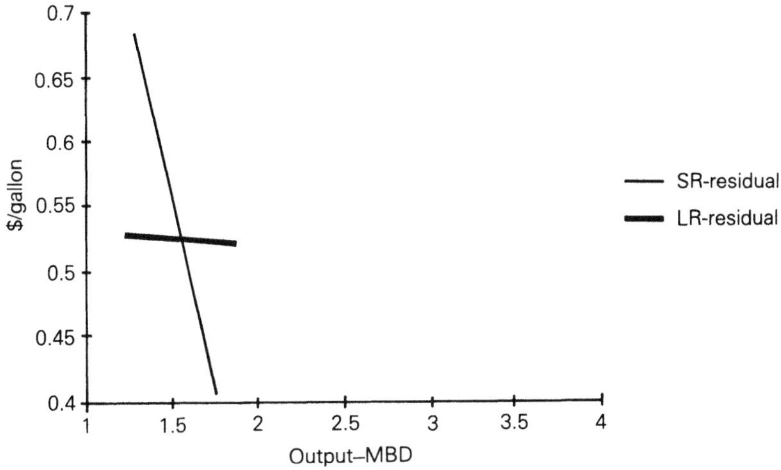

extent to which this adjustment affects quantity and price depends on the elasticities of demand *and* the cross elasticities of supply.

Table 9.2 illustrates the effect of varying crude throughput on the equilibria in the three product markets for the base case and for two other scenarios, for quantities and for prices. All values in the table are deviations from the base case for each product, with 13.5 mbd of crude throughput, shown with zeros in the third column. On either side are shown the effects of running the model with a crude throughput 0.5 mbd greater and 0.5 mbd smaller. As expected, an increase in throughput increases product supplies and reduces product prices, and a decrease in throughput has the opposite effect. Also as expected, the price adjustment is greater in the short run than in the long run since demand elasticities are lower in the short run than in the long run. The quantity adjustments, however, are not always larger in the long run than in the short run as one would expect them to be. For example, in the base case when crude throughput increases by 0.5 mbd, distillates increase by only 0.105 mbd in the long run even though they increase by 0.126 mbd in the short run. To see why this can happen, refer to Fig. 9.5.

Figure 9.5 illustrates how an increase in crude throughput affects the market for distillates in our model. The figure gives both the long-run demand schedule D_L and the short-run demand schedule D_s for the product as well as the supply schedule S. The increase in throughput shifts S to the right, but the shift is greater in the short run (S_s) than in the long run (S_L). S_s lies to the right of S_L because of the refiner's decision of how to allocate between the three joint products in response to their changed relative prices. Hence, even though equilibrium P_s is below equilibrium P_L (and below the original price P), the equilibrium Q_s is *above* equilibrium Q_L.

Table 9.2 Product market response to a change in crude throughput

	Base			*Income increased by 5%*			*Capacity reduced by 10%*		
Crude (mbd)	*−0.5*	*0*	*0.5*	*−0.5*	*0*	*0.5*	*−0.5*	*0*	*0.5*
Output (mbd)									
Gasoline									
Short run	−0.196	0.000	0.195	−0.203	−0.005	0.192	−0.336	−0.097	0.100
Long run	−0.195	0.000	0.190	−0.197	−0.005	0.184	−0.355	−0.124	0.073
Distillates									
Short run	−0.129	0.000	0.126	−0.123	0.005	0.136	−0.192	−0.035	0.091
Long run	−0.126	0.000	0.105	−0.109	0.005	0.126	−0.196	−0.053	0.070
Residual									
Short run	−0.129	0.000	0.138	−0.126	0.000	0.122	0.029	0.105	0.239
Long run	−0.138	0.000	0.190	−0.169	0.000	0.165	0.030	0.215	0.353
Price ($/gal)									
Gasoline									
Short run	0.149	0.000	−0.148	0.195	0.045	−0.105	0.255	0.073	−0.076
Long run	0.025	0.000	−0.025	0.066	0.042	0.017	0.045	0.015	−0.010
Distillates									
Short run	0.109	0.000	−0.105	0.201	0.093	−0.017	0.162	0.030	−0.076
Long run	0.047	0.000	−0.039	0.138	0.096	0.051	0.073	0.020	−0.026
Residual									
Short run	0.079	0.000	−0.085	0.107	0.030	−0.045	0.018	−0.065	−0.146
Long run	0.001	0.000	−0.002	0.031	0.030	0.028	−0.001	−0.002	−0.004

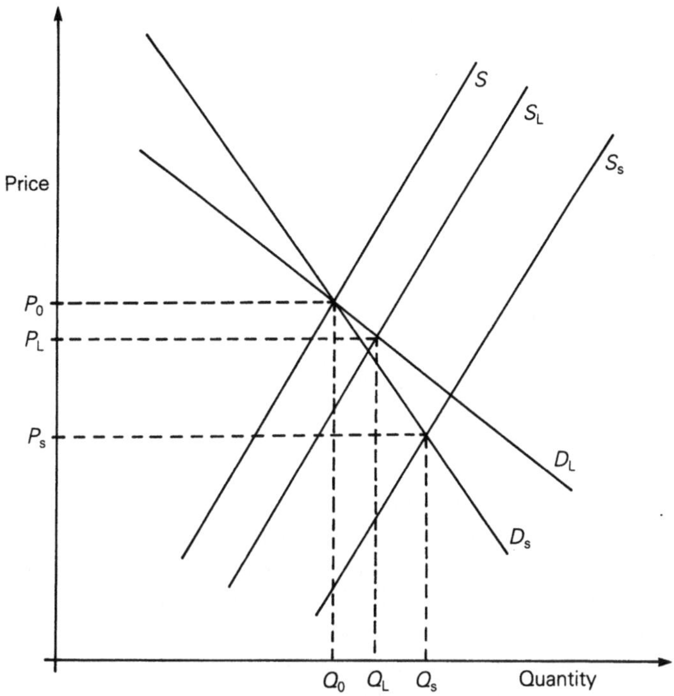

Fig. 9.5 Distillate market: effects of increases in refinery crude throughput on product supply and demand.

Note from Table 9.2 that increased crude throughput results in a change in the price of distillates *relative* to the other product prices, but that the direction of this relative price change is different in the short run from the direction of the price change in the long run. Although all product prices fall, in the short run the distillate price falls less than the gasoline price—10.5 cents versus 14.8 cents per gallon. However, in the long run the distillate price falls more than the gasoline price—3.9 cents versus 2.5 cents. This means that in the short run the product mix will move towards distillates, but in the long run the mix will move away from distillates in favour of the other products.

In the central columns of Table 9.2 are shown the effects of increasing personal disposable income of energy consumers by 5%. All changes are relative to the base case with 13.5 mbd of crude throughput. Note that since we have used a stock adjustment model here, the long-term effects reflect a long-term adjustment of products demand that is greater than the short-term adjustment because of both the income change and the resulting price changes. The base solution (with crude input of 13.5 mbd) shows that most impact is on price with little on quantity.

The right-hand side of Table 9.2 gives the effects of a 10% across the board reduction of refining capacity. Remember that all changes are relative to the base case and base crude level. Capacity reduction has the greatest effect on gasoline and other high fractions. Output of these products falls and prices rise as expected for the base level of crude throughput. But, increasing throughput by 0.5 mbd reverses these effects (and reducing throughput exacerbates them). Note that capacity restriction shifts the product mix toward residuals for which there is excess refining capacity.

9.4.3 The derived demand for crude oil

This modelling system can be operated to set up a derived demand for crude oil. To do this, we fix the product demand curves and systematically vary the available quantity of crude oil throughput, thus allowing the model to compute the refiners' netback and establish a derived demand for crude.

The netback per barrel of crude oil throughput can be understood as the weighted sum of product values and is a measure of the implicit value of crude oil. Figure 9.6 shows this derived demand for crude oil, the combination of quantity and netback for crude at various volumes of crude throughput as obtained by the model. These results are shown both for short-run demand elasticities and for long-run elasticities. The demand curves are almost linear over the range considered here. The difference between the elasticities of derived demand for crude oil is striking, as shown in Table 9.3. For short-run product demand curves the derived demand for crude is price inelastic (−0.264), but for long-run product demands the derived demand for crude oil is elastic, at −1.286.

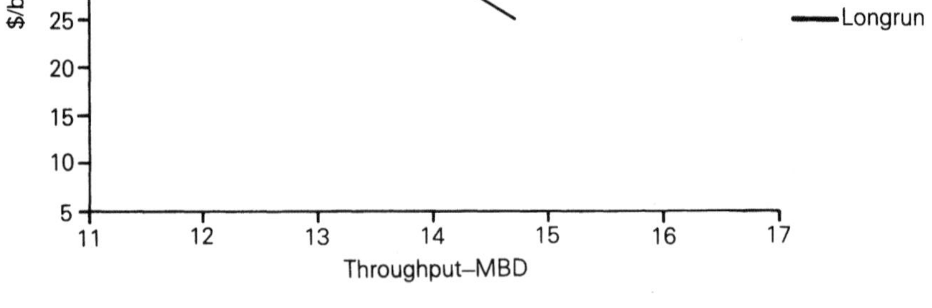

Fig. 9.6 Derived demand for crude.

The effects of a 5% increase in the disposable income of refined product consumers and of a 10% reduction in capacity across all refining units are shown in Figure 7(a) and 7(b). The increase in consumer income increases the value of crude oil, as shown by an upward shift in the derived demand schedule in both cases. Comparison of these cases is complicated by the fact that in comparison to the short term solution, the long run case has greater income effects pushing in one direction offset by greater price effects which operate in the opposite direction.

A reduction in refining capacity tends to shift the derived demand curve

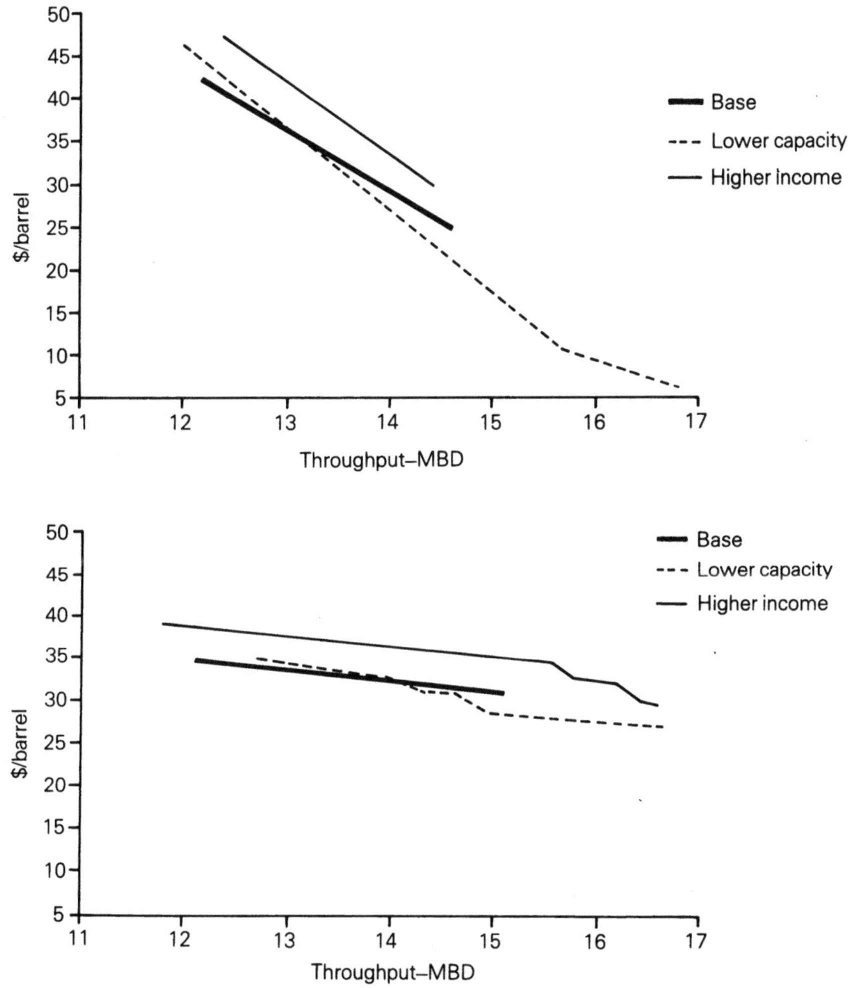

Fig. 9.7 (a) Derived demand for crude, short-run; (b) derived demand for crude, long-run.

Table 9.3 Price elasticities

Demand elasticities	*Base*	*Income increased by 5%*	*Capacity reduced by 10% across all refinery units*
Short run	−0.264	−0.281	−0.271
Long run	−1.286	−1.377	−1.308

for crude oil only very little, and again to the right. On changing refining capacity by 10% across the board, downstream processes which are nearest to capacity operation are the most affected. Gasoline production is constrained. Gasoline will reach equilibrium at a higher price, resulting in a higher value of crude. The derived demand elasticity for crude oil is not greatly affected by the alternative income and capacity assumptions, as shown in Table 9.3. These elasticities were computed over the same range of throughput for each of the three scenarios. An increase in income causes the crude oil price elasticity to increase in the short run and in the long run. Lower refining capacity increases the elasticities slightly.

9.4.4 Concluding remarks

We have laid out a model system for linking the market of products and crude in a consistent fashion. Clearly this is a highly simplified model and it is still in an early stage of construction. The applications presented are illustrative. Specifically, we show the demand curve for crude derived from equilibrium in product markets and refinery maximization in an LP framework. This demand curve can be confronted with a supply curve which maximizes producer revenues to provide an equilibrium solution for both the products and the crude oil markets.

There are many opportunities for elaborating this system further in order to provide more realistic applications.

REFERENCES AND FURTHER READING

Adams, F. G. and Griffin, S. (1972) An econometric linear programming model of the U.S. refining industry. *Journal of the American Statistical Association*, 67, (339). 542–51

Alt, C., Bopp, A. and Lady, G. (1976) Short term forecast of energy supply and demand, in *Econometric Dimensions of Energy Demand* (eds A. B. Askin and J. Kraft), Lexington Books, Lexington, MA.

Baltagi, B. H. and Griffin, J. M. (1983) Gasoline demand in the OECD: an application of pooling and testing procedures. *European Economic Review*, 22 117–37.

Bohi, R. (1981) *Analyzing Demand Behaviour* Johns Hopkins University Press for Resources for the Future, Baltimore, MD.

Dahl, C (1982) Do gasoline demand elasticities vary? *Land Economics*, 58, (3). 373–82

Dahl, C. A. (1986), Gasoline demand survey. *The Energy Journal*, 7 (1). 67–82

Edgar, M. D. (1974) New look at cat cracking LSR. *Oil and Gas Journal*, 3 June, 166–9.

Energy Information Administration, *Petroleum Supply Annuals*, (1987) Vol. 1, pp. 62–7.

Gary, J. H. and Handwerk, G. E. (1975) *Petroleum Refining*, Marcel Dekker, New York.

Gately, D. and Rappoport, P. (1988) The adjustment of U.S. oil demand to the price increases of the 1970s. *The Energy Journal*, April, **9** (2). 93–107.

Griffin, J. M. (1985) OPEC behaviour: a test of alternative hypotheses. *American Economic Review*, December, **75** (5). 954–63.

Hariu, O. H. and Sage, R. C. (1969) Crude split figured by computer. *Hydrocarbon Processing*, April, 143–8.

Houthakker, H. S., Verleger, P. K. and Sheehan, D. P. (1974) Dynamic demand analyses for gasoline and residential electricity. *American Journal of Agricultural Economics*, **56** (2). 412–18

Hudson, E. A. and Jorgenson, D. W. (1974) US energy policy and economic growth, 1975–2000, *Bell Journal of Economics and Management Science*, **5** (2), 461–514.

Kennedy, M. (1974) An econometric model of the world oil market. *Bell Journal of Economics and Management Science*, **5** 540–76.

Kroch, E. and Didziulis, V. (1988) The demand for petroleum products in the OECD: some new time-series evidence, ERU Working Paper 233, Department of Economics, University of Pennsylvania.

Leuenberger, E. L. (1988) Optimum FCC conditions give maximum gasoline and octane, *Oil and Gas Journal*, 21 March, 45–50.

Manne, A. S. (1958) A linear programming model of the US petroleum refining industry, *Econometrica*, **26** (1)

Marshack, T. A. (1966) A spatial model of US refining, in *Studies in Process Analysis* (eds Manne A. S. and Markowitz, H. M), Wiley, New York.

Moorman, J. W. (1954) How conversion level affects product distribution, *Oil and Gas Journal*, 4 January, 76–7.

Nelson, W. L. (1958) *Petroleum Refining Industry*, 4th edn, McGraw-Hill, New York.

Summers, R. and Heston, A. (1984) Improved international comparisons of real product and its composition, *Review of Income and Wealth*, **30** (2), 207–62.

Verleger, P. K. and Sheehan, D. P. (1975) The demand for distillate fuel oil and residual oil: a cross-section time-series study, Report prepared for the US Council on Environmental Quality, July.

10

Modelling the international natural gas market: the case of the Western European natural gas market

JEAN-BAPTISTE LESOURD, JACQUES PERCEBOIS

AND JEAN-MICHEL RUIZ

10.1 INTRODUCTION

The purpose of this chapter is to present a simple model of international gas contract behaviour on the Western European market. We can consider that the international gas market is broken down into three main segments: a North American segment, a European segment and an Asian segment. In each segment we have specific types of behaviour. We first present an outline of the European gas market before building a model aimed at explaining its historical evolution since the first oil shock.

10.2 THE EUROPEAN NATURAL GAS MARKET AND ITS RECENT EVOLUTION

10.2.1 The role of natural gas in Europe

Over the last 20 years natural gas has penetrated the global energy balance rapidly in Western Europe. However, the penetration has been variable from one country to another. To expand, the natural gas market needs a conjunction of several favourable factors:

1. a critical bulk of potential users;
2. an end-user price which is relatively competitive (in fact natural gas suffers from a major handicap in that it has no captive market—hence

gas penetration will largely depend on the prevailing economic conditions, i.e. on its relative price with respect to other energy sources);
3. a selling price profitable enough to justify the expenditures made all along the gas chain.

At the world level most of the natural gas produced and marketed is today marketed in producing countries. Only a small fraction of the production (from 12% to 13%) is traded internationally. The main natural gas importers are OECD countries and Eastern European countries.

Roughly speaking, the structure of the North American market corresponds to perfect competition (with a large part of natural gas sold at spot prices). The Asian market is totally dominated by Japan which is the main buyer. In Europe the rate of gas penetration in the primary energy balance is lower than in the USA and higher than in Japan (15% on average) and the part played by imports is much more important than in the USA. In Europe about 50% of the gas consumed crosses a border, coming either from another European country or from a non-European country (Tables 10.1 and 10.2). Western Europe is and will remain a net importing area. The main outside suppliers for Europe as a whole are the USSR and Algeria. Inside Europe, there are the Netherlands and Norway. The main importers of this gas sold on the international market are West Germany, France, Italy and the United Kingdom.

10.2.2 Natural gas price: a fragile compromise between contradictory requirements

Since the beginning of the 1970s the evolution of gas prices has been characterized by, on the one hand, a strong dependence on the price of oil (Percebois, 1986; Lesourd *et al.,* 1987) and, on the other hand, fluctuations in the relative price of gas depending on the situation prevailing on the market, as a result of the seller's and buyer's bargaining powers.

Two consistent pricing methods are encountered on the international natural gas market: the 'parity method', and the 'net-back method' (Percebois, 1986). According to the second approach it is necessary to start from the average price of gas substitutes (for end-users) and, going back up the gas chain, to deduct the costs of distribution, storage, eventual regasification and international transport to obtain the good level of the free-on-board (FOB) price which can be paid to the exporter. For this reason oil prices and gas prices are generally on the same scale but not necessarily on the same level of this scale.

Each seller on the European market owns winner cards: large reserves and low prices for the USSR, guaranteed supplies for the Netherlands and Norway, quite large reserves and historical relationships with Western Europe for Algeria.

Table 10.1 Natural gas production, consumption and trade at the international level ($\times 10^9$ m^3)

Zones	1950 P	1950 C	1960 P	1960 C	1970 P	1970 C	1980 P	1980 C	1986 P	1986 C	Reserves (1 Jan 1987)
North America	170.6	170.0	358.6	369.3	651.8	651.3	622.6	625.7	530.94	529.64	7970
Latin America	3.7	4.3	13.0	12.1	34.5	33.7	65.9	63.0	77.51	77.56	6581
Western Europe	0.9	0.9	11.9	11.9	77.5	80.0	192.5	226.0	190.01	251.18	5554
Eastern Europe[a]	9.0	9.0	56.7	56.7	235.5	238.0	492.7	469.6	749.42	711.52	41748
Africa	—	—	—	—	3.4	1.9	20.5	12.3	56.25	35.38	7109
Middle East	0.1	0.1	2.7	2.7	20.1	19.2	43.0	40.1	75.68	72.68	26203
Far East	0.3	0.3	3.6	3.6	25.3	24.0	80.1	80.6	127.48	129.33	8399
Total	184.6		446.5		1048.1		1517.3		1807.29		103564
International trade	0.83		5.35		45.05		199.52		229.49		
Part of international trade in marketed production (%)	0.4		1.2		4.3		13.1		12.6		
Part of liquefied natural gas (LNG) in international trade (%)	—		—		6.1		16.3		22.4		

[a] Including the USSR.
P, marketed production; C, consumption.
Source: CEDIGAZ.

Table 10.2 Estimate of international gas trade by pipeline and LNG tanker for Western Europe in 1986 ($\times 10^9$ m^3)

	Countries exporting by pipeline							*LNG*			
Importing countries	*Denmark*	*Nether-lands*	*Norway*	*West Germany*	*USSR*	*Algeria*	*Total, by pipe*	*Algeria*	*Libya*	*Total LNG*	*Total imports*
Austria				0.15	3.90		4.05				4.05
Belgium		4.34	1.60				5.94	2.59		2.59	8.53
Finland					1.24		1.24				1.24
France		5.47	3.66		8.88		18.01	7.67		7.67	25.68
Italy		4.31			7.65	8.01	19.97				19.97
Luxembourg		0.42					0.42				0.42
Netherlands			1.68				1.68				1.68
Spain							—	1.62	0.86	2.48	2.48
Sweden	0.23						0.23				0.23
Switzerland		0.60		0.96			1.56				1.56
United Kingdom			12.68				12.68				12.68
West Germany	0.38	19.80	6.43		14.73		41.34	0.12		0.12	41.46
Yugoslavia					3.90		3.90				3.90
Total	0.61	34.94	26.05	1.11	40.30	8.01	111.02	12.00	0.86	12.86	123.88

Source: CEDIGAZ.

The capability for natural gas to influence energy prices is low. Thus the natural gas market remains vulnerable and dependent on the oil market. In a saturated market and owing to a large uncertainty concerning the future price for oil, many factors must be taken into account (in particular macroeconomic factors). Therefore aversion to risk is an important element of the behaviour of sellers and buyers.

10.3 MODELLING GAS CONTRACT BEHAVIOUR

10.3.1 International gas contracts as protection against oil market risks

Although the natural gas market has been extensively studied, few authors have studied natural gas contracts, except perhaps for the recent work of Boucher *et al.* (1986), Golombek *et al.* (1987), Boucher and Smeers (1988), and Estrada *et al.* (1988).

The economic theory of contracts, however, is well developed and has been the object of numerous studies since the pathbreaking articles by Nash (1953) on two-person bargaining. Applications of this theory have been found in various fields (see, for example, Johnson, 1960; Black and Scholes, 1972; Levasseur and Simon, 1980; Tirole, 1986). In all these applications a contract is interpreted as a protection against some economic risk at a price which is paid by one (or both) of the two economic agents involved in the contract.

For instance, futures markets contracts involve insurance against the risk of a stochastic price variation of some commodity or some financial asset, which is embodied in a premium over the spot price of this commodity or asset; this premium is positive (thus supported by the buyer) if the market's expectations are that the price will rise during the contract's time of application, and negative (thus supported by the seller) if the market's expectations are that the price will decrease.

The observation of the Western European[1] international natural gas market leads us to formulate our model in terms of just one gas consumer (because importing countries often group in consortiums to negotiate contracts with exporting countries) against four gas exporters (namely, Algeria, the Netherlands, Norway and the USSR, which account for practically all the internationally traded natural gas in the European OECD area). We consider that the gas consumer is a consortium of gas utility companies which are engaged in natural gas distribution, and which will be referred to as 'the consortium'.

It is reasonable to assume that, if the consortium of buyers bought its gas on a spot market, its prices would be equal (except for small differences that reflect specific technical characteristics of gas and the local situation of

[1]Defined as the European OECD area.

the various exporters) to those of oil substitutes on spot markets. This means that, up to a conversion coefficient k that ensures technical (and, in particular, calorific) parity, the spot price of gas at time t would be

$$\pi_g^s(t) = k\pi_p^s(t) + v_i \tag{10.1}$$

where $\pi_p^s(t)$ is the spot price of the competing petroleum product. Thus, for some time interval $[0, T]$ $(T > 0)$, the *ex post* cost C_i of the consortium's gas purchases in quantity x_{gi} from supplier i would be

$$C_i = \pi_{gi}^s x_{gi} = k \int_0^T \pi_p^s(t)\dot{x}_{gi}dt + v_i x_{gi} \tag{10.2}$$

where $x_{gi} = \int_0^T \dot{x}_{gi}dt$ and π_{gi}^s is the average price of gas under this scheme.

Equation 10.2 describes the behaviour of the consortium if its gas purchase were at spot prices. The fact that gas contracts are actually used is linked to the fact that both gas buyers and sellers are risk averse, because Equation 10.2 would lead to undesirable cost instabilities that would reflect the instabilities on the underlying markets for competing oil products. To be protected against these instabilities, the consortium and the gas exporting countries will sign a contract in which they will agree on an *ex ante* price π_{gi}^C that results from an equation analogous to 10.2, in which the spot price of competing petroleum products will be replaced by an implicit futures' price $\pi_p^F(t)$, whereby

$$C_i = \pi_{gi}^C x_{gi} = k \int_0^T \pi_p^F(t)\dot{x}_{gi}dt + v_i x_{gi} \tag{10.3}$$

Integrating with the condition that, at time 0, $\pi_p^F(t)$ will be equal to the spot price $\pi_p^s(0)$ and that the gas purchase flux $\dot{x}_{gi}$ will be constant (so that $x_{gi} = T\dot{x}_{gi}$) we obtain

$$\pi_{gi}^C = k\left\{\pi_p^s(0) + \frac{1}{T}\int_0^T [\pi_p^F(t) - \pi_p^s(0)]dt\right\} + v_i \tag{10.4}$$

Thus national buyers and sellers that wish to hedge themselves against price instabilities will be led into entering a contract that includes the following.

1. Gas prices are indexed on competing oil products—hence the term $k\pi_p^s(0)$;
2. The price differential is equal to the average value of the price differential between an implicit futures price for competing oil products and the spot price for those products, with indexing factor k—hence the presence of the integral

$$\sigma_p = k\frac{1}{T}\int_0^T [\pi_p^F(t) - \pi_p^s(0)]dt$$

$\sigma_p > 0$ denotes increasing price expectations, while $\sigma_p < 0$ denotes decreasing price expectations.

3. Finally, another price differential v_i, specific to the particular contract signed, takes into account some particular cost situation, and/or some particular political situation.

It is worth noting that, while components 1 and 2 are in fact exogenous, v_i in our model will be endogenous as it is the result of supply and demand equilibrium.

At equilibrium, if natural gas was a perfect substitute of competing oil products and was traded on a single and perfectly fluid market, all prices should be equal to $k\pi_p^s(0) + \sigma_p$, and this exogenous value should be equal to the marginal utility of the consumer u_c' and to the common value of the marginal costs C_i' of all sellers:

$$\pi_{gi}^C = k\pi_p^s(0) + \sigma_p = U_C' = C_i' \qquad (i = 1, \ldots, n) \tag{10.5}$$

However, natural gas markets are not entirely tied to oil markets and may be discussed independently of oil markets, to a certain extent at least.

In effect, if, strictly speaking, 10.5 were to apply, there could be no natural gas market, since π_{gi}^C would not be determined endogenously as the solution of a supply–demand equilibrium equation. Actually, there are differences in utility between natural gas and competing oil products, reflecting specific technical differences and differences in market behaviours.

10.3.2 Specification of the econometric model

We now develop a model, along the lines of section 10.3.1, concerning internationally traded natural gas in Western Europe (OECD Europe). According to 10.4 the price of natural gas for a contract with supplier i should depend on the spot price of competing oil products, as well as on the average of price expectations, as measured by the average of some implicit futures price, and on some price increment v_i, which leads us to the following specification for the price of gas π_{git}^c from supplier i:

$$\ln \pi_{git}^c = a \ln \pi_{gt-1}^c + b\pi_{pt}^s + c(\pi_{pt}^s - \pi_{pt-1}^s) + \ln v_i \tag{10.6}$$

In this equation, according to the theoretical analysis that results in Equation 10.4, the price of natural gas from supplier i is separable into a term depending on oil price and another term $\ln v_i$; this second term is assumed to depend on quantities demanded at time t and on quantities demanded and the gross domestic product (GDP) of OECD European countries lagged once, with a log-linear specification:

$$\ln v_i = \alpha_i \ln D_{git} + \beta_i \ln D_{git-1} + \gamma_i \ln \mathrm{GDP}_{t-1} + \delta_i \tag{10.7}$$

All prices have been deflated with a GDP deflator for OECD countries. It may be shown that, to a first approximation, 10.6 and 10.7 lead to an expression for π_{git}^c similar to 10.4.

As will be shown below, we have introduced two main suppliers: USSR, on the one hand, and all the other suppliers (essentially the Netherlands, Norway and Algeria), on the other hand, while total OECD European demand is also assumed to obey 10.6 and 10.7.

We assume that all α_i are equal for all i to their counterparts in the total demand equation. This specification means that the consumer's cost function for imported gas is of the constant elasticity of substitution form.

To be well behaved, the above demand functions should be such that $\beta_i \geqslant 1$ (because β_i reflects memory of past variables, which should be decreasing with time), and that $\alpha_i \leqslant 0$ (a general property of demand functions).

We specify the logarithms of supply functions for each group of suppliers as defined above as depending on real lagged contract prices, real oil prices, lagged total oil and natural gas consumption for OECD Europe, lagged supply for the competitor and time. The supply equations thus determined are reduced equations, so that instrumental variables for quantities equal to estimated supplies, through the supply equals demand identities ($\ln D_{it} = \ln \widehat{S_{git}}$), can be used for the estimation of 10.7 (Malinvaud, 1983).

10.3.3 Empirical results and discussion

A preliminary principal component analysis was carried out with the following variables: D_t is the global demand at time t, D_{t-1} is the global

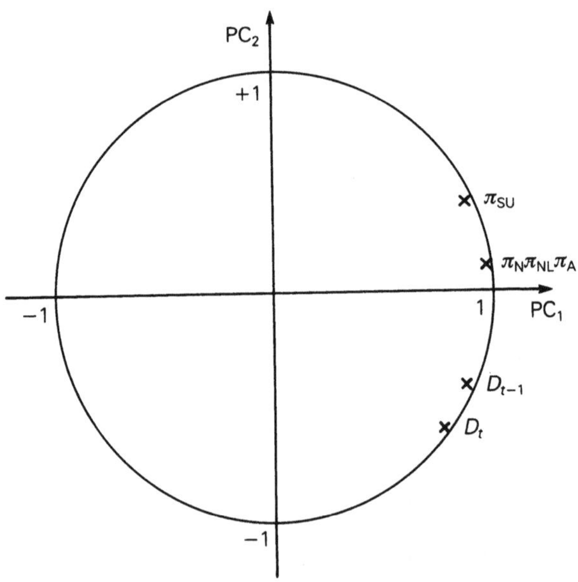

Fig. 10.1

demand at time $t-1$ and π_{SUt}, π_{Nt}, π_{NLt}, π_{At} are real contract prices at time t.

Two principal components (Fig. 10.1) appear to account for 96.5% of variance. Furthermore, the price of Soviet gas supplies π_{SU} is significantly different from the three other prices, which led us to specify our model with two main supply areas (USSR, and the other suppliers), as mentioned previously (section 10.3.2).

As shown in Tables 10.3 and 10.4, the regressions on supplies and demands concerning the two supply areas defined above are quite satisfactory. We observe correlation coefficients near to unity in all cases, and the elasticities are of the expected signs.

The specific behaviour of the USSR may be interpreted as the behaviour of a supplier with a large capability of adjusting to prevailing market conditions (because its exports to the OECD area are only a small share of its production and because of its quite important reserves). This specific behaviour also reflects systematically low prices, because of the USSR's need for foreign currencies.

10.4 CONCLUSION

Our model shows that there exists a well-defined market for natural gas in Western Europe, although gas prices are largely correlated with oil prices. Another interesting point is that suppliers may be classified into two main groups: the USSR and all other suppliers. However, we cannot reject the hypothesis that the market is competitive. It would be necessary to

Table 10.3 Supply equations

$$\ln(S_{SU})_t = \underset{(7.671)}{16.4565} + \underset{(5.365)}{0.26408}\ln(\pi_{SU})_{t-2} - \underset{(-4.942)}{0.02887}(\pi_{oil})_{t-1} - \underset{(-4.938)}{0.0573}(\pi_{oil})_{t-2}$$

$$\underset{(-9.003)}{-3.0812}\ln(D_{og})_{t-1} + \underset{(3.050)}{0.6175}\ln(D_{nsu})_{t-1} + \underset{(7.620)}{1.4662}\ln(D_{nsu})_{t-2}$$

$\bar{R}^2 = 0.995$ DW $= 2.217$ SEE $= 0.5\%$

D_{og}, total oil and gas consumption for OECD Europe
D_{nsu}, total international gas demand to countries other than USSR for OECD Europe

$$\ln(S_{nsu})_t = \underset{(0.379)}{0.4197} + \underset{(1.511)}{0.5442}\ln(S_T)_{t-1} - \underset{(-0.231)}{0.03844}\ln(S_{SU})_{t-1} + \underset{(1.479)}{0.34785}t$$

$$-\ 0.03812t^2 + 0.001266t^3$$

$\bar{R}^2 = 0.972$ DW $= 2.195$ SEE $= 5.8\%$

S_T, total gas supply (production) for area

DW, Durbin–Watson statistic; SEE, standard error estimate.

Table 10.4 Price and demand equations

Country or area concerned	*Constant*	*Endogenous demand-related part of the price*			*Quasi-exogenous oil-indexed part of the price*			*Statistics*		
		$\ln Q_t$	$\ln Q_{t-1}$	$\ln (GDP)_{t-1}$	$\pi_{oil,t}$	$\dot{\pi}_{oil,t}$	$\ln(\pi_{gt-1})$	$\bar{R}^2$	DW	SEE (%)
USSR	−10.380 (1.988)	−0.349 (1.87)	0.031 (0.269)	2.156 (1.938)	0.0624 (3.498)	0 (as constrained)	0.169 (2.865)	0.863	1.472	23.9
Countries other than USSR: Netherlands Norway Algeria Germany Denmark	−11.357 (−3.502)	−0.349 (4.11)	0.288 (3.08)	2.307 (3.373)	0.0799 (3.282)	0.0050 (0.217)	0.2346 (1.258)	0.963	1.815	14.9
Total demand	−10.170 (−3.295)	−0.349 (4.11)	0.235 (2.558)	2.080 (3.121)	0.0827 (4.209)	0.0050 (0.284)	0.2355 (1.526)	0.961	1.729	14.3
Expected sign	Undetermined	−	+	+	+	+	+	n.a.	n.a.	n.a.

introduce exogenous, non-economic factors to describe the European gas market more completely.

ACKNOWLEDGEMENTS

Financial support for this work from the Programme Interdisciplinaire de Recherche sur l'Energie et les Matières Premières du Centre National de la Recherche Scientifique (PIRSEM – CNRS, Paris, France) is gratefully acknowledged.

REFERENCES AND FURTHER READING

Black, F. and Scholes, M. (1972) The valuation of option contracts and a test of market efficiency. *Journal of Finance,* **27**, 399–418.

Boucher, J. and Smeers, Y. (1988) Economic forces in the European gas market. *Energy Economics,* **9**, 2–16.

Boucher, J., Hefting, J. and Smeers, Y. (1986) Economic analysis of natural gas contracts, Communication at the IAEE Conference, Bergen.

Estrada, J., Bergesen, H. O., Moe, A. and Sydnes, A. K. (1988) *Natural Gas in Europe: Markets Organization and Politics,* Pinter Publishers, London and New York.

Golombek, R., Hoel, M. and Vislie, J. (1987) *Natural Gas Markets and Contracts,* North-Holland, Amsterdam.

Johnson, L. L. (1960) The theory of hedging and speculation in commodity futures. *Review of Economic Studies,* **3**, 139–51.

Lesourd, J. B., Percebois, J. and Ruiz, J. M. (1987) Equilibre et déséquilibre sur le marché international du gaz naturel, in *Pétrole: marchés et stratégies* (eds A. Ayoub et J. Percebois), Economica, Paris.

Levasseur, M. and Simon, Y. (1980) *Marchés de capitaux: options et nouveaux contrats à terme,* Dalloz, Paris.

Malinvaud, E. (1983) *Méthodes statistiques de l'économétrie,* Dunod, Paris.

Nash, T. F. (1953) Two-person cooperative games. *Econometrica,* **21**, 128–40.

Percebois, J. (1986) Gas market prospects and relationships with oil prices. *Energy Policy,* **14** (4), 329–46.

Percebois, J. (1989) *Economie de l'Energie,* Economica, Paris.

Tirole, J. (1986) Procurement and renegotiation. *Journal of Political Economy,* **94**, April, 235–59.

PART THREE

Application of New Methodologies to Commodity Futures Markets

11

Dynamic welfare analysis and commodity futures markets overshooting

GORDON C. RAUSSER

AND NICHOLAS WALRAVEN

11.1 INTRODUCTION

Much of the analysis conducted on commodity futures markets focuses on partial equilibrium frameworks (Stein, 1981). However, linkages among markets implied by general equilibrium representations show that such analyses can suffer from serious limitations. In particular, studies of futures market efficiency which search for single series martingale or random walk processes cannot be expected to classify markets correctly (Rausser and Carter, 1983).

Linkages among markets mean that inefficiencies in one market may be transmitted to related markets. Nowhere is this more likely to be evident than in commodity futures markets. Since these markets reflect price expectations, differential information flows in the various markets will generally result in varying speeds of adjustment to causal forces.

Varying speeds of market adjustment have been used by Dornbusch (1976) and others to show that exchange rates can overshoot as a result of such market behaviour. In this work, and the subsequent work by Frankel (1979), exchange rates over-react to a monetary shock in order to compensate for the disequilibrium arising in a more slowly adjusting goods market. In the Dornbusch formulation, the long-run steady state remains unchanged while the exchange rate equates (temporarily) demand and supply in both the exchange and goods markets. For an expansionary monetary shock, the exchange rate moves to a level higher than that implied by the new long-run equilibrium and falls gradually as the sticky goods market adjusts. Prices in the efficient market overshoot the eventual equilibrium levels in order to clear the relatively inefficient goods market.

In addition to disequilibrium conditions arising in all sectors from monetary shocks, similar behaviour can arise from other types of shocks. In the case of commodity markets, market-specific shocks frequently can result from droughts and other weather-related phenomena. In the case of exchange rate markets, political instability is one frequent source of unexpected shocks. Furthermore, attempts on the part of central banks to manage the value of their country's currency can lead to disequilibria in exchange rate markets which, in turn, spill over to related markets. In fact, in many agricultural commodity markets, interest rate markets and exchange rate markets, unanticipated government policy is a likely source for shocks that arise in a specific market.

The basic Dornbusch model will be extended in this paper to examine the linkages among three groups of markets: interest rates, exchange rates and commodity markets. Price expectations, as represented by futures markets, will be emphasized. Unlike the Dornbusch model, in which all goods prices except exchange rates are presumed to be sticky, interest rates and agricultural prices will be allowed to be 'flexible'. Futures contracts for agricultural commodities are homogeneous, frequently traded and (mostly) storable and thus are presumed to be flexible and governed by instantaneous arbitrage.

Allowing flexibility in some markets with less flexibility in other markets admits the possibility of overshooting in spot markets for interest rates, exchange rates and commodity markets. To the extent that this overshooting exists, does it carry over to the formation of expectations as reflected by futures markets? If resource allocation decisions are based on these expectations, what are the welfare implications of overshooting? These questions are investigated by quantifying the dynamic linkages among US treasury bills (T-bills), the British pound, the Canadian dollar, the German mark, the Japanese yen and three agricultural commodity markets—corn, cotton and wheat.

A vector autoregressive moving average (ARMA) model is empirically estimated for a specific period of tight monetary policy, namely 1980 through the spring of 1982. Based on the dynamic adjustment paths, pricing efficiency is examined, accuracy and speed of convergence measures are calculated and dynamic welfare measures are computed. The accuracy measure is the total absolute deviation from final equilibrium levels of each price series during the adjustment period. The speed measure is the number of trading days required for some percentage of the total deviation to occur.

11.2 BASIC SPECIFICATION

The existence of differential responses to monetary shocks among markets can lead to 'overshooting'. More specifically, price stickiness in some markets is a necessary but not a sufficient condition for overshooting. Hence, whether pricing inefficiencies in one market lead to overshooting and allocative inefficiencies in another market is an empirical question. To demonstrate this result, a basic model specification linking interest rates, exchange rates, sticky price and flexible price markets is advanced followed by two alternative formulations. Both formulations are based on an open economy, one of which imposes fixed output whilst the other allows the possibility of endogenous output responses to dynamic price adjustment paths.

11.2.1 Basic model specification

For the case of fixed output, the manufactured good sector is represented by two basic equations. The first specifies output supply and the second excess demand:

$$y_{\rm m}^{\rm s} - \bar{y}_{\rm m}^{\rm p} = 0 \qquad (11.1)$$

$$y_{\rm m}^{\rm d} - \bar{y}_{\rm m}^{\rm p} = a_0(p_{\rm c} - p_{\rm m}) - a_1(i - u - \bar{r}) + a_3(p_{\rm m}^* + e - p_{\rm m}) \qquad (11.2)$$

where $y_{\rm m}^{\rm e}$ denotes output supply, $\bar{y}_{\rm m}^{\rm p}$ defines potential output, $y_{\rm m}^{\rm d}$ denotes demand for manufactured goods, $p_{\rm c}$ denotes the price of commodities, $p_{\rm m}$ denotes the price of manufactured goods, i denotes the nominal rate of interest, u denotes rate of inflation, $\bar{r}$ denotes the long-run equilibrium real rate of interest and e denotes the exchange rate. All variables are defined in logarithms. The rational expectations for commodity prices are generated by an arbitrage condition

$$\dot{p}_{\rm c}^{\rm e} = i + {\rm sc} \qquad (11.3)$$

where sc represents the storage costs associated with withholding the flexible commodity stock from one period to another. For the money market, a standard money demand equation is assumed with the following equilibrium condition:

$$m - p = \phi y - \lambda i \qquad (11.4)$$

where m is the logarithm of the money supply, p is the logarithm of the general price index, y is the logarithm of income and λ is the interest rate semi-elasticity of demand for real balances. This specification presumes that the money market clears at each moment in time.

Under the above specifications, manufactured goods are produced and imported; agricultural commodities are exported but not imported, with agricultural exports normalized to be zero; domestic and foreign manufactured goods are perfect substitutes; and for the agricultural commodity, representing the flex-price good, purchasing power parity is assumed to hold in the long run. However, deviations from purchasing power parity occur in the short run because of differential adjustment speeds in exchange rates and manufactured goods prices. Note also that the small country assumption is presumed to hold.

For the price index p, the underlying utility functions are assumed to be Cobb–Douglas so that the prices are weighted by their expenditure shares, i.e.

$$p = \alpha_1 p_m + \alpha_2 p_c + (1 - \alpha_1 - \alpha_2)(p_m^* + e) \tag{11.5}$$

In the short run, uncovered interest parity is presumed to hold:

$$i - i^* = \dot{e}^e \tag{11.6}$$

where i^* is the foreign short-term nominal interest rate and $\dot{e}^e$ is the expected rate of depreciation or appreciation of domestic currency. Since the small-country assumption is imposed, the nominal interest rate adjusted for expected appreciation is always equal to the (given) foreign rate. Implicitly, the specification assumes perfect substitutability between domestic and foreign interest-bearing instruments (a one bond world), absence of risk premia and perfect capital mobility.

Since the manufactured goods market is presumed to be sticky in the short run, the price adjustment process is specified to be some distributed lag of excess demand, i.e.,

$$\dot{p}_m = \Pi(y_m^d - \bar{y}_m^p) + u \tag{11.7}$$

For the flexible price commodity market, expectations are presumed to be rational:

$$\dot{p}_c^e = \dot{p}_c \tag{11.8}$$

Finally, for the case of exchange rate expectations, two alternative specifications are evaluated. One is based on rational expectations,

$$\dot{e}^e = e \tag{11.9a}$$

while the other is based on regressive expectations,

$$\dot{e}^e = -\delta(e - \bar{e}) \tag{11.9b}$$

For the second specification (11.9b), the expected rate of appreciation or depreciation is assumed to be proportional to the gap between the exchange rate and its long-run equilibrium value $\bar{e}$. Hence, if the spot rate exceeds its long-run value, these investors expect the rate gradually to appreciate at a speed of adjustment equal to δ. Finally, the normalization imposed is $\bar{p}_m = \bar{p}_c$, $\bar{y}_c = 0$, where y_c is the level of commodity output.

11.2.2 Fixed output and overshooting

The excess demand equation (11.2) does not admit output responses to alternative dynamic price paths. Combining equations in the basic model specification, the following price dynamic equations are obtained for the case of rational expectations (11.9a):

$$\dot{p}_c = \frac{\alpha_1}{\lambda}(p_m - \bar{p}_m) + \frac{\alpha_2}{\lambda}(p_c - \bar{p}_c) + \frac{1 - \alpha_1 - \alpha_2}{\lambda} \times (e - \bar{e}) + u + \bar{r} + \mathrm{sc} - \frac{1}{\lambda}\varepsilon \tag{11.10}$$

$$\dot{e} = \frac{\alpha_1}{\lambda}(p_m - \bar{p}_m) + \frac{\alpha_2}{\lambda}(p_c - \bar{p}_c) + \frac{1 - \alpha_1 - \alpha_2}{\lambda} \times (e - \bar{e}) + u + \bar{r} - i^* - \frac{1}{\lambda}\varepsilon \tag{11.11}$$

$$\dot{p}_m = -\Pi\left(a_0 + \frac{\alpha_1 a_1}{\lambda}\right)(p_m - \bar{p}_m) + \Pi\left(a_0 - \frac{a_1\alpha_2}{\lambda}\right)(p_c - \bar{p}_c) - \frac{\Pi a_1(1 - \alpha_1 - \alpha_2)}{\lambda}(e - \bar{e}) + \Pi a_3(p_m^* + e - p_m) + u + \frac{\Pi}{\lambda}\varepsilon \tag{11.12}$$

These dynamic equations (or equivalently those resulting from the regressive expectation specifications on exchange rates) can be solved to obtain the following commodity price determination equations:

$$p_c = \bar{m} - \phi\bar{y} + \lambda(\bar{r} + u) - \frac{1}{\theta_1}(i - u - \bar{r}) \tag{11.13}$$

$$\Delta p_c = \frac{1 + \lambda\theta_1}{1 - \alpha_1 + \lambda\theta_1}\Delta m + \frac{\lambda(1 - \lambda\theta_1)}{1 - \alpha_1 + \lambda\theta}\Delta u \tag{11.14}$$

where $-\theta_1$ is the negative root of the solutions which were obtained by solving Equations (11.10), (11.11) and (11.12).[1] The results for regressive and rational expectations of exchange rates collapse to the same outcome when $\delta = \theta_1$.

It is transparent from Equation (11.14) that overshooting of agricultural commodity prices follows immediately from the assumption of no output response. Since all parameters in the basic model specification are positive, the coefficient on a change in money supply Δm is greater than unity. This result shows that the transmission of price disequilibrium in the manufactured good market to deviations of commodity market prices is higher the larger is α_1, for a given rate of adjustment and interest semi-elasticity of money demand. Note also that the degree of overshooting is decreasing in

[1]Note that $\Delta \mathrm{sc} = \Delta\bar{r} = \Delta\bar{y}m$ ($= \Delta\bar{y} = \Delta p_m^* = 0$) and $m = \bar{m}$ are assumed in these derivations.

the responsiveness of money demand. This is simply because more of a shock to the money supply will be absorbed within the money market the larger is the value of λ. Finally, the length of time that the overshooting will last is a negative function θ_1 or equivalently, when regressive expectations are rational, δ.

The driving force behind the overshooting result of Equation (11.14) is the sticky price market represented by Equations (11.1) and (11.7) in combination with the fixed output specification, Equation (11.2). The arbitrage, Equation (11.3), combined with short-run disequilibrium between nominal interest rates and the long-run real rate of interest generates short-run disequilibrium in agricultural commodity markets. For example, to compensate the holders of grain inventories for forgoing present consumption, the grain price must rise at the nominal interest rate between harvests once convenience yields, storage costs and risk premia are properly incorporated. If money growth occurs so that a liquidity effect causes a fall in the nominal interest rates and in the short run the real interest rate due to sticky manufactured goods prices, a better return is available for storing grain than dollars and investors compete to hold grain inventories. This causes an immediate jump in the price of grain so that the condition for asset market equilibrium of equal rates of return is restored. All commodity prices jump and then are expected to rise at the new lower interest rates.

11.2.3 Endogenous output, overshooting and undershooting

The fixed output specification can be relaxed by replacing Equations 11.1 and 11.2 by the following three equations:

$$y_{\mathrm{m}}^{\theta} = y_{\mathrm{m}}^{\mathrm{d}} \tag{11.1a}$$

$$\bar{y}_{\mathrm{m}}^{\mathrm{d}} = a_0(p_{\mathrm{c}} - p_{\mathrm{m}}) - a_1(i - u) + a_2 y_{\mathrm{m}}^{\mathrm{s}} + a_3(p_{\mathrm{m}}^{*} + e - p_{\mathrm{m}}) \tag{11.2a}$$

$$\bar{y}_{\mathrm{m}}^{\mathrm{p}} = \frac{a_0}{1 - a_2}(\bar{p}_{\mathrm{c}} - \bar{p}_{\mathrm{m}}) - \frac{a_1}{1 - a_2}\bar{r} \tag{11.2b}$$

where $0 < a_2 < 1$.

Replacing Equations 11.1 and 11.2 by 11.1a, 11.2a and 11.2b and leaving Equations 11.3–11.9b unchanged leads to the following price dynamic equations in place of Equations 11.10, 11.11 and 11.12:

$$\dot{p}_{\mathrm{c}} = \frac{\alpha_1 - b_0\phi - b_3\phi}{\lambda + b_1\phi}(p_{\mathrm{m}} - \bar{p}_{\mathrm{m}}) + \frac{\alpha_2 + b_0\phi}{\lambda + b_1\phi}(p_{\mathrm{c}} - \bar{p}_{\mathrm{c}}) + \frac{1 - \alpha_1 - \alpha_2 + b_3\phi}{\lambda + b_1\phi}(e - \bar{e}) + u + \bar{r} + \mathrm{sc} - \frac{1}{\lambda + b_1\phi}\varepsilon \tag{11.10a}$$

$$\dot{e} = \frac{\alpha_1 - b_0\phi - b_3\phi}{\lambda + b_1\phi}(p_m - \bar{p}_m) + \frac{\alpha_2 + b_0\phi}{\lambda + b_1\phi}(p_c - \bar{p}_c)$$
$$+ \frac{1 - \alpha_1 - \alpha_2 + b_3\phi}{\lambda + b_1\phi}(e - \bar{e}) + u + \bar{r} - i^* - \frac{1}{\lambda + b_1\phi}\varepsilon \tag{11.11a}$$

$$\dot{p}_m = \Pi\left[b_0 - \frac{b_1(\alpha_2 + b_0\phi)}{\lambda + b_1\phi}\right](p_c - \bar{p}_c)$$
$$- \Pi\left[b_0 + b_3 + \frac{b_1(\alpha_1 - b_0\phi - b_3\phi)}{\lambda + b_1\phi}\right](p_m - \bar{p}_m)$$
$$+ \Pi\left[b_3 - \frac{b_1(1 - \alpha_1 - \alpha_2 + b_3\phi)}{\lambda + b_1\phi}\right](e - \bar{e}) + u + \frac{\Pi b_1}{\lambda + b_1\phi}\varepsilon \tag{11.12a}$$

where

$$b_0 = \frac{a_0}{1 - a_2} \qquad b_1 = \frac{a_1}{1 - a_2} \qquad b_3 = \frac{a_3}{1 - a_2}$$

Commodity price determination equations can be analytically derived from the above price dynamic equations. Equation 11.13 for endogenous output remains the same but Equation 11.14 becomes

$$\Delta p_c = \frac{1 + \theta_2(\lambda + \phi b_1)}{1 - \alpha_1 + \theta_2\lambda + \phi(b_0 + b_3 + \theta_2 b_1)}\Delta m$$
$$+ \frac{\lambda + \lambda\theta_2(\lambda + \phi b_1)}{1 - \alpha_1 + \theta_2\lambda + \phi(b_0 + b_3 + \theta_2 b_1)}\Delta u \tag{11.14a}$$

where $-\theta_2$ is the negative root of the solution obtained by solving Equations 11.10a, 11.11a and 11.12a. Note that the results for the regressive and rational expectations of exchange rates collapse to the same outcome when $\delta = \theta_2$.

For Equation 11.14a overshooting is only one of three possible outcomes. This can be seen from

$$\frac{\Delta p_c}{\Delta m} = 1 + \frac{\alpha_1 - \phi(b_0 + b_3)}{1 - \alpha_1 + \theta_2\lambda + \phi(b_0 + b_3 + \theta_2 b_1)} \tag{11.15}$$

Hence, given that the denominator of the second term is positive, whether overshooting, undershooting or neutrality results from a particular monetary shock depends upon the numerator of the second term in Equation (11.15).

More precisely,

$$\alpha_1 \gtreqless \frac{\phi(a_0 + a_3)}{1 - a_2} \begin{cases} > \text{overshooting} \\ < \text{undershooting} \\ = \text{neutral} \end{cases} \tag{11.16}$$

In contrast with the result obtained for fixed output (Equation 11.14), the mixed results of 11.16 reflect the possibility that the output and demand reponses to prices ($a_0 + a_3$), appropriately modified for the effect of output on money demand (ϕ) and the dynamic output adjustment ($1 - a_2$), can swamp the initial moves in flexible price commodity markets that would otherwise result. Hence, it depends empirically on the relative size of the expenditure share of the sticky price markets (α_1) and the responsiveness of output to prices and, in turn, the responsiveness of demand for real money balances to changes in output levels.

If the overshooting result occurs, its degree will depend upon four key parameters: the relative share of the sticky price markets in the domestic economy (α_1); the responsiveness of money demand (λ); the effect of output response on money demand; and the speed of adjustment (θ). The length of time overshooting will last is a negative function of θ. This parameter plays a major role in the speed of convergence to any new long-run equilibrium resulting from a monetary shock. The remaining parameters dictate the magnitude of Δp_c during any particular period. The accuracy of any particular level of price relative to the long-run equilibrium level Δp_c and the speed of convergence θ are the major components of the appropriate dynamic welfare measure.

11.3 A DYNAMIC MULTICOMMODITY WELFARE MEASURE

The allocative efficiency loss during the joint adjustment of all prices to some new stationary state following a shock requires the development of general equilibrium welfare measures (Rausser and Just, 1981). Conventional dynamic and simultaneous supply and demand equations, where futures prices p^f representing expectations enter the supply equations, will be presumed. As shown in numerous places (Zellner and Palm, 1974; Rausser and Carter, 1983), if a set of endogenous variables is generated by a dynamic simultaneous equation model, then it is often possible to solve for the transfer function of individual endogenous variables (such as exchange rates, interest rates etc.) through algebraic manipulation. In essence, each endogenous variable in a structural form model has associated with it an explicit and unique transfer function equation which expresses the endogenous variable as a linear combination of current and past values of the exogenous variables and an ARMA error term. Similarly, given that each exogenous variable can be expressed in terms of an ARMA process, it is possible to re-specify the transfer function equation as an ARMA process for each endogenous variable. Accordingly, it is possible to represent the basic model presented in section 11.2 as a multivariate time series model as long as the relevant error terms in each exogenous variable can be represented as an ARMA process.

If the vector ARMA process represents adequately the endogenous price variables, then this framework approximates a rational expectations formulation, with the error in the approximation caused by transactions costs, risk aversion of agents etc. (Rausser and Carter, 1983). The resulting price path is a function of structural supply S and demand D parameters and expected (futures) prices. More specifically, assume that an initial steady state level of prices $\hat{p}$ exists and there is a shock Z_0 at time 0. Since $\hat{p}$ represents the net effect of all past adjustments, the future expected prices p_t^{f} for any time after time 0 are

$$p_t^{\mathrm{f}} = \hat{p} = Q_i Z_0 \tag{11.17}$$

where

$$Q_i Z_0 = \sum_{j=0}^{j-1} [A^{-1}(L)\, B(L)]_j Z_0 \tag{11.18}$$

$A(L)$ and $B(L)$ represent polynomials in the lag operator that reflect the dynamic interactions of the endogenous prices. These lag polynomials are assumed to be stationary and invertible. If there are no subsequent shocks, the effect of Z_0 at any time may be expressed as the initial steady state $\hat{p}$ plus the net effect of Z_0 to that time. The deviation of prices from the eventual long-run equilibrium changes over time; therefore the amount of welfare loss also changes over time.

To determine the welfare loss in a market at any time during the adjustment period, consider the static welfare analysis depicted in Fig. 11.1. The long-run steady state is $\bar{p}$ following the shock, and $\bar{y}$ is the corresponding quantity. Given a stable and invertible model, $A(L)p_\infty = B(L)Z_\infty \to \bar{p}$; i.e. given no other shocks, the price path converges to its new equilibrium level. The expected level of prices for t periods following the shock is $Ep_t = Q_t Z_0 + \hat{p}$.

It is clear that the welfare loss at p_t relative to $\bar{p}$ is given by triangle abc. This area is given algebraically by halving the product of the base and height of abc:

$$\mathrm{WL}_t = \tfrac{1}{2}[S(p^f) - S(\bar{p})]\{p^{\mathrm{f}} - D^{-1}[S(p^{\mathrm{f}})]\} \tag{11.19}$$

Adding and subtracting $S(\bar{p}) = D(\bar{p})$ and taking a first-order Taylor series expansion about $\bar{p}$, the welfare loss at any time t is

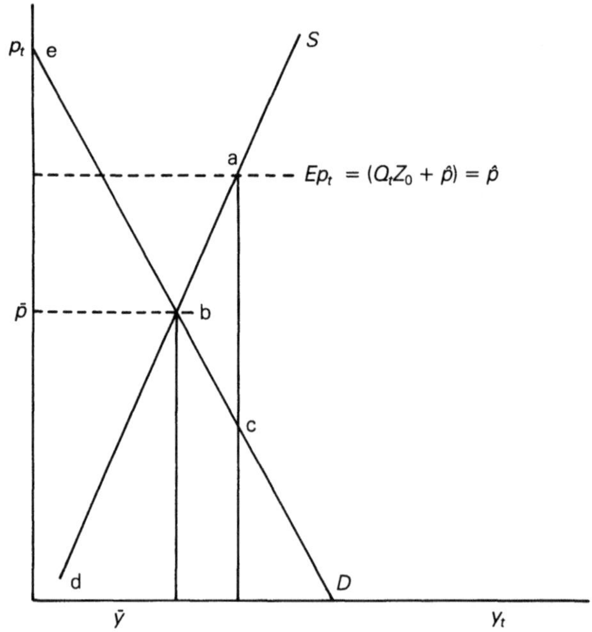

Fig. 11.1 Static welfare analysis.

$$\mathrm{WL}_t = \frac{1}{2}\frac{A}{B}(B + A)(p^{\mathrm{f}} - \bar{p})^2 \tag{11.20}$$

where A represents the slopes of the supply equations and B the slopes of the demand equations. Note that this measure depends on the squared deviation of prices from the eventual equilibrium level and slope parameters of the supply and demand functions. This result is similar to the partial equilibrium measure of Stein for welfare losses due to futures market price inaccuracies. Because A and B represent $\partial S/\partial p$ and $-(\partial D/\partial p)$ respectively, the welfare loss expressed in terms of the supply and demand elasticities at $(\bar{p}, \bar{y})$ is

$$\mathrm{WL}_t = \frac{1}{2}\frac{\bar{y}}{\bar{p}}\frac{\eta_{\mathrm{s}}}{\eta_{\mathrm{d}}}(\eta_{\mathrm{d}} - \eta_{\mathrm{s}})(p^{\mathrm{f}} - \bar{p})^2 \tag{11.21}$$

where η_{s}, η_{d} define the supply and demand elasticities respectively. The total consumer and producer surplus at $(\bar{p}, \bar{y})$ is given by the area of triangle deb in Fig. 11.1 which, when expressed in terms of elasticities, is

$$\mathrm{TS} = \frac{1}{2}\bar{p}\bar{y}\left[\frac{A_1\bar{p}/\bar{y} + B_1\bar{p}/\bar{y}}{B_1(\bar{p}/\bar{y})A_1\bar{p}\bar{y}}\right] = \frac{1}{2}\bar{p}\bar{y}\left(\frac{\eta_{\mathrm{d}} - \eta_{\mathrm{s}}}{\eta_{\mathrm{s}}\eta_{\mathrm{d}}}\right) \tag{11.22}$$

Dividing Equation (11.21) by Equation (11.22) gives the percentage of total surplus lost as a result of the deviation of futures prices, i.e.

$$\frac{\mathrm{WL}}{\mathrm{TS}} = \eta_s^2 \left(\frac{1}{\bar{p}}\right)^2 (p^f - \bar{p})^2 \tag{11.23}$$

Representation 11.23 has several advantages. First, the expression is solely in terms of the elasticity of supply, the new steady state price and the squared deviation of prices during the adjustment period. The last two variables are known for each market; therefore the percentage welfare loss may be expressed in terms of one parameter, the elasticity of supply. Another advantage is the lack of scale for this loss measure. This allows various markets to be compared regardless of their size.

The form of this welfare measure depends upon several assumptions. First, the results hold only to the extent that the linear representation approximates the supply and demand relationships. Note that, in this measure, futures prices affect welfare through expectations, and the critical parameters are those appearing in the spot market supply and demand relationships. The dependence of 11.23 on the supply elasticity rather than on both supply and demand elasticities occurs because of the linear structure and because p^f in the supply equation determines the quantity in any period. Once one component of the supply and demand relationships determining the surplus measure is fixed (i.e. the supply elasticity at $(\bar{p}, \bar{y})$); any change in the other component has offsetting effects on total surplus and welfare loss. That is, rotating the demand curve clockwise around $(\bar{p}, \bar{y})$ in Fig. 11.1 proportionately increases both the welfare loss 11.21 and the total surplus 11.22, leaving their ratio unchanged. The supply elasticity becomes the scaling factor in 11.23 because the expected price determines the quantity via supply.

Of course, to obtain the total welfare loss for the adjustment period, one should discount losses at future dates by some discount rate. Specifying the number of periods for an arbitrary amount of the total adjustment to occur, the total welfare loss due to the deviation of prices is a function of the discount rate, the elasticities of the supply and demand functions, the number of periods for the adjustment to occur, the new steady state level of prices and quantities and the squared deviations of futures prices from the new steady state following a shock. The first three parameters are assumed to be constant over the adjustment period, and so welfare loss may be viewed as a function of the dynamic adjustment path of prices. In other words, the welfare loss depends on the accuracy (squared deviations) and the speed of the convergence of the price series.

A crucial assumption for this view of the efficiency of observed price series is the nature of the long-run equilibrium in an environment of slowly adjusting prices. Futures markets clear each day, and allocative decisions involving futures prices may be made during the adjustment of prices to their new equilibrium levels. If the final equilibrium is affected by the series of temporary equilibria of all related markets, any measure of welfare loss based upon the final equilibrium which it obtains must be incorrect. The long-run equilibrium which would exist in the absence of lagged adjustment of prices is unobservable; thus the new efficiency measures developed in this paper, while correcting for some of the possible misspecification of previous studies, may remain only partial evaluations of total efficiency.

Although the loss measure in Equation 11.23 provides a convenient comparison of the relative efficiency of various markets, the absolute levels of welfare loss remain informative. The measurement problem that arises for the absolute welfare loss 11.22 is the absence of observations of $\bar{y}$, the equilibrium quantity. Although some approximations of $\bar{y}$ are used in the empirical section, the approximations may be crude. A loss measure consisting of the forecast error weighted by the particular market's importance to the economy is both easily determined from available data and useful in assessing the total welfare loss in each market.

Dividing both sides of equation 11.21 by $(\bar{p}, \bar{y})$ yields

$$\frac{\mathrm{WL}}{\bar{y}\bar{p}} = \frac{1}{2}\frac{\eta_s}{\eta_d}(\eta_d - \eta_s)\left(\frac{p^f - \bar{p}}{\bar{p}}\right)^2 \qquad (11.24)$$

This value expresses the welfare loss scaled by the total sales in the market or, in other words, the welfare loss per unit of revenue. An attraction of this quantity is the absence of unobserved equilibrium quantities on the right-hand side of the expression. The scaled welfare measure depends only on the elasticities of supply and demand, the steady state price level and the squared deviation of prices.

11.4 METHODOLOGY, DATA AND EMPIRICAL RESULTS

To capture empirically the dynamic price linked paths, a multivariate time series model is specified for an eight-market system. This model incorporates the relationships of T-bills and exchange rates (British pound, Canadian dollar, German mark and Japanese yen—all in cents per unit of foreign currency) with corn, cotton and wheat prices. As mentioned earlier, the dynamic interactions of the price series depend on the structural relationships of the underlying supply and demand functions. The existence of these interactions among agricultural commodities, as well as the relationship of agricultural commodities to interest and exchange rates, has been documented in numerous studies (Rausser, 1985).

The accuracy of the estimated efficiency for each market and for the system depends critically upon expressing the dynamic interactions adequately by identifying a suitable time series representation. We chose vector

ARMA representations because of their parsimony relative to the more widely used vector autoregressive models. Nevertheless, large-vector ARMA models still fall into the general class of overparameterized models, implying that some sort of restrictions other than identifying the order of the autoregressive and moving-average polynomials may become necessary (Sims, 1980).[2]

The data used to estimate the vector ARMA consist of 205 observations for the March 1981 delivery contracts of the eight variables mentioned above and 195 observations for March 1982. These data span the period beginning in the spring of 1980 through the spring of 1982, a period during which financial markets adjusted to the new Federal Reserve (Fed) policy of targeting the money supply rather than interest rates. The two sets of observations provide estimates of pricing and allocative efficiency immediately following the Fed policy change (1981 delivery contracts) and much later (1982 delivery contracts).

The choice of the steady state vector of prices used in the dynamic analysis influences all the subsequent results. The welfare measures developed in section 11.3 depend on the steady state level of prices. In addition, since the models all have autoregressive terms, the new equilibrium price levels, as well as the dynamic adjustment paths, depend upon the initial steady state chosen.

The initial equilibrium of each series is obtained by forecasting from the end of the time series until no further change in the variables is observed. This approach, of course, provides only one of many possible steady state levels for the vector of series. At any point in the sample, one could assume that there are no further shocks and find a different steady state level. Any one of these steady states is preferable to some *ad hoc* level, such as the mean for each series, because the simultaneous observation of all series at their mean level may be highly unlikely. Since there is no definite trend in the price series, the choice of the last observation rather than another simultaneously observed set of prices will not affect the results.

The degree of correlation among the series in this study may be highly positive or negative, and so a simultaneously observed set of prices was chosen—the last observation vector. Then, assuming no futher disturbance, the estimated parameters are used to compute successive forecasts until there is no change in the forecast price. The estimated models are stable and

[2]One option is to use t tests to set individual elements of parameter matrices to zero (Tiao and Box, 1981), reducing the degree of overparameterization. The undesirable decrease in statistical power due to the extra coefficients may therefore be reduced by constraining particular values to zero. The increase in power is achieved, of course, at the risk of biasing the remaining parameter estimates. Since there is little prior information concerning parameter values used in the vector ARMA models (in particular, whether or not to include variables in certain equations), the possibility of biased parameter estimates is high. One should therefore avoid selecting extremely low significance levels for any tests. The major concern in this study is not hypothesis testing but in reflecting as much of the dynamic adjustment as possible.

invertible, and so the forecasts converged to the equilibrium level $\bar{p}$—usually in about 25 periods.

The proper type of shock to consider when calculating the multipliers is an interesting question relating to the selection of the initial steady state. Most authors who construct vector autoregressive models analyse the dynamic properties of their models by using one standard deviation of a single series as a shock. In other words, one element of the shock vector is the standard deviation of the corresponding error series, and all other elements are zero. The probability of observing this particular shock may be extremely low, especially when the residual error terms are correlated. For example, given positive expected correlation of interest rate and the costs of storage, one might not expect to see large positive shocks both in T-bills and all commodity prices. Even if all the series are positively correlated, adjustment by the correlation matrix will reflect the relationships among errors.

The plausibility of the shock is very important since the resulting dynamic patterns are used to construct empirical measures. As mentioned earlier, positive shocks in one market or set of markets might be associated with a particular type of shock in a related market. An arbitrary choice of a particular shock might obscure this empirical relationship in the efficiency measure.

Consideration of a large positive shock in one leading market and none in another should yield a different adjustment pattern from that of a simultaneous shock in several markets. If some particular type of shock rarely occurred and therefore hardly affected the estimated relationships among the price series, then one should not use it to calculate the efficiency measure for the entire sample. In other words, the most likely shock during the sample period should be used to summarize the relative efficiency of the markets. Accordingly, a multimarket shock is employed here. It is generated by a one-standard-deviation vector of errors from the fitted ARMA model multiplied by the empirical correlation matrix of the errors. This procedure yields the best estimate of the signs and relative magnitudes of the elements of the shock vector given the observed data. Multiplication by the correlation matrix adjusts each standard deviation by its correlation with all other series multiplied by the standard deviation of that particular series. This perturbation helps to demonstrate the dynamic interactions of the series and allows calculation of the empirical efficiency measures resulting from the single shock.

Fitting an (AR(1)MA(6,13,18)) model to the differenced 1981 data and an (AR(1)MA(5,6,8,14,15)) model to the differenced 1982 data yields the parameter estimates necessary for constructing the efficiency measures. Graphs of eight markets in terms of these efficiency measures, accuracy and speed are given in Figs 11.2 and 11.3. Figure 11.2 effectively summarizes the previous comparison of the dynamic behaviour of each market from the perspective of the absolute deviation measure of accuracy. Figure 11.3 displays a similar summary for a total deviation measure. In both graphs, the furthermost point on the left for each series represents the number of periods required following a shock for 50% of the total response to occur. Similarly, the furthermost point on the right represents the number of

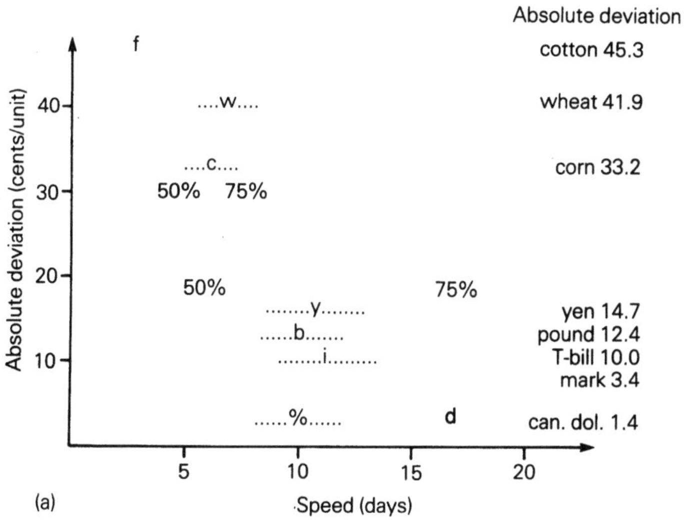

(a)

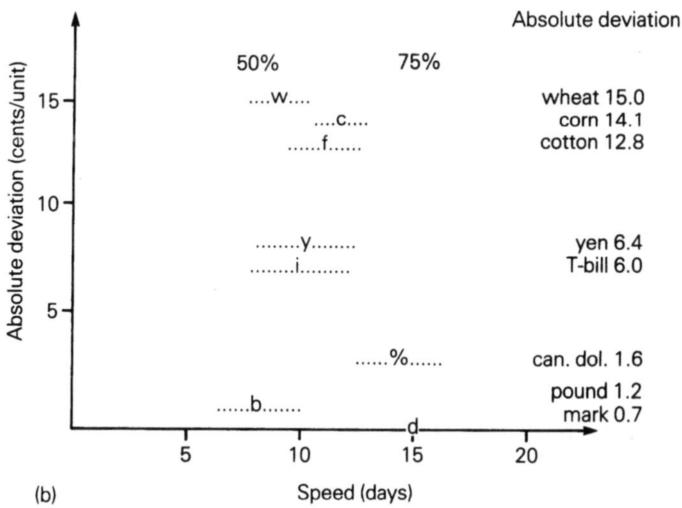

(b)

Fig. 11.2 Accuracy in terms of total absolute deviations and speed: (a) 1981; (b) 1982 (i, T-bills; b, British pound; %, Canadian dollar; d, German mark; y, Japanese yen; c, corn; f, cotton; w, wheat).

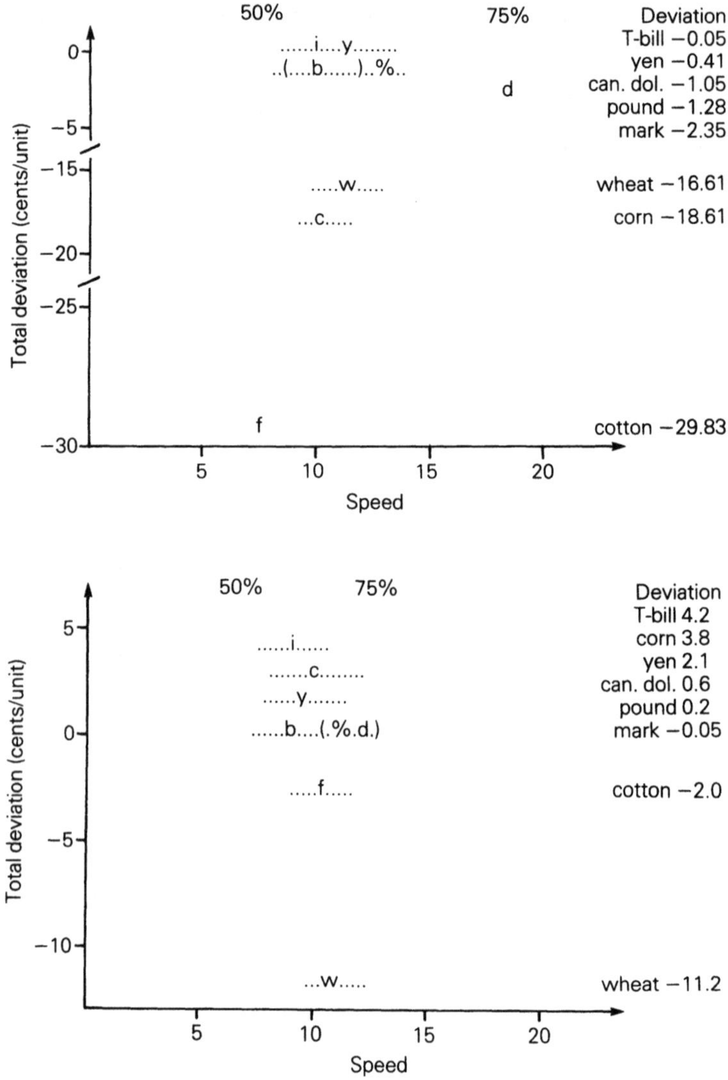

Fig. 11.3 Accuracy in terms of total deviation and speed: (a) 1981; (b) 1982.

periods required for 75% of the total fluctuations to occur. The 1981 Canadian dollar series in Fig. 11.2, for example, achieves 50% of the total adjustment by period 7 and 75% by period 17. The German mark, in contrast, achieves both 50% and 75% of its adjustment at period 18. When the deviation measures for the two series overlap, parentheses indicate differing values of the speed measure. To illustrate, in Fig. 11.3 the British pound and Canadian dollar have total deviation values of −1.28 and −1.05, respectively, which are too close to distinguish on the graph. The parentheses surrounding the symbol for the pound indicate that its speeds for 50% and 75% adjustment were 9 and 15 days respectively. The vertical axes for Figs 11.2 and 11.3 measure the total absolute deviation and the total deviation respectively. The values of these deviations are printed along the right-hand side of each group.

These results indicate that agricultural markets for 1981 delivery tended to adjust more quickly and to deviate more than either interest rate or exchange rate markets. As shown in Fig. 11.2, the agricultural markets achieved 50% of the total absolute deviation by period 7 in 1981. Futhermore, 75% of the total absolute adjustment occurred by period 7 for cotton, period 10 for corn and period 14 for wheat. In contrast, both interest rates and exchange rates generally took much longer to reach either 50% or 75% adjustment.

The total deviation values in Fig. 11.3 show that the agricultural series dropped to their final levels while the other markets oscillated about their initial levels. The combination of large negative total deviations and large positive absolute deviations of the agricultural markets suggests that they generally fell after the period 0 reaction to the shock. The agricultural markets overshoot to the greatest degree, followed by exchange rates. The empirical results indicate that agricultural markets exhibited both greater net overshooting and a faster speed of convergence in 1980–1.

For March 1982 delivery contracts, the German mark continued to show little deviation and to adjust slowly. The agricultural markets continued to deviate substantially more in absolute value than exchange rates or interest rates, but the speed was much more similar to the financial markets than previously. Indeed, the British pound achieved both 50% and 75% of its total adjustment faster than any other series, while the speed for the agricultural series was similar to that of the Japanese yen and T-bills.

The total deviation measures in 1982 are markedly different from those in 1981. In contrast with the −0.05 to −30 range for 1981, the total deviation varied from 4.2 for T-bills to −11.2 for wheat; and the distinct difference between agricultural and financial contracts observed in 1981 in less clear.

The empirical dynamic adjustment paths can be used to compute the welfare measures developed in section 11.3. Figures 11.4 and 11.5 express the percentage of total welfare lost as a result of the dynamic path of prices for 1981 and 1982 respectively. The percentage of total welfare lost as a function of squared elasticity of supply is represented along the vertical axis in these figures.

The figures show clearly that a much greater percentage of total welfare in the cotton and Japanese yen markets is lost in both 1981 and 1982. The other series have values markedly smaller suggesting that, relative to total trade of the commodities, the cotton and Japanese yen markets were

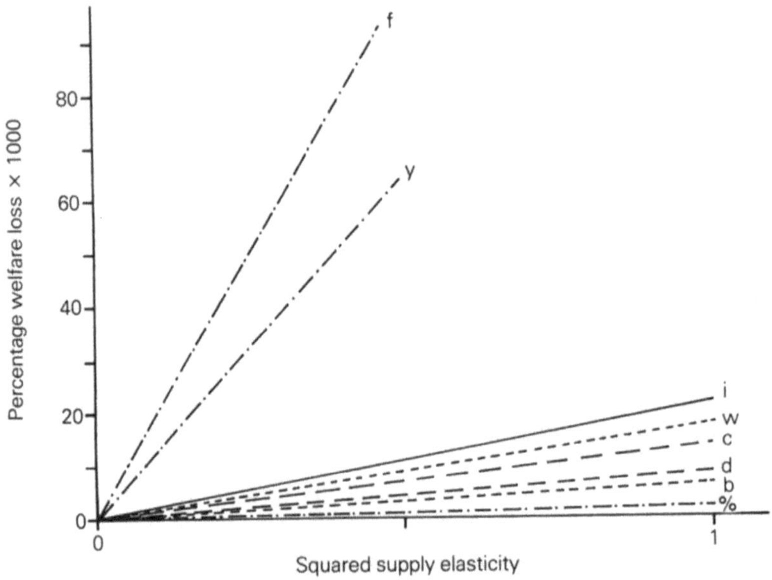

Fig. 11.4 1981 Percentage welfare loss.

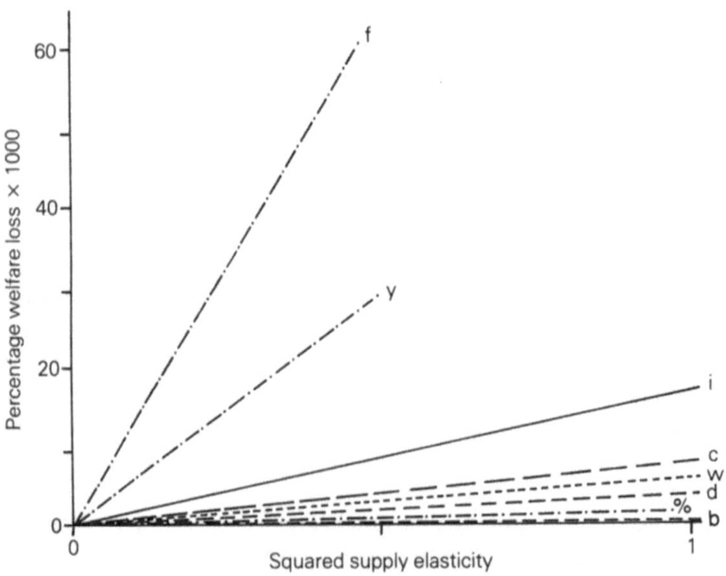

Fig. 11.5 1982 Percentage welfare loss.

allocatively inefficient. Of course, this observation depends upon an assumption that the elasticity of supply for yen and cotton is non-zero and not drastically different from the those for other exchange rates and agricultural commodities respectively.

The elasticity of supply for both yen and cotton in 1981 must be approximately $\sqrt{(0.25)}$ while the elasticity of supply for the other commodities must be approximately $\sqrt{3}$ for there to be a roughly comparable percentage welfare loss in all the markets. Although there are no empirical estimates of daily supply and demand elasticities, general assumptions regarding relative elasticities can be made. It seems reasonable to assume that there is some degree of price responsiveness, even on a daily basis. The difference in the loss measure becomes more pronounced if all series have roughly similar elasticities of supply. The similarity of the relative values in 1981 and 1982 suggests that greater welfare loss may be endemic to the yen and cotton markets.

Daily quantities supplied or demanded are necessary to calculate the welfare loss. Quantities supplied are readily available for agricultural markets, and the *Federal Reserve Bulletin* reports average daily trade in T-bills. The only source of data regarding volume of spot market currency transactions seems to be a sampling done by the New York Federal Reserve Bank every 3 years. The last available data, sampled for the month of April 1983, are given in Table 11.1. These monthly trade volume numbers provide an estimate of the daily volume which may be used to estimate the total welfare loss in the exchange markets.[3]

The welfare measure can be rearranged to be an expression involving two multiplicands, specifically:

$$\mathrm{WL}_t = \left[\frac{\eta_s}{\eta_d}(\eta_d - \eta_s)\right]\left[\frac{\bar{p}\,\bar{y}}{2}\left(\frac{p^f - \bar{p}}{\bar{p}}\right)^2\right] \tag{11.25}$$

The first term, involving elasticities of supply and demand, will be denoted subsequently the elasticity multiplicand. The second term, hereafter called the deviation multiplicand, consists of the squared forecast error, constants, market revenues and the squared equilibrium price.

The deviation multiplicand and its components are given in Table 11.1. The deviation multiplicand indicates that the loss due to deviations in the agricultural markets is minuscule compared with the losses in the T-bill, German mark and Japanese yen markets if the elasticity multiplicands are of similar magnitudes across markets. That is, if the elasticity multiplicand $\eta_s/\eta_d(\eta_d - \eta_s)$ is roughly comparable, the welfare loss is much less in the agricultural markets. The greater deviations of prices for agricultural series are more than counterbalanced by the large volume of trade in the financial markets.

[3]The exchange rate volumes presented in Table 11.1 may overstate actual volume by up to 25% as a result of double counting.

Table 11.1 Welfare loss measure

Series	*Squared deviation ($)*		*Yearly volume*[a]		*Steady state ($)*		*Yearly value ($ billion)*		*Deviation multiplicand ($ million)*	
	1981	1982	1981	1982	1981	1982	1981	1982	1981	1982
US treasury bills	0.27	0.56	3707.0	4618.0	0.851	0.869	3155.0	4013.0	0.515	0.634
British pound	0.64	0.04	[b]		2.21	1.84	861.0	861.0	0.060	0.003
Canadian dollar	0.02	0.04			0.838	0.810	318.0	318.0	0.004	0.006
German mark	0.04	0.02			0.468	0.416	1929.0	1929.0	0.147	0.078
Japanese yen	1.22	0.93			0.469	0.421	1006.0	1006.0	3.14	1.56
Corn	2.71	1.85	8.3	9.2	3.50	2.54	29.0	23.5	0.002	0.001
Cotton	4.11	4.52	10.7	15.2	0.847	0.628	4.3	4.6	0.010	0.008
Wheat	6.61	2.80	3.8	4.0	4.40	3.51	16.7	14.0	0.001	0.0005

[a]Dollars for treasury bills; million bushels for corn and wheat; thousand bales for cotton.
[b]Blanks indicate no data available.

The deviation multiplicand indicates that the Japanese yen, followed by T-bills and the German mark, should exhibit the greatest daily welfare loss due to slowly adjusting prices. The relatively small squared deviations of T-bill prices are offset by its enormous volume, so that any deviation of prices causes a great welfare loss. The Japanese yen exhibits the highest squared deviations among the financial markets, and its relatively large volume gives it a large welfare loss. Agricultural markets have squared deviations about ten times greater than those of the financial markets, but their trade value makes them have a relatively low welfare loss.

Figure 11.6 depicts the trade-off between elasticities of supply and demand for the elasticity multiplicand. For welfare losses in agricultural markets to be as large as in the T-bill market, the elasticity multiplicand (*k* in Fig. 11.6) must be roughly 100 times greater for agricultural markets than for T-bills.

11.5 CONCLUSION

Allowing for varying flexibility among exchange rates, interest rates and commodity markets and dynamic linkages among these various markets, overshooting is revealed as a common empirical phenomenon. For the eight

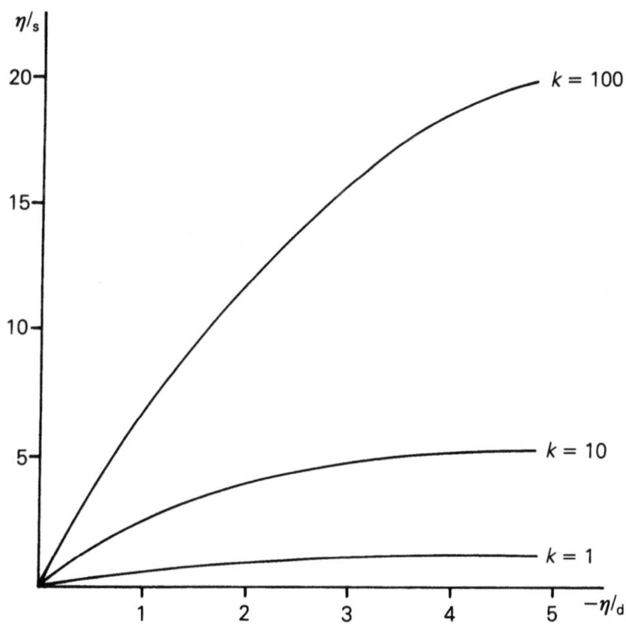

Fig. 11.6 Elasticity multiplicand trade-offs.

futures markets investigated (T-bills, the British pound, the Canadian dollar, the German mark, the Japanese yen, corn, cotton and wheat), overshooting occurs in the formation of expectations for each market.

Although interest rate, exchange rate and commodity markets are all shown by the estimated vector ARMA model to overeact to an initial shock, commodity markets (corn, cotton and wheat) do so to a much greater degree than either exchange rate or short-term interest rate markets. However, the period length of this overeaction, for a major portion of the degree of disequilibrium, is much shorter for the agricultural commodity markets. In the context of resource allocation decisions, the dynamic welfare measures reported suggest that the cotton and yen markets have the greatest loss as a proportion of the total consumer and producer surplus in each. For comparable elasticities of supply and demand, the total welfare losses are found to be the largest in the short-term interest and Japanese yen exchange rate markets.

REFERENCES

Dornbusch, R. (1976) Expectations and exchange rate dynamics. *Journal of Political Economy*, **84**, 1161–76.

Frankel, J. (1979) On the mark: a theory of floating exchange rates based on real interest differentials. *American Economic Review*, **69**, 610–22.

Rausser, G. C. (1985) Macroeconomics and U.S. agricultural policy, in *U.S. Agricultural Policy: 1985 Farm Legislation* (ed. B. L. Gardner), American Enterprise Institute for Public Policy Research, Washington DC, pp. 207–52.

Rausser, G. C. and Carter C. (1983) Futures market efficiency in the soybean complex. *Review of Economics and Statistics*, **65**(3), August, 469–78.

Rausser, G. C. and Just, R. E. (1981) Principles of policy modeling in agriculture, in *Modeling Agriculture for Policy Analysis in the 1980s*, Federal Reserve Bank of Kansas City, Kansas City, September, pp. 139–74.

Sims, C. (1980) Macroeconomics and reality. *Econometrica*, **48**(1), 1–48.

Stein, J. (1981) Speculative price: economic welfare and the idiot of chance. *Review of Economics and Statistics*, **63**, May, 223–32.

Tiao, G. C. and Box, G. E. P. (1981) Modelling multiple time series with applications. *Journal of the American Statistical Association*, **76** (376), 802–16.

Zellner, A. F. and Palm, F. (1974) Time series analysis and simultaneous equation econometric models. *Journal of Econometrics*, **2** April, 17–54.

12

When does the creation of a futures market destabilize spot prices?

PATRICK ARTUS

12.1 INTRODUCTION

The question of whether spot prices are stabilized or not by the creation of a futures market has been examined in many theoretical studies. The earliest of these conclude that the futures market should logically exert a stabilizing effect (Turnovsky, 1979, 1983; Kawai, 1983; Turnovsky and Campbell, 1985; Peck, 1985; Artus, 1987). The arguments advanced for this are (a) the increase in speculative activity and (b) the fact that shocks occurring in the spot market are diluted between two markets. These findings run counter to the conventional wisdom and to certain empirical studies. More recent research has therefore consisted in altering the basic model so as to produce the opposite result. The present chapter seeks to incorporate these new ideas into a unified model where their value can be assessed and their effects compared.

Although we chiefly had in mind financial futures markets, our model can be applied to any futures market. Accordingly, we shall sometimes describe market hedgers as 'producers', in reference to futures contracts on raw material and farm commodity output. In the case of financial markets, the hedgers can be portfolio managers protecting themselves from capital gains or losses due to exchange rate movements, or they could be firms with the skills to hedge against a rise in long-term rates. When we speak of 'production techniques', a clear enough concept in raw materials markets, we are therefore designating the process used to determine the size of the portfolio that requires actual hedging. A mutual-fund manager, for example, must hedge the securities that he plans to sell to cover dividend payments or redemptions. The latter may be random and dependent on his management or on market swings.

Starting from the basic model, which allows us to describe the functioning of any futures market (e.g. for tangible commodities or financial assets), we shall successively examine the following.

The effect of liquidity constraints, which, in particular, limit the activity of arbitragers: it has often been claimed that arbitragers are sometimes prevented from undertaking profitable transactions by factors such as restrictions on their indebtedness and the need to hedge losses on other markets. We incorporate this phenomenon by introducing an arbitrage inertia or 'viscosity' that permits a divergence between spot and futures prices.

The effect of erroneous expectations about futures prices: Stein (1986) introduces a random bias in speculators' price forecasts. He sees the introduction of the futures market as exerting a stabilizing effect through the variance of this bias for 'professional' speculators and 'amateur' speculators—whose forecasts are less accurate. But the origin of the bias is not clearly stated in his model, which also poses feedback problems. We shall assume here that, when the futures market opens, speculators are not immediately familiar with the new interactions of the spot and futures markets and require a learning period. This means that they continue to form their expectations according to the pattern that was rational when the spot market alone existed.

In the long run, however, speculators eventually learn how the markets work. We distinguish between two cases: (a) only their expectations about price levels become rational again, or (b) their price volatility expectations become rational again too. In the second case, which implies perfect variability, the effects of changes in price variability are amplified. For example, any price fall will intensify speculative activity—normally a stabilizing factor—and will further reduce variability.

We shall examine in what conditions speculators—starting from mistaken expectations about price levels and volatility after the opening of the futures market—can gradually converge toward a rational mode of expectations formation simply by observing market prices and volatility and periodically revising their expectations patterns.

The case where the production technique is random, this disturbance being correlated with the one affecting the spot market (Newbery, 1987): one could model a random morality behaviour of market participants. We would assume they hedge because the existence of a futures markets incites them to choose a riskier production technique, resulting in higher price variability. We show that the introduction of the futures market can destabilize spot prices through a slightly different mechanism. Without a futures market, hedgers rely on the correlation between production risk and market disturbance to reduce their income volatility. This stabilizing mechanism disappears when the futures market opens.

The effects of an imperfection in the signal about future price variations: Hart and Kreps (1986) have shown that, in this case, speculation could be destabilizing since speculators reduce their exposure when the signalled price rise does not occur and the signal disappears. This resale depresses prices. Taking an example of their model, we introduce an imperfect signal, observed by speculators, of future swings in spot demand. We show that the new futures market can be destabilizing here owing to the induced volatility of the exposure.

Our study is therefore organized as follows. We present the basic model (section 12.2) and analyse the effect of the futures market introduction according to different modes of price and volatility expectation formation (section 12.3). We examine the possibility of converging toward rational expectations (section 12.3.4). We conclude with two extensions of the basic model: the first embodies a random production technique (section 12.4) and the second a signal of future price variability (section 12.5).

12.2 HOW THE MARKET WORKS WITHOUT A FUTURES MARKET

12.2.1 Participant behaviour

We assume a one-period horizon for all market participants, which also corresponds to the duration of the futures contract. Net spot demand (excess of non-professional public demand over supply) is a simple function of the spot price. This prevents intertemporal price-destabilizing behaviour on the part of speculators as described by Hart (1977).

Producers (e.g. portofolio managers, firms) make a spot sale of an exogenous quantity K in period t. In section 12.4 we introduce random production.

Net spot demand (excess of demand by the public over supply) is given by

$$D_t = a_0 - a_1 C_t + \varepsilon_t \tag{12.1}$$

where ε is a non-autocorrelated disturbance with a variance of σ_ε^2. It will be useful to suppose a normal distribution of ε so that we can write the utility functions in expectation–variance form.

Demand decreases ($a_1 > 0$) with price. This demand does not come from speculators. Rather, in the case of financial markets, we should interpret it as demand from investors with a long horizon, interested in long-term yields and not in short-term capital gains. It should be noted that introducing 12.1 is also equivalent to making supply increase with price.

Speculators maximize the utility of their profit—a utility expressed in expectation–variance form[1]

$$\max_{x_t} E_t^s(\pi_{t+1}^s) - \frac{\rho^s}{2} \operatorname{var}(\pi_{t+1}^s) \tag{12.2}$$

where E^s and var^s are the expectations and variances used by speculators, π^s is their profits, X their speculative position and ρ^s their risk-aversion coefficient. We obtain

$$\pi_{t+1}^s = X_t(C_{t+1} - C_t) \tag{12.3}$$

For simplicity we assume a zero discount rate.

We also assume that the spot market has been functioning for a long time and that speculators have had time to become rational. It follows that

$$X_t = \frac{E_t^s C_{t+1} - C_t}{\rho^s \operatorname{var}_t^s C_{t+1}} \tag{12.4}$$

The speculative position increases in proportion to the excess of the expected price over the spot price.

There are N^p hedger-producers (suppliers who will use the forward market to hedge) and N^s speculators. Market equilibrium in period t is therefore written.

$$N^S \frac{E_t^s C_{t+1} - C_t}{\rho^s \operatorname{var}_t^s C_{t+1}} + a_0 - a_1 C_t + \varepsilon_t = N^p k + \frac{N^s(E_{t-1}^s C_t - C_{t-1})}{\rho^s \operatorname{var}_{t-1}^s C_t} \tag{12.5}$$

Speculators sell off in t the speculative position built up in $t - 1$.

In the long-term deterministic stationary equilibrium ($\varepsilon = 0$), we posit a standard $C = 1$. We therefore take $a_0 - a_1 = N^P K$, or $a_0 = a_1 + N^p K$ ($a_1 > 0$). Let

$$S_t = C_{t-1}$$

$$M^s = \frac{N^s}{\rho^s \operatorname{var}_t^s C_{t+1}}$$

$$(\operatorname{var}_t^s C_{t+1} = \operatorname{var}_{t+1}^s C_t).$$

We obtain

$$M^s E_t^s S_{t+1} - M^s E_{t-1} S_t - (M^s + a_1) S_t + M^s S_{t-1} + \varepsilon_t = 0 \tag{12.6}$$

We then look for solutions to 12.6 such that $\operatorname{var}_t C_{t+1} = \operatorname{var}_{t-1} C_t$ on account of the solution's stationary character.

The stationary solution to this differential equation with rational expectations is

[1]We assume a one-period investment horizon for all types of behaviour. Introducing an intertemporal behaviour creates vast analytical complications when we need to resolve the dynamic model with rational expectations.

$$S_t = \lambda S_{t-1} \frac{\varepsilon_t}{M^s(1 - \lambda) + a_1} \tag{12.7}$$

where λ is the stable root of the characteristic polynomial of 12.8 taken in expectation form in $t - 1$:

$$\lambda = \frac{2M^s + a_1 - (a_1^2 + 4M^s a_1)^{1/2}}{2M^s} \tag{12.8}$$

The solution method is as follows. We take the mathematical expectation of 12.6 conditional on the information available in $t - 1$. This gives a second-order differential equation for the expectations in $E_{t-1}S_{t+1}$, $E_{t-1}S_t$, and $E_{t-1}S_{t-1}$, which we solve by keeping only the stable solution. This yields the relation between $E_{t-1}S_t$ and S_{t-1}, as well as E_tS_{t+1} and S_t, which we later identify in 12.6 to find the general solution.

The conditional variance of the price is given by

$$\sigma^2 = \text{var}_{t-1}C_t = \left[\frac{1}{M^s(1 - \lambda) + a_1}\right]^2 \sigma_\varepsilon^2 \tag{12.9}$$

which is an implicit equation in σ^2 since $M^s = N^s/\rho^s\sigma^2$. We obtain

$$M^s(1 - \lambda) + a_1 = \frac{a_1}{2} + \frac{1}{2}(a_1^2 + 4M^s a_1)^{1/2}$$

The conditional variance decreases with speculative activity and with the price sensitivity to spot demand, since a high value for either of these implies that a small price swing in the case of shock is enough to restore market equilibrium. We naturally have

$$E_{t-1}S_t = \lambda S_{t-1} \tag{12.10}$$

The solution chosen excludes 'bubble' paths that would show the unstable root λ' of the characteristic polynomial and would be distinguished by the fact that $E_tS_{t+k} \to \infty$.

12.3 INTRODUCING THE FUTURES MARKET

In the futures market, participants can buy or sell contracts for the delivery of an asset in $t + 1$. Arbitragers can buy spot (at price C_t) and sell forward (at price F_t). Speculators can buy forward to sell spot in $t + 1$ (at price C_{t+1}).

12.3.1 Consequences during the learning period

Introducing the futures market has the following consequences.

1. It prompts the emergence of arbitragers operating between the spot and futures market on the basis of price spreads (we continue to assume a

zero discount value). They can thus buy spot, invest (here, at a zero rate) and sell forward. This means that their profit is linked to the $F - C$ spread between forward and spot prices. Their behaviour is

$$\max_{A_t} E_t^{\mathrm{A}}(\pi_t^{\mathrm{A}}) - \frac{\theta^{\mathrm{A}}}{2} A_t^2 \tag{12.11}$$

π^{A} is their profit, A is the value of arbitrages completed (on a spot purchase/forward sale basis), and θ^{A} is a friction coefficient representing factors such as response slowness, liquidity constraints and transaction costs. Arbitrages are not perfect unless θ^{A} equals zero. Since $\pi^{\mathrm{A}} = A(F - C)$, where F_t is the forward price in t for the futures contract maturing in $t + 1$, we have

$$A_t = \frac{F_t - C_t}{\theta^{\mathrm{A}}} \tag{12.12}$$

2. Speculators begin to operate in the futures market, since the lack of an initial funding requirement makes speculation easier.[2] We assume that this causes the number of speculators to increase from N^{s} to $\bar{N}^{\mathrm{s}} > N^{\mathrm{s}}$. Since the value in $t + 1$ of the futures contract bought in t is C_{t+1}, the speculative position of each player is given by

$$X_t^2 = \frac{E_t^{\mathrm{s}} C_{t+1} - F_t}{\rho^{\mathrm{s}} \operatorname{var}_t^{\mathrm{s}} C_{t+1}} \tag{12.13}$$

3. Participants are not very familiar with the workings of the new market. We assume that they initially resort to the same expectations formation mechanisms—as regards both prices and volatility—that they applied when the spot market alone was in operation. $E_t^{\mathrm{s}} C_{t+1}$ is given by 12.10 and $\sigma^2 = \operatorname{var}_t C_{t+1}$ by 12.9. Expectations are thus initially biased.
4. Producers can hedge their risk. Let T_t be the forward sales of hedger-producers. They have a strict hedging behaviour and minimize the conditional variance of their profit:

$$\begin{aligned} \min_{T_t} \operatorname{var}_t (\pi_{t+1}^{\mathrm{p}}) &= \min_{T_t} \operatorname{var}_t [K C_{t+1} - T_t (C_{t+1} - F_t)] \\ &= \min_{T_t} [(K - T_t)^2 \operatorname{var}_t C_{t+1}] \end{aligned} \tag{12.14}$$

   Hence

$$T_t = K \tag{12.15}$$

   Only perfect hedging exists in this basic model. It we introduced a non-infinite risk aversion for producers, part of their futures demand would be strictly identical with that of the speculators and would therefore have no effect on the model.

[2] We use this intuitive (umodelled) argument to put speculators in the futures market. Actually, because of arbitrage, it makes no difference in theory whether speculators are assumed to operate on the spot market or the futures market.

12.3.2 Market equilibrium

On the spot market we have

$$D_t + N^A A_t = N^p(K - T_{t-1})$$

(net demand + purchases by arbitragers = net sales by producers). On the futures market

$$\bar{N}^s X_t^s = N^p T_t + N^A A_t + \bar{N}^s X_{t-1}^s$$

(change in speculative position = forward sales by producers and arbitragers). N^A is the number of arbitragers. Producers sell forward at date t part of their production at date $t + 1$ and deliver the corresponding commodities in $t + 1$. Hence

$$\bar{N}_s(X_t^s - X_{t-1}^s) - N_p(T_t - T_{t-1}) + D_A - N^p K = 0$$

or, in a different form,

$$\bar{M}^s[(E_t^s C_{t+1} - F_t) - (F_{t-1}^s C_t - F_{t-1})] - a_1 C_t + \varepsilon_t + a_1 = 0 \quad (12.16)$$

where $\bar{M}^s = \bar{N}^s/\rho^s\sigma^2$ and σ^2 is again given by 12.9.

The arbitrage condition 12.12 is written

$$F_t - C_t = \theta^A A_t = \frac{\theta^A}{N^A}[N^p(K - T_{t-1}) - D_t]$$

or

$$F_t = C_t + \frac{1}{M^A}(-N^p K - a_1 + a_1 C_t + \varepsilon_t) \quad (12.17)$$

($M^A = N^A/\theta^A$). The term in (12.17) represents excess spot supply. If it is positive, arbitragers must buy spot and sell forward. As a result, if the arbitrages are imperfect ($\theta^A \neq 0$), the forward price must exceed the spot price. Substituting into 12.16 the final expression of market equilibrium obtained by setting $S_t = C_{t-1}$ as before, we obtain

$$\bar{M}^s F_t^s S_{t+1} - \bar{M}^s F_{t-1}^s S_t - \left[\bar{M}^s + a_1 + a_1 \frac{\bar{M}^s}{M^A}\right] S_t + \left[\bar{M}^s + \frac{\bar{M}^s}{M^A} a_1\right] S_{t-1}$$
$$+ \varepsilon_t\left(1 + \frac{\bar{M}^s}{M^A}\right) - \frac{\bar{M}^S}{M^A \varepsilon_{t-1}} = 0$$
$$(12.18)$$

The differences from condition 12.5 obtained in the case of the spot market as sole market are the following.

1. Speculators' expectations are given by 12.10 and are not rational.

2. Arbitrages are imperfect ($M^A \neq +\infty$).
3. The number of speculators is greater ($\bar{M}^s > M^s$).

Identifying 12.10 we obtain

$$S_t\left[\bar{M}^s(1-\lambda) + a_1 + a_1\frac{\bar{M}^s}{M^A}\right] = S_{t-1}\left[\bar{M}^s(1-\lambda) + \frac{\bar{M}^s}{M^A}a_1\right] + \varepsilon_t\left(1+\frac{\bar{M}^s}{M^A}\right) - \frac{\bar{M}^s}{M^A}\varepsilon_{t-1} \tag{12.19}$$

We can readily verify that, if $M^A = +\infty$, then $\bar{M}^s = M^s$. This is the earlier solution 12.7 for the model without a futures market: with expectations now correct, perfect arbitrage means that $C_t = F_t$. In other words—and most significantly—futures speculation and spot speculation are equivalent here (compare 12.4 and 12.13).

In the general case, we see that 12.19 leads to

$$\sigma'^2 = \mathrm{var}_{t-1}\, C_t = \left[\frac{1+\bar{M}^s/M^A}{\bar{M}^s(1-\lambda) + a_1 + a_1\bar{M}^s/M^A}\right]^2 \sigma_\varepsilon^2 \tag{12.20}$$

12.3.3 Stabilizing or destabilizing effect of creation of the futures market during the learning phase

The conditional variance of the price exceeds that obtained without the futures market 12.9 if

$$\left[\left(1+\frac{\bar{M}^s}{M^A}\right)^2[\bar{M}^s(1-\lambda)+a_1)^2 - \left(\bar{M}^s(1-\lambda) + a_1 + a_1\frac{\bar{M}^s}{M^A}\right)^2\right]\sigma_\varepsilon^2 > 0 \tag{12.21}$$

i.e.

$$M^s - \bar{M}^s + \frac{M^s\bar{M}^s}{M^A} > 0 \tag{12.21'}$$

Spot prices will thus be destabilized when we introduce a futures market without a change in the expectations formation mechanism if the increase in the number of speculators ($\bar{M}^s > M^s$) does not offset the possible imperfection of arbitrages ($M^A \neq \infty$).

The explanation is as follows. First, the increase in the number of speculators ($\bar{M}^s > M^s$) enables the market to absorb demand shocks ε_t with narrower price swings. As each participant's speculative position depends on $E_tS_{t+1} - S_t$, i.e. on $-(1-\lambda)S_t$, an unexpected rise in demand ε will drive up prices by $\varepsilon_t/(1-\lambda)M^s$, all other things being equal. Imperfect arbitrages also have a favourable effect shown in the denominator on the right-hand side of

12.20. The forward speculative position depends on $E_{t+1} - F_t$, i.e. on $\lambda S_t - S_t(1 + a_1/M^A)$. A given variation (e.g. a positive variation) in price S_t will entail a fall in spot demand, and therefore a rise in excess spot supply. Arbitragers will have to buy spot and sell forward, generating a positive spread between the forward price F_t and the spot price C_t. This spread means that speculators sell forward (the speculative position depends on $E_t S_{t+1} - F_t$). A positive demand shock ($\varepsilon > 0$) will be offset by arbitragers' spot purchases and forward sales. The incentive for speculators to sell forward therefore has a stabilizing effect.

This favourable influence, however, is dominated by the fact that supply or demand shocks, as we can see in 12.18, have a stronger initial effect—before prices even begin to move—on the forward price, and therefore on the market equilibrium.

If the arbitrages are very efficient, and arbitragers face minimal liquidity constraints or security shortages, then the introduction of the futures market is stabilizing from the very start, even if expectations do not adjust.

Equation 12.19 implies that

$$E_{t-1}S_t = \frac{1}{\bar{M}^s(1-\lambda) + a_1 + a_1\bar{M}^s/M^A}\left\{\left[\bar{M}^s(1-\lambda) + \frac{\bar{M}^s}{M^A}a_1\right]S_{t-1} - \frac{\bar{M}^s}{M^A}\varepsilon_{t-1}\right] \tag{12.22}$$

whereas speculators expect

$$E^s_{t-1}S_t = \lambda S_{t-1} \tag{12.23}$$

Hence, for the expectation error,

$$E^s_{t-1}S_t - E_{t-1}S_t = S_{t-1}\left[-\frac{\bar{M}^s(1-\lambda) + \bar{M}^s/M^A a_1}{\bar{M}^s(1-\lambda) + a_1 + a_1\bar{M}^s/M^A} + \frac{M^s(1-\lambda)}{M^s(1-\lambda) + a_1}\right] + \frac{\bar{M}^s/M^A}{\bar{M}^s(1-\lambda) + a_1 + a_1\bar{M}^s/M^A}\varepsilon_{t-1} \tag{12.24}$$

12.3.4 Rational expectations: long-term consequences of the introduction

In the long run, expectations become rational again once the futures market opens. We must therefore solve 12.18 on a rational expectations basis. We set

$$\tilde{S}_t = S_t - \frac{\bar{M}^s/M^A}{\bar{M}^s + a_1\bar{M}^s/M^A}\varepsilon_t$$

Equation 12.18 is

$$\bar{M}^{s}E_t\tilde{S}_{t+1} - \bar{M}^{s}F_{t-1}\tilde{S}_t - \left(\bar{M}^{s} + a_1 + a_1\frac{\bar{M}^{s}}{M^{A}}\right)\tilde{S}_t + \left(\bar{M}^{s} + \frac{\bar{M}^{s}}{M^{A}}a_1\right)\tilde{S}_{t-1} + \varepsilon_t\frac{\bar{M}^{s}}{\bar{M}^{s} + \bar{M}^{S}/M^{A}a_1} = 0 \qquad (12.18')$$

Setting

$$\tilde{\lambda} = \frac{2\bar{M}^{s} + a_1 + a_1\dfrac{\bar{M}^{s}}{M^{A}} - [a_1^2(1 + \bar{M}^{s}/M^{A})^2 + 4\bar{M}^{s}a_1]^{1/2}}{2\bar{M}^{s}} \qquad (12.25)$$

we obtain the solution

$$\tilde{S}_t = \tilde{\lambda}\tilde{S}_{t-1} + \varepsilon_t\bar{M}^{s}\left[\frac{\bar{M}^{s} + \bar{M}^{s}/M^{A}a_1}{\bar{M}^{s}(1 - \tilde{\lambda}) + a_1 + a_1\bar{M}^{s}/M^{A}}\right]^{-1} \qquad (12.26)$$

or, expressed differently,

$$\tilde{S}_t = \tilde{\lambda}\tilde{S}_{t-1} - \tilde{\lambda}\frac{\bar{M}^{s}/M^{A}}{\bar{M}^{s} + \bar{M}^{s}/M^{A}a_1}\left[\frac{\bar{M}^{s}}{\bar{M}^{s}(1 - \tilde{\lambda}) + a_1 + a_1\bar{M}^{s}/M^{A}} + \frac{\bar{M}^{s}}{M^{A}}\right] \qquad (12.26')$$

Hence

$$E_{t-1}S_t = \tilde{\lambda}S_{t-1} - \tilde{\lambda}\frac{\bar{M}^{s}/M^{A}}{\bar{M}^{s} + \bar{M}^{s}/M^{A}a_1}\varepsilon_{t-1}$$

$$\mathrm{var}_{t-1}S_t = \tilde{\sigma}^2 = \sigma_\varepsilon^2\left[\frac{\bar{M}^{s}}{\bar{M}^{s}(1 - \tilde{\lambda}) + a_1 + a_1\bar{M}^{s}/M^{A}} + \frac{\bar{M}^{s}}{M^{A}}\right]^2\left(\bar{M}^{s} + \frac{\bar{M}^{s}}{M^{A}}a_1\right)^{-1} \qquad (12.27)$$

This implicitly defines $\tilde{\sigma}^2$ since $\bar{M}^{s} = N^{s}/\rho\tilde{\sigma}^2$ if the expectations of second-degree moments also are rational in the long run and if speculators use the true conditional variance to calculate their positions. The solutions to 12.27 are analysed below.

12.3.5 Stabilizing or destabilizing effect in the long run

Let us assume for the moment that expectations are rational about future price ($E_{t-1}S_t$) but not about conditional variance. It may be objected that this abstract hypothesis is not very realistic, since the price's conditional expectation is partly determined by the second-order moments. If market participants underwent econometric training, they would learn about both mathematical expectation (via the regression coefficient) and variance (via the estimated standard deviation).

In this case we have $\bar{M}^s = \bar{N}^s/\rho^s\sigma^2$, with σ^2 again defined by 12.9, producing volatility if no futures market exists.

The conditional variance $\tilde{\sigma}^2$ with rational expectations is inferior to that obtained without a futures markets (σ^2 in 12.9) when

$$-1 + \frac{(a_1^2 + 4M^s a_1)^{1/2} + a_1}{a_1 + a_1\bar{M}^s/M^A + [a_1^2(1 + \bar{M}^s/M^A)^2 + 4\bar{M}^s a_1]^{1/2}}$$

$$+ \frac{(a_1^2 + 4\bar{M}^s a_1 - a_1)^{1/2}}{2M^A} < 0 \qquad (12.28)$$

This is true when M^A is very large (since $\bar{M}^s > M^a$, the second term is always smaller than unity), but is false when M^A is small. When rational expectations are formed for future price levels only, the introduction of a futures market is stabilizing if the arbitrages are sufficiently good, and destabilizing if they are imperfect.

When rational expectations also apply to second-order moments, we must look for the σ^2 solution of 12.9 and the σ^2 solution of 12.17, taking into account the expression for λ and M^s. We obtain

$$\sigma = \frac{\sigma_\varepsilon^2 - N^s a_1/\rho^s}{a_1\sigma_\varepsilon} \qquad (12.9')$$

for σ^2 (spot market alone) and

$$\tilde{\sigma} = \frac{1}{2\sigma_\varepsilon a_1}\left(\sigma_\varepsilon^2 - \frac{\bar{N}^s a_1(1 + a_1/M^A)}{\rho^s} + \left\{\left[\sigma_\varepsilon^2 - \frac{\bar{N}^s a_1(1 + a_1/M^A)^2}{\rho^s}\right]^2 + \frac{4\sigma_\varepsilon^2 a_1^2 \bar{N}^s}{\rho^s M^A}\right\}^{1/2}\right) \qquad (12.27')$$

for $\tilde{\sigma}^2$ (futures market). $\tilde{\sigma}$ is always positive (not generally the case in this type of model: see McCafferty and Driskill (1980)) and we therefore have $\tilde{\sigma} > \sigma$ if

$$(\bar{N}^s - N^s)\left(\sigma_\varepsilon^2 + \frac{2\bar{N}^s - N^s}{\rho^s} a_1 + \frac{\bar{N}^s a_1^2}{M^A\rho^s}\right) - \frac{(\bar{N}^s)^2 a_1^2(1 + a_1/M^A)}{M^A\rho^s} > 0 \qquad (12.28')$$

If $\bar{N}^s = N^s$ (same number of speculators in both cases), $\tilde{\sigma} > \sigma$ as soon as $M^A \neq \infty$: even if the future price volatility expectation is rational, the introduction of a futures market becomes destabilizing as soon as the increase in the number of speculators does not offset a possible arbitrage imperfection.

The earlier result (unchanged conditional variance) therefore remains valid even if a decrease in the conditional variance here has a positive stabilizing effect on speculator activity (measured by $\bar{M}^s = \bar{N}^s/\rho^s\sigma^2$) and further reduces the conditional variance. Let us assume, for simplicity, that $M^A = +\infty$ (perfect arbitrage). When the conditional variance used by speculators remains the same, the creation of the futures market causes volatility to shift from

$$\sigma_\varepsilon^2 \frac{1}{[N^s/\rho^s\sigma^2(1-\lambda)+a_1]^2}$$

to

$$\sigma_\varepsilon^2 \frac{1}{[\bar{N}^s/\rho^s\sigma^2(1-\tilde{\lambda})+a_1]^2}$$

or, to put it differently, from

$$\sigma_\varepsilon^2 \frac{1}{[a_{1/2}+\frac{1}{2}(a_1^2+4N^s a_1/\rho_s\sigma^2)^{1/2}]^2}$$

to

$$\sigma_\varepsilon^2 \frac{1}{[a_{1/2}+\frac{1}{2}(a_1^2+4\bar{N}^s a_1/\rho^s\sigma^2)^{1/2}]^2}$$

or, when the volatility expectation is rational, from

$$\left(\frac{\sigma_\varepsilon^2\bar{N}^s - N^s a_1/\rho^s}{a_1\sigma_\varepsilon}\right)^2$$

to

$$\left(\frac{\sigma_\varepsilon^2 - \bar{N}^s a_1/\rho^s}{a_1\sigma_\varepsilon}\right)^2$$

In the first case, the volatility's derivative with respect to the number of speculators is $-(1-\lambda)/\rho^s\sigma_\varepsilon$; in the second case it is $-1/\rho^s\sigma_\varepsilon$. The fact that volatility is endogenous and rationally expected increases the effect of the introduction of the futures market, since the higher speculation reduces volatility, which in turn stimulates speculation.

12.3.6 Learning

Let us now assume that speculators do not know how the market really works. After the futures market is set up they continue to form their expectations just as they did when the spot market was the only market. Once they become aware that their expectations are biased, they begin to learn the workings of the market. For this purpose, they adopt an analytic form of the expectation mechanism that is accurate in principle, since they know the determining variables; its parameters are erroneous, however, and are regularly updated to reflect observed values.

(a) *Learning about price expectations*

Initially, therefore, speculators form their expectations according to 12.23, which explains why the market follows 12.19. Observing their expectation errors, speculators revise their expectations mechanism and adopt 12.22, which corresponds to the market's actual functioning.

Let us suppose for the moment that volatility (σ^2 in 12.9) is constant and that, at a given moment, speculators expect the future price to follow

$$E_t S_{t+1} = \lambda_k S_t + \mu_k \varepsilon_t \tag{12.29}$$

The speculators' choice of expectations mechanism 12.29 naturally influences the equilibrium price. Speculators observe that the price conforms not to 12.29 but to another equation with the same explanatory variables but other λ and μ parameters. The speculators accordingly revise these, e.g. by using an econometric method.

Substituting 12.29 into the market equilibrium condition, we see that the new market functions according to

$$E_t S_{t+1} = \lambda_{k+1} S_t + \mu_{k+1} \varepsilon_t$$

$$\lambda_{k+1} = \frac{\bar{M}^s(1-\lambda_k) + \bar{M}^s/M^A a_1}{\bar{M}^s(1-\lambda_k) + a_1 \bar{M}^s/M^A a_1}$$

$$\mu_{k=1} = -\frac{\bar{M}^s/M^A + M^s \mu_k}{\bar{M}^s(1-\lambda_k) + a_1 + a_1 \bar{M}^s/M^A} \tag{12.30}$$

As speculators regularly revise their expectations they adopt λ_{k+1}, μ_{k+1} and then revise them again. The stationary solutions in λ, μ are

$$\tilde{\lambda}, -\tilde{\lambda}\frac{\bar{M}^s/M^A}{\bar{M}^s + \bar{M}^s/M^A a_1}$$

i.e. the parameters of the rational expectations pattern of 12.27.

The learning curve for λ can be represented as in Fig. 12.1. If $\lambda_k = 1$, the slope of the $\lambda_{k+1}(\lambda_k)$ curve in absolute value is equal to $\bar{M}^s/a_1(1 + \bar{M}^s/M^A)^2$ and is not necessarily smaller than unity. If it is not, let us call $\hat{\lambda}$ the value between $\tilde{\lambda}$ and 1 where the curve's slope is equal to -1.

At the fixed points A and C, the slope of $\lambda_{k+1}(\lambda_k)$ is smaller than unity in absolute value. Thus, if we start from λ_0, $0 \leqslant \lambda_0 \leqslant \hat{\lambda}$, there will always be convergence towards the stationary solution. The initial choice for this parameter, therefore must not be too close to unity.

When $\tilde{\lambda} = \lambda$, we obtain the following relationship for parameter μ:

$$\mu_{k+1} = \frac{-(\bar{M}^s/M^A + \bar{M}^s\mu_k)}{a_{1/2} + a_{1/2}\dfrac{\bar{M}^s}{M^A_A} + \dfrac{1}{2}\sqrt{a_1^2\left(1 + \dfrac{\bar{M}^s}{M^A}\right)^2 + 4\bar{M}^s a_1}} \tag{12.31}$$

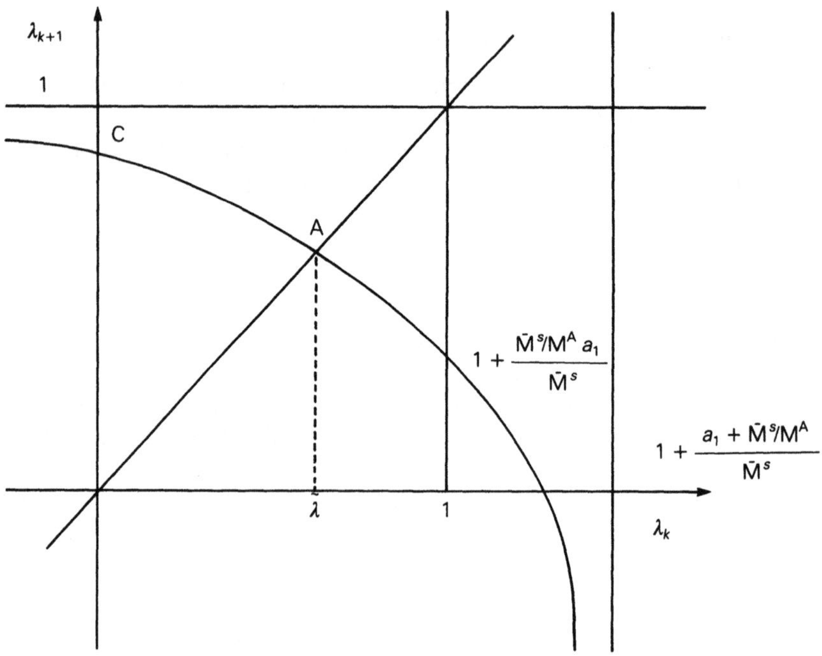

Fig. 12.1

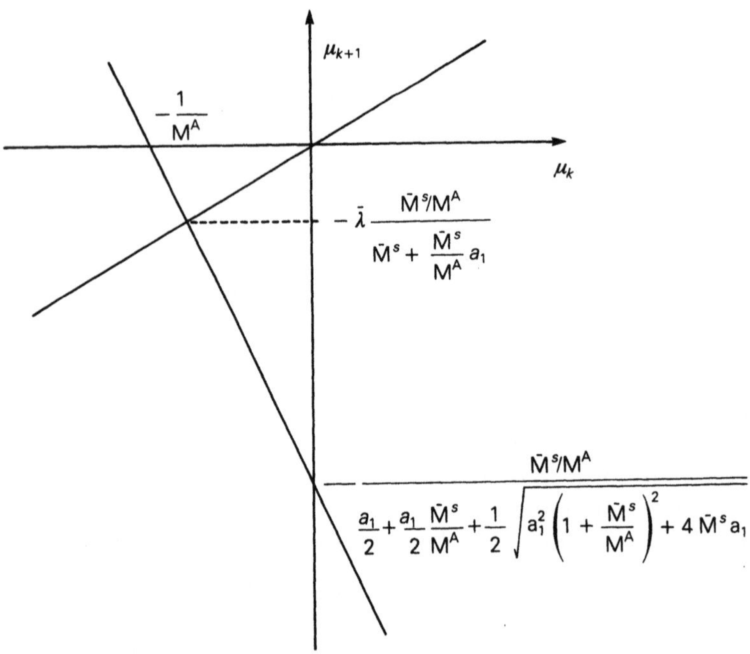

Fig. 12.2

which is shown graphically in Fig. 12.2. We can see that convergence towards the stationary point is not inevitable. The convergence will occur if

$$\bar{M}^s\left\{\frac{a_1}{2}+\frac{a_1}{2}\frac{\bar{M}^s}{M^A}+\frac{1}{2}\left[a_1^2\left(1+\frac{\bar{M}^s}{M^A}\right)^2+4\bar{M}^s a_1\right]^{1/2}\right\}^{-1}<1$$

i.e. if speculator activity (measured by M^s is not too intense, whatever the arbitrage quality. If $M^s \to \infty$ (very active speculators), $\tilde{\lambda} \to 1$ and we see in 12.30 that the μ learning process is divergent.

(b) *Learning about volatility*

Let us suppose that the convergence toward a rational expectation about futures prices has effectively taken place, i.e. that 12.17 is fulfilled. We must then examine the price volatility learning process. We have seen the importance of this issue, as learning amplifies the initial volatility response.

Equation 12.27 shows that if, at a given instant, speculators use the σ_k volatility, the price will exhibit the conditional variance

$$\sigma_{k+1}=\frac{\sigma_\varepsilon}{1+a_1/M^A}\left(\left\{\frac{a_1}{2}+\frac{a_1}{2}\frac{\bar{N}^s}{M^A\rho^s\sigma_k^2}+\frac{1}{2}\left[a_1^2\left(1+\frac{\bar{N}^s/M^A}{\rho^s\sigma_k^2}\right)\right.\right.\right.$$
$$\left.\left.\left.+\frac{4\bar{N}^s a_1}{\rho^s\sigma_k^2}\right]^{1/2}\right\}^{-1}+\frac{1}{M^A}\right) \tag{12.32}$$

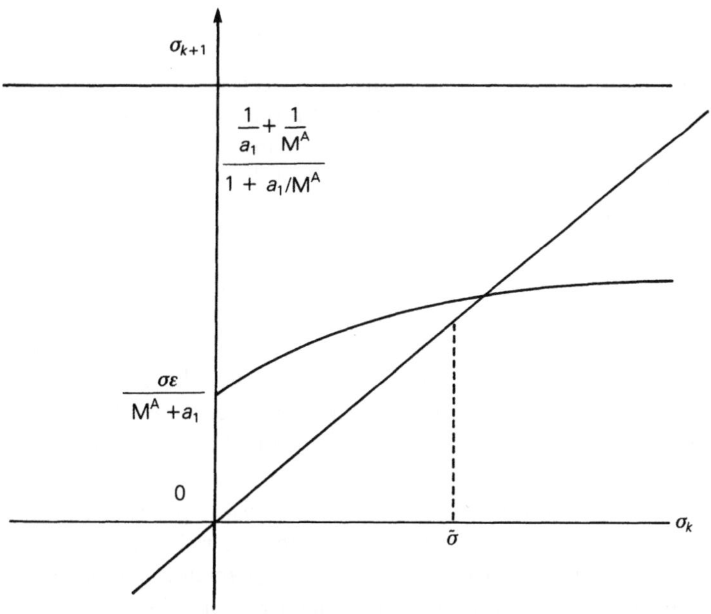

Fig. 12.3

as shown in Fig. 12.3. We can clearly observe the convergence toward the stationary solution $\tilde{\sigma}$.

12.4 FIRST EXTENSION: RANDOM HEDGING AMOUNT

So far, we have assumed that the amount to be hedged is fixed and equal to K. We now suppose that producers (hedgers) have the choice between two solutions:

1. choosing in t to put a specified quantity on the market in $t + 1$;
2. choosing a random 'production technique' that allows them to sell $1 + \varphi_{t+1}$ spot in $t + 1$ per currency unit invested in t, where φ_{t+1} is a zero-mean disturbance that is non-autocorrelated but correlated with the demand disturbance $(\text{cov}(\varphi_{t+1}, \varepsilon_{t+1}) = \sigma^2_{\varepsilon\varphi})$.

We can give a few examples of such choices. A fund manager hedges the bonds he will have to sell to meet bond-coupon or fund-redemption payments. If he adopts a risky management technique (by 'sensitizing' his portfolio), he may face large withdrawals if prices fall. In the case of farm commodity markets, producers may actually have the choice between more or less random yield-producing technologies. The hedger-producer's profit is therefore

$$\pi^{\text{P}}_{t+1} = [x_t K + (1 - x_t)(K + K\varphi_{t+1})]C_{t+1} - T_t(C_{t+1} - F_t) \quad (12.33)$$

We maintain the assumption that the producer is simply hedging and not speculating. If he chooses the random technique, it is to reduce his risk and not to increase his expected profits. Minimizing the conditional variance of π^{P}_{t+1} with respect to T_t and x_t we obtain the following.

Without a futures market $(T_t = 0)$

$$x_t = 1 + \frac{\text{cov}_t(C_{t+1}\varphi_{t+1}C_{t+1})}{\text{var}_t(\varphi_{t+1}C_{t+1})} \quad (12.34)$$

If φC and C are negatively correlated, a rise in prices C_{t+1} coincides on average with a fall in production obtained with the risky technique. To minimize variance, a profit-boosting price rise can be offset by a fall in profits due to the presence of risky production—hence $x_t < 1$ (see Anderson and Danthine, 1980, 1981, 1983).

With a futures market we get $x_t = 1$ (no risky production) and $T = K$ (perfect hedging), which completely eliminates the risk. In this case, we have exactly the same market mechanism as in the preceding paragraph. Without a futures market, production becomes stochastic, but the analysis in the first paragraph is very easily generalized since the disturbance in 12.6 becomes

$$\varepsilon_t + \gamma K\varphi_t N^{\mathrm{P}}, \text{ with } \gamma = \frac{\operatorname{cov}_{t-1}(C_t, \varphi_t C_t)}{\operatorname{var}_{t-1}(\varphi_t C_t)}$$

(If φ_t grows and $\gamma < 0$, the $-\gamma$ fraction of production is achieved with the random technique. Spot sales therefore include the term $-\gamma K\varphi_t N^{\mathrm{P}}$ since there are N_{p} producers.) The solution therefore becomes

$$S_t = \lambda S_{t-1} + \frac{\varepsilon_t + \gamma K\varphi_t N^{\mathrm{P}}}{M^{\mathrm{s}}(1-\lambda) + a_1} \tag{12.35}$$

with the price's conditional variance

$$\sigma''^2 = \operatorname{var}_{t-1} C_t = \frac{1}{M^{\mathrm{s}}(1-\lambda) + a_1} (\sigma_\varepsilon^2 + \gamma^2 K^2 N^2 p^2 \sigma_\varphi^2 + 2\gamma K N^{\mathrm{P}} \sigma_{\varepsilon\varphi}^2) \tag{12.36}$$

In addition, and assuming an approximately symmetrical distribution, 12.35 entails

$$\begin{cases} \operatorname{Cov}_{t-1}(C_t, \varphi_t C_t) \cong \dfrac{1 + \lambda S_{t-1}}{M^s(1-\lambda) + a_1}\, \sigma_{\varepsilon\varphi}^2 + \dfrac{\gamma K N^p (1 + \lambda S_{t-1})}{M^s(1-\lambda) + a_1}\, \sigma_\varphi^2 \\ \operatorname{Var}_{t-1}(\varphi_t C_t) \cong (1 + \lambda S_{t-1})^2 \sigma_\varphi^2 + \dfrac{1}{(M^s(1-\lambda) + a_1)^2} \\ \qquad (E(\varphi^2 \varepsilon^2) + (\gamma k N^p)^2 E(\varphi^4)) \end{cases} \tag{12.37}$$

which implicitly defines γ. The expectations formation mechanism remains $E_{t-1}S_t = \lambda S_{t-1}$. This is identical with that of the first paragraph, where we assumed a deterministic production technique.

We can thus begin by comparing 12.36 (volatility of the spot market as the only market with the possibility of random production) with 12.20 (volatility with futures market when expectations are not changed by the latter's introduction). We observe that, even without introducing imperfect arbitrage ($M^{\mathrm{A}} = \infty$), the creation of a futures market may well prove destabilizing when hedgers have a risky production technique (investment choice) at their disposal. This requires

$$\gamma^2 KNP\sigma_\varphi^2 + 2\gamma\sigma_{\varepsilon\varphi}^2 < 0 \tag{12.38}$$

Equation 12.37 shows that γ has the same sign as

$$\sigma_{\varepsilon\varphi}^2 + \gamma K N^{\mathrm{P}} \sigma_\varphi^2$$

Let

$$\gamma = \frac{\sigma_{\varepsilon\varphi}^2 + \gamma K N^{\mathrm{P}} \sigma_\varphi^2}{\Delta} \qquad \Delta > 0$$

or, to put it differently,

$$\gamma = \frac{\sigma^2_{\varepsilon\varphi}}{\Delta - KN^{\mathrm{P}}\sigma^2_{\varphi}}$$

Equation 12.38 becomes

$$2\Delta - KN^{\mathrm{P}}\sigma^2_{\varphi} < 0 \tag{12.38'}$$

If the uncertainty about the production (or investment) technique is very great, γ is negative if $\sigma^2_{\varepsilon\varphi} > 0$ and positive if $\sigma^2_{\varepsilon\varphi} < 0$, which stabilizes the price disturbance $\varepsilon_t + \gamma K\varphi_t N^{\mathrm{P}}$ in 12.35. Choosing a hedger-protective production technique therefore stabilizes the spot price, as the swings in the random share of production offset swings in demand. We can also compare 12.36 (spot market with random output) with 12.27 (futures market with rational price expectations). We observe that, in the same circumstances as above, the opening of the futures market can be destabilizing.

12.5 SECOND EXTENSION: SPECULATORS WITH ACCESS TO A PRICE VARIATION SIGNAL

We continue using the same basic model, assuming for simplicity that arbitrages are perfect, i.e. that $M^{\mathrm{A}} = +\infty$, and that the production technique is not stochastic. However, we introduce an imperfect signal, available at date t, of the demand disturbance ε_{t+1} occurring in $t+1$. Speculative expectations are based on this signal and are rational. More specifically, we suppose that, at date t, speculators will observe μ_t given by

$$\mu_t = E_{t+1} + K_t \tag{12.39}$$

To permit the calculation of the conditional moments, all the disturbances are assumed to have normal distributions.

ε and k are independent and non-autocorrelated. The signal μ_t is therefore imperfect, since it is disturbed by the random factor k_t, unobserved in t. This gives

$$E_{t-1}(\varepsilon_t) = \mu_t \frac{\sigma^2_{\varepsilon}}{\sigma^2_{\varepsilon} + \sigma^2_{k}} \qquad E_{t-1}(k_{t-1}) = \mu_t \frac{\sigma^2_{k}}{\sigma^2_{\varepsilon}\sigma^2_{k}}$$

$$\mathrm{var}_{t-1}(\varepsilon_t) = \sigma^2_{\varepsilon} - \frac{\sigma^4_{\varepsilon}}{\sigma^2_{\varepsilon} + \sigma^2_{k}} = \mathrm{var}_{t-1}(k_{t-1})$$

$$E_{t-1}(\mu_t) + 0 \qquad \mathrm{var}_{t-1}(\mu_t) = \sigma^2_{k} + \sigma^2_{\varepsilon} \tag{12.40}$$

The only reason why this changes the model seen earlier is that the expectation of ε_t, conditional on the information held in $t-1$, is no longer equal to zero. Solving the market equilibrium equation in the same manner as before, we finally obtain

$$S_t = \lambda^* S_{t-1} + \frac{1}{M_s^*(1 - \lambda^*) + a_1} \left[\varepsilon_t + \frac{\lambda^* \sigma_\varepsilon^2}{\sigma_k^2 + \sigma_\varepsilon^2} (\mu_t - \mu_{t-1}) \right] \quad (12.41)$$

where $\lambda^* = \lambda$ (12.8) and $M_s^* = M^s$ without a futures market and $\lambda^* = \tilde{\lambda}$ (12.25) and $M_s^* = \bar{M}^s$ with a futures market. We see that

$$\text{var}_{t-1} S_t = \frac{1}{[M_s^*(1 - \lambda^*) + a_1]^2} - \frac{\sigma_k^2 \sigma_\varepsilon^2 + (\lambda^* \sigma_\varepsilon^2)^2}{\sigma_k^2 + \sigma_\varepsilon^2} \quad (12.42)$$

Taking the case where the expected variance remains the same (rationality confined to price levels), we can see that the variability with a futures market will be greater than the variability without a futures market if

$$\frac{2M^s}{(a_1 + 4M^s)^{1/2}[a_1 + (a_1^2 + 4M^s a_1)^{1/2}]} < \frac{\sigma_\varepsilon^2}{\sigma_k^2 + \sigma_\varepsilon^2} \quad (12.43)$$

If the left-hand side is smaller than unity, this condition will be met if the variability of k_t is small, i.e. if the signal is good enough. (When no signal exists, $\sigma_k^2 = +\infty$ and we see that 12.43 never obtains.) If speculators have access to a fairly accurate signal of future demand swings, the introduction of a futures market can be destabilizing.

The rational expectations about future price are given by

$$E_t S_{t+1} = \lambda^* S_t + \lambda^* \frac{\sigma_\varepsilon^2}{M_s^*(\sigma_\varepsilon^2 + \sigma_k^2)} \mu_t \quad (12.44)$$

The inertia λ^* grows with the number of speculators (if $M^s \to \infty$, the spot price becomes constant). The random signal μ_t therefore leads to a random position-taking linked to $E_t S_{t+1} - S_t$, whose size increases with the number of speculators. (It is readily seen that μ is weighted by λ in 12.41: this can more than offset the stabilizing effect of heightened speculation in response to common demand shocks.)

12.6 CONCLUSION

In recent literature, different ideas have been advanced concerning the potentially destabilizing effects of the creation of a futures market. We have succeeded in combining these ideas in a unified framework. If we complicate the basic model by introducing elements such as imperfect arbitrages, a randomness of the hedging quantity and a signal of future swings, we can fairly easily obtain a result that is contrary to the standard findings, i.e. we shall observe higher volatility after the establishment of the futures market.

Arbitrage imperfection seems to be a relatively serious shortcoming since it leads to greater price volatility after the opening of the futures market if speculation does not increase sufficiently.

ACKNOWLEDGEMENTS

A French version of this work was previously published in *Revue Economique* of January 1990 whose permission to publish this new English version is gratefully acknowledged.

REFERENCES AND FURTHER READING

Anderson, R. and Danthine, J. P. (1980) Hedging and joint production: theory and illustration. *Journal of Finance*, March, 487–98.

Anderson, R. and Danthine, J. P. (1981) Cross-hedging. *Journal of Political Economy*, December, 1182–96.

Anderson, R. and Danthine, J. P. (1983) Hedger diversity in futures markets. *Economic Journal*, June, 370–89.

Artus, P. (1987) Marché à terme, options et stabilité du marché au comptant des taux d'intérêt. *Finance*, **8** (2), 25–54.

Gilbert, G. (1985) Futures trading and the welfare evaluation of commodity price stabilization. *Economic Journal*, September, 637–60.

Hart, O. (1977) On the profitability of speculation. *Quarterly Journal of Economics*, October, 579–97.

Hart, O. and Kreps, D. (1986) Price destabilizing speculation. *Journal of Political Economy*, October, 927–52.

Kawai, M. (1983) Price volatility of storable commodities under rational expectations in spot and futures markets. *International Economic Review*, **24**, 43–54.

McCafferty, S. and Driskill, R. (1980) Problems of existence and uniqueness in non-linear rational expectation models. *Econometrica*, **48**, 1313–17.

Newbery, D. (1987) When do futures prices destabilize spot prices? *International Economic Review*, June, 291–8.

Peck, A. (1985) *Futures markets: their economic role,* American Enterprise, Institute for Public Policy Research, Washington, DC.

Stein, J. (1981) Speculative priceeconomic welfare and the idiot of chance. *Review of Economics and Statistics*, **63**, 223–32.

Stein, J. (1986) Real effects of futures speculation: asymptotically rational expectations. *Economica*, **53** (210), 159–80.

Turnovsky, S. (1979) Futures markets, private storage and price stabilisation. *Journal of Public Economics*, **12**, 301–27.

Turnovsky, S. (1983) The determination of spot and futures prices of storable commodities. *Econometrica*, September, 1363–87.

Turnovsky, S. and Campbell, S. (1985) The stabilising and welfare properties of futures markets: a simulation approach. *International Economic Review*, June, 277–303.

13

The producer and futures markets

JEAN PIERRE DALOZ

13.1 INTRODUCTION

Ever since the work of McKinnon (1967), considerable attention has been paid to the problem of producers of agricultural commodities and their strategies on futures markets. Use is generally made of the so-called J–S model, named after Johnson (1960) and Stein (1961). In this model the producer is given a financial type of behaviour; it is assumed that he will follow a utility function whereby he tries to maximize his expected yield whilst reducing the associated risk. Contributors to this approach include Ederington (1979), Rolfo (1980) and Gemmill (1983).

The present contribution is based on the belief that this 'financial' model may not be the most appropriate. This is particularly the case if the producer takes the form of a marketing board (Caisse de Stabilisation or Office, in French-speaking countries) responsible for the marketing of tropical products such as coffee and cocoa in subsaharan Africa.

13.2 THE FINANCIAL MODEL AND ITS RESULTS

The financial approach defines two main strategies once it has been agreed that 'traditional hedging' is not the best way to use futures markets. This follows from demonstrations by Working (e.g. Working, 1961). On this basis, it is then assumed that the cornerstone is the covariance between prices on the cash market and prices on the futures markets. The analogy with the financial model of portfolio diversification is evident. The first result with this approach is sometimes called the 'pure hedging' strategy.

It is unnecessary to explore in detail the formalization of the model giving the pure hedging solution. Our producer can ignore the futures markets and stay solely on the spot market. He is, in fact, speculating on inventories. The producer's results are:

$$E(R) = SE(\mathrm{PS}_2 - \mathrm{PS}_1) \tag{13.1}$$

$$\mathrm{var}(R) = S\,\mathrm{var}(\mathrm{PS}_2 - \mathrm{PS}_1) \tag{13.2}$$

where R is the return, E is the expected value, PS is the price on the spot market (in periods 2 and 1) and S is the position on the spot market.

Our producer can also choose another strategy: he can combine positions on the spot market and positions on the futures market, thereby practising a hedge. This strategy gives the following results:

$$E(R) = SE(\mathrm{PS}_2 - \mathrm{PS}_1) + FE(\mathrm{PF}_2 - \mathrm{PF}_1) \tag{13.3}$$

$$\mathrm{var}(R) = S^2\,\mathrm{var}(S) + F^2\,\mathrm{var}(F) + 2\,\mathrm{cov}(sf)\,SF \tag{13.4}$$

where F is the position on the futures market and PF is the price on the futures market.

He can easily compute his minimal risk position as the ratio $F°$ of the futures position over the spot position on inventories.[1]

$$\frac{F°}{S} = -\frac{\mathrm{cov}\,(sf)}{\mathrm{var}\,F} = b° \tag{13.5}$$

Parameter $b°$ is the 'pure hedge' ratio. This follows from demonstrations by Ederington (1979) and Ntamatungiro (1988).

The model can be developed further if we have some information about the degree of risk aversion of our producer. This new step was introduced by Ederington (1979), Anderson and Danthine (1980, 1981) and Gemmill (1983). In this case a utility function for the producer can be expressed as follows:

$$U = E(\mathrm{RH}) - \lambda\,\mathrm{var}(\mathrm{RH}) \tag{13.7}$$

where RH is the return of the hedging strategy. Substituting into 13.4 and 13.7, we obtain

$$F^* = \frac{0.5 \times \lambda^{-1} E(\mathrm{PF}_2 - \mathrm{PF}_1)}{\mathrm{var}(F)} - S\,\frac{\mathrm{cov}(sf)}{\mathrm{var}(F)} \tag{13.8}$$

where F^* is the optimal hedging ratio.

The result 13.8 shows that our producer adopts a strategy which can be divided into pure hedging and pure speculative elements. If his utility function shows very high risk aversion, the speculative element shrinks towards a zero value and the entire strategy returns to the pure hedge.

While this is undoubtedly an elegant result, analysis of actual situations in

[1]It is assumed that the producer has no choice as far as the volume of his inventories is concerned, because purchase of the entire crop is obligatory.

this manner encounters difficulties. Empirical work for evaluation of the pure hedging ratio gives values which differ not only according to the product but also, and more puzzling perhaps, for the same product from different countries. Ntamatungiro, for example, computed a ratio of 70% for cocoa in Brazil versus 29% for the same product in the Ivory Coast and Cameroun.

However, the heart of the problem is the widely acknowledged distortion between academic results and the actual strategy of marketing boards. Academic results seem to indicate that positive advantages can be obtained by using futures markets and adopting a hedging strategy in order to reduce risk. Yet it is well known that marketing boards shy away from these futures markets. Hope was expressed that the introduction of risk aversion into the picture might provide a solution.

Results vary considerably from one contribution to another. Gemmill expressed a rather skeptical point of view (1983) whereas, for Ntamatungiro, practising hedging strategies with futures markets is of definite interest.

These considerations leave largely unanswered the question raised above: why do marketing boards and caisses or Offices refuse to cash in the advantages offered by the risk reduction effect of hedging strategies?

13.3 THE UTILITY FUNCTION OF MARKETING BOARDS

In this chapter we postulate that our difficulties are due to inappropriate definition of the utility function.[2] The 'producer' in the form of a marketing board does not show the 'financial' behaviour assumed in the studies quoted above. Both the producer's objective and his attitude towards risk are completely different.

The marketing board is obliged to buy the crop from the farmers at a specified price. This price, sometimes called the intervention price (PA) is determined by a political decision, where market considerations play a minor role (differences can thus exist between countries in this respect). The price varies (actually more or less increasing) over time. After purchase, the board must pay the technical expenses (TE) involved in moving the product from the field to the embarkment bank. Costs of marketing and embarking the product (D) are a third item of expense (Daloz and Badillo (1985), Ch. IV). This can even include tax payment. The sum of the inevitable expenses of the board is therefore

$$\mathrm{EXP} = \mathrm{PA} + \mathrm{TE} + D$$

This is sometimes called the 'theoretical' Free-On-Board (FOB) price.

[2]It is assumed that marketing boards have a utility function, revealed by their behaviour. In this case, the utility function differs from the usual function used in financial analysis.

On the other side of the marketing operation, the product is negotiated on the 'spot' market at a CIF (Cost, Insurance and Freight) price in the consumption areas. Assuming that the difference between FOB and CIF is known, it is easy to compare the following two prices: EXP + (FOB to CIF) and CIF. Let us call them CIFTH and CIFAC respectively. If equality prevails, the board has zero return. However, it is easy to understand that the board hopes to achieve a positive result. In the 'golden years' such a surplus could be enjoyed. These sums were the most important source of investment financing in the producer countries. The funds were invested through public investment programmes as some studies have shown (Daloz, 1980).

The real risk for the board is that it may run short of funds if the CIF price on the spot market is lower than the 'theoretical' CIF price of the board. When this happens, the board must draw on its reserves—if there are any.

These considerations about the way boards operate suggest a utility function that differs from behaviour consistent with portfolio theory. Our producer is going to maximize his expected return, defined as the difference between the theoretical CIF and the actual CIF price on the spot market:

$$E(R) = QE(\text{CIFAC} - \text{CIFTH})$$

For our producer, the risk is defined as the probability of being obliged to sell at a true CIF price (CIFAC), which is lower than the theoretical CIF price CIFTH (Martin and Hope, 1984): CIFAC < CIFTH. This is what I have termed the 'loss risk'.

The utility function can thus be expressed as follows:

$$U = E(R) - \lambda P(\text{CIFAC} - \text{CIFTH})$$

One more specification is in order. Board performance is assessed on a yearly basis because of the links, suggested above, between the return of the board and planning of public expenses over the fiscal year. Each performance is judged individually rather than as an average over a more or less set period of time. Moreover, no attempt is made here to evaluate the risk-aversion coefficient—an almost hopeless task. Instead, we concentrate solely on the evaluation of the loss risk in a way consistent with the utility function suggested above.

The last step at this point of our analysis concerns the method of computing the loss risk probability. This can be done simply by using a t coefficient. The corresponding values given by a probability table one estimates of this loss probability. Let RAV be the average return obtained by the board and RAV the standard deviation on this return. The probability at any moment can be measured using the t ratio

$$t = \frac{\text{RAV} - \text{CIFTH}}{\sigma_{\text{rav}}}$$

It must be noted that the CIFTH variable is exogenous to the board because it is essentially a political price. It differs from one period to another according to government policy.

The remainder of this chapter is a preliminary attempt to determine whether strategies using futures markets could give more efficient results than a simple strategy of marketing the product solely on the spot markets. The experiment described uses data on sales of cocoa by the Ivory Coast from 1982 to 1988.

13.4 AN EMPIRICAL ATTEMPT TO MEASURE THE RISK OF THE BOARD

Empirical tests encounter two major problems. The first difficulty is an almost unsolvable problem of data. Completion of our experiment required three kinds of data; unfortunately these were not available and second-best solutions had to be relied on instead.

Reliable information about the intervention price and the various costs called the 'internal differential' was not too difficult to obtain because the level of the intervention price is public. The only difficulty was to pass from campaign prices to annual prices. This can be done with a small margin of estimation error.

Information about the CIF price is nearly impossible to obtain. We have no reliable data about spot prices. The only good sources are the invoices signed by the officials of the Caisse. Unfortunately, this information is secret. To circumvent this obstacle, prices on the futures markets (on the nearest position), were used as substitutes for the spot prices. This prevents examination of many other possible strategies by the Caisse, but there was no other solution, at least for the moment.

The difference between the CIF prices and the FOB prices is another problem. Precise knowledge of the difference 'from FOB to CIF' is not only difficult but actually almost impossible. We were thus fortunate to have the results of a recent study on the subject (Descamps, 1988). Even so, some errors undoubtedly remain.

Having more or less solved these three difficulties, a fourth problem arose: for simplicity, the futures prices (nearest position) were taken from documents on the London Market (Gill and Dufus, various years). We then had to compute an exchange rate because the prices for the Caisse in the Ivory Coast are expressed in CFA francs.

Once these estimations were made, it was possible to compare CIFAC prices expressed in sterling and theoretical (CIFTH) prices estimated on the basis of the various information or estimations presented above and converted into this same currency.

The second set of difficulties concerns the wide variety of strategies available for the Caisse, using the futures markets or not, once it is admitted that the portfolio strategy is not the most adequate. There is a definite loss of simplicity if we compare the present situation with analysis based on the J–S model. With the J–S model, we had the advantage of a well-established scientific body. If we modify the utility function as described, we enter unexplored territory. We must advance without the guidelines of any well-known theoretical apparatus. For this reason, the present study has a more empirical character.

In an initial step in this new domain we limited our analysis to comparison of five different strategies. The first three do not use futures markets.

1. The first strategy (the reference strategy), calls for sales on the spot market on a monthly basis. Each month, one-twelth of the crop is sold at the ruling spot price. The average performance is not very different from the average price observed over the year.
2. The second strategy calls for sale of the entire crop (supposedly known at the time of the sale) a few months before the usual time of the main sales. We have assumed that selling activity takes place in September, October and November at the ruling spot prices of each of these three months.
3. The third strategy is the same as the second except that the sales take place in January, February and March.

Two other alternative strategies make use of the futures markets.

4. The first strategy is rather simple, even naïve, but could easily be made more complex. The inventories are hedged in totality (a one-to-one hedge) in January. The hedge is placed on a single rather distant position (March of the following year) on the futures market. (We have excluded the possibility of hedging on various positions, a solution which, of course, should present great interest.) Inventories are then disposed of month after month (one-twelth at each sale). Simultaneously, the hedge is lifted for the same quantity. The result of each of these twelve operations is computed when the hedge is lifted. The new set of marketing results thus obtained can easily be compared with what happens with any of the strategies avoiding the futures market.
5. The last strategy simulated in this chapter is a discretionary hedge (more precisely a 'carrying charge' hedge). It is assumed that the Caisse watches the evolution of the basis in January and hedges only if the situation seems favourable (a weak basis). If the basis is weak enough the Caisse goes on the futures market and places a one-to-one hedge; if the basis is too strong, the inventories are kept unhedged for the remainder of the year. Any hedges are lifted in the same way as in strategy 4.

The discretionary hedge strategy led to no hedging in 1984 and 1985, when we had an inverted market. Thus the true strategy of the Caisse during the whole of the period studied came out as a combination of strategy 1 and strategy 3.

At this stage of the experiment, we left aside the cost of storage. This cost item does not discriminate between strategies 1, 4 and 5 because inventories are the same in these three alternative solutions. However, it does make a difference between strategies 3 and 2 versus the other strategies (as well as between 2 and 3 but to a lesser extent).

The following variables were computed for all five scenarios:

1. the average yield for the period;
2. the *t* parameter providing information about the producer's risk of being unable to meet his obligations.

Both these variables change over time under the effect of the decisions regarding the intervention price in the producing countries. This applies at the annual level. At the intra-annual level, fluctuations in the exchange rate between sterling and CFA are an important modifying factor because all prices are converted into sterling equivalent.

Despite the relative simplicity of the experiment, preliminary findings are interesting. A preliminary remark is in order. The situation on the cocoa market differed considerably from one year to another over the period 1982–8. 1982 saw a down-trend movement in prices; the basis was well behaved, with weakening in the middle of the year. In 1983, the prices went up and the basis which was normally weak at the beginning of the year went up at the end. Prices remained stable over 1984 and 1985, with an inverted basis. In 1986–8 prices went down, with a fluctuating basis.

Analysis of the average selling prices for the various strategies envisaged failed to reveal any marked discrepancy between results. The average selling prices over the period 1982–8 (sterling) were as follows: strategy 1, 1425; strategy 2, 1454; strategy 3, 1770; strategy 4, 1512; strategy 5, 1484.

Strategy 3 (selling in January, February and March) gives the best result, but we have to make sure that such a decision is realistic. If the producer is large enough, such a possibility is more theoretical than real.

This is why we concentrated our attention on two other 'best' strategies: 1 (reference strategy) and 5. Figure 13.1 illustrates the situation if values of the *t* parameter are taken into account. It is immediately clear that the strategy with discretionary hedging is slightly better than the reference strategy. The second curve almost always lies above the curve of the simple spot selling strategy. It shows eight points above the line, representing a probability of 7%; the spot selling strategy shows only five points, marginally above this same line. Use of the futures markets to hedge inventories appears to make the danger of running short of funds less probable than a 'spot selling only' practice.

A complete picture of the results of the experiment can be gained by consulting the Appendix. Here again, strategy 3 gives the best results. The *t*

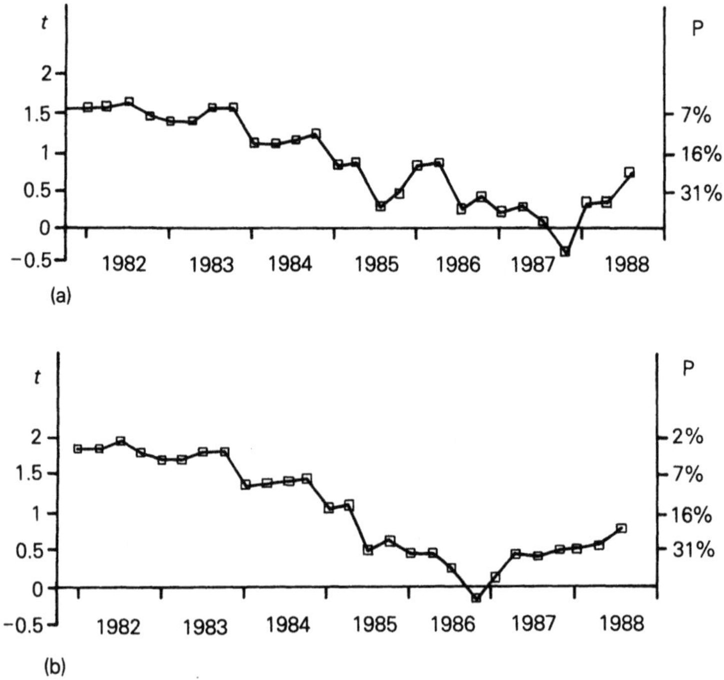

Fig. 13.1 The two best strategies in perspective: quarterly values of the *t* parameters from 1982 to 1988. (a) Strategy 1: selling monthly on the spot market. (b) Strategy 5: discretionary hedging (carrying charges).

parameter is always the highest of the entire array of strategies. Even at the end of the period when the cocoa prices were rather depressed, this parameter still gave a respectable 1.30 on average (corresponding to a 10% probability of running out of funds).

This initial study is only a first step; findings must now be verified and computed in different ways before becoming the basis for recommendations. For example, *t* should be computed not only over the entire survey period but also for shorter periods of time. Another analytical approach would involve deriving a strategy for 'selective hedging'. Decisions to hedge or to stay away from the futures market could be based on a model anticipating the future course of prices. Such a model, of course, remains to be defined.

Finally, nothing has been said about the various costs of using these 'fine-tuned' strategies: these include not only financial cost (interest on deposits on margins) but also various costs related to permanent access to information and processing of the various data required for definitions of reasonable marketing decisions (Daloz, 1985).

If these elements are all incorporated into the analysis, they can raise the cost (in terms of money or in the risks that our predictive models will make mistakes) of operations using the futures markets. This is probably a very powerful brake limiting more general use of futures markets by 'producer countries'.

ACKNOWLEDGEMENTS

Financial support for this work from the Programme Interdisciplinaire de Recherche sur l'Energie et les Matières Premières du Centre National de la Recherche Scientifique (PIRSEM – CNRS, Paris, France) is gratefully acknowledged.

REFERENCES AND FURTHER READING

Anderson, R. W. and Danthine, J. P. (1980) *Journal of Finance,* 35, 487–49.

Anderson, R. W. and Danthine, J. P. (1981) Cross hedging. *Journal of Political Economy,* **89**, 1182–96.

Bra Kanon, D. (1985) *Developpement et Appauvrissement,* Vol. 1, Economica, Paris.

Daloz, J. P. (1980) Le secteur public et l'extèrieur dans le financement des investissements en Côte d'Ivoire 1974–1978. *Revue Banque,* 27–35.

Daloz, J. P. and Badillo, D. (1985) *Marché, Speculation, Stabilisation,* Vol. 1, Economica, Paris.

Daloz, J. P. (1985) Dans quelle mesure les Caisses de Stabilisation peuvent-elles intervenir sur les marchés à terme? *Matières Premieres et Echanges Internationaux* (eds Cl. Mouton and Ph. Chalmin), Vol. 1, Economica, Paris, pp. 75–80.

Descamps, A. (1988) Rapport de Stage, Ecole Polytechnique et Caisse Centrale de Coopération Economique, September.

Ederington, L. H. (1979) The hedging performance of new futures markets. *Journal of Finance,* **34**, 164–70.

Gemmill, G. (1983) Optimal hedging on futures markets for commodity exporting nations, Working Paper, City of London Business School.

Gemmill, G. (1984) Forward Contracts or International Buffer Stocks. City of London Business School.

Gemmill, G. (1985) Producer welfare and synthetic long-term futures contracts. City of London Business School. IFCI Conference, Geneva.

Gill and Dufus, *Cocoa Report,* London, various years.

Johnson, L. L. (1960) The theory of hedging and speculation in commodity futures. *Review of Economic Studies,* **74**, 139–51.

Martin, L. J. and Hope, D. (1984) Risk and returns from alternative marketing strategies for corn producers. *Journal of Futures Markets,* 513–30.

McKinnon, R. I. (1967) Futures markets, buffer stocks and income stability for primary producers. *Journal of Political Economy,* 844–61.

Ntamatungiro, J. (1988) *Stabilisation des Recettes d'Exportation: Stock Regulateur, Contrats à Terme et Options,* Vol. 1, Economica, Paris.

Overdahl, J. A. (1987) The use of crude oil futures by the government of oil producing states. *Journal of Futures Markets*, **7**, 603–17.

Rolfo, J. (1980) Optimal hedging under price and quantity uncertainty. The case of a cocoa producer? *Journal of Political Economy,* **88**, 100–16.

Stein, J. (1961) The simultaneous determination of spot and futures prices. *American Economic Review,* **51**, 1012–25.

Working, H. (1961) New concepts concerning futures markets and prices. *American Economic Review,* **51**, (*Papers and Proceedings*), 160–3.

Appendix: *t* values according to different strategies

	I/82	II/82	III/82	IV/82
1	1.57	1.57	1.66	1.47
2	1.81	1.82	1.91	1.71
3	2.41	2.41	2.49	2.32
4	1.90	1.90	1.99	1.80
5	1.78	1.79	1.87	1.68
	I/83	II/83	III/83	IV/83
1	1.43	1.44	1.57	1.55
2	1.65	1.67	1.81	1.80
3	2.27	2.29	2.41	2.40
4	1.75	1.76	1.89	1.88
5	1.64	1.65	1.78	1.77
	I/84	II/84	III/84	IV/84
1	1.10	1.14	1.17	1.22
2	1.29	1.34	1.37	1.43
3	1.96	2.00	2.03	2.07
4	1.40	1.45	1.48	1.53
5	1.30	1.34	1.37	1.42
	I/85	II/85	III/85	IV/85
1	0.80	0.87	0.28	0.41
2	0.97	1.04	0.39	0.54
3	1.68	1.74	1.18	1.31
4	1.09	1.16	0.54	0.68
5	0.99	1.06	0.45	0.59
	I/86	II/86	III/86	IV/86
1	0.19	0.25	0.05	−0.38
2	0.30	0.36	0.15	−0.33
3	1.10	1.15	0.97	0.55
4	0.45	0.52	0.31	−0.15
5	0.36	0.43	0.22	−0.23
	I/87	II/87	III/87	IV/87
1	−0.13	0.21	0.15	0.28
2	−0.05	0.32	0.26	0.40
3	0.79	1.11	1.06	1.19
4	0.12	0.47	0.41	0.55
5	0.04	0.38	0.33	0.46
	I/88	II/88	III/88	IV/88
1	0.31	0.35	0.56	
2	0.43	0.47	0.70	
3	1.21	1.25	1.45	
4	0.58	0.62	0.84	
5	0.48	0.53	0.74	

14

Futures prices and hidden stocks of refined oil products

MARK NEWTON LOWRY

Short-run analysis of petroleum markets is complicated by poor data on the size of distributor and end-user stocks of refinery products. In OECD countries, data are collected frequently only for primary sector stocks. These are stocks held at refineries, pipeline facilities and large bulk terminals. Data on the size of stocks for the secondary (distribution) and tertiary (end-use) sectors are collected infrequently or not at all. In the USA, for example, good data on these stocks are collected only for one date in every 5 years. We can therefore think of secondary–tertiary stocks as being 'hidden' from the view of market analysts.

Lacking data on secondary–tertiary stocks, we also lack data on the volume of refinery product consumption. The best consumption proxy, termed 'product supplied' in the USA, is actually the domestic disappearance of refinery products from the primary sector. Hence it is the amount of primary sector supplies demanded by distributors and end-users for use in storage and consumption.

Most quantitative work on refinery product demand ignores this complication by treating product supplied as consumption. Yet adjustments in the level of hidden stocks may occasionally alter the disappearance volume significantly. If so, structural econometric models of disappearance and price formation should account for hidden stock adjustments.

Casual empiricism suggests that the problem is potentially important. The periodic US estimates of secondary and tertiary stocks of fuel oils, for example, are larger than those for primary sector stocks. Fuel oils account for over 25% of US refinery product disappearance currently and for a much larger share overseas. Fluctuations in primary sector fuel oil stocks are considered by market analysts to be an important determinant of short-run price movements. The theory of competitive storage suggests that the adjustment in hidden stocks is a function of price expectations and other variables. Given a good proxy for the price expected in the near future such as the published price provisions of futures contracts, it is then possible to enhance econometric models of refinery product demand. To illustrate the potential usefulness of the approach, this chapter presents results of an econometric model of the disappearance of distillate fuel oil in the eastern US market. New York Mercantile Exchange (NYMEX) futures prices are used as a proxy for expected future spot prices.

The plan for the chapter is as follows. In the next section, the theoretical basis of the econometric model is first discussed. There follows a discussion of the empirical work and a presentation of results. The chapter concludes with a brief summary and evaluation of the research.

14.1 MODELLING HIDDEN STOCK ADJUSTMENTS

Suppose that the amount X_t of a refinery product consumed in a given period is explained by the equation

$$X_t = X_t(P_t, \mathrm{ZX}_t) \tag{14.1}$$

where P_t is the spot price of the product and ZX_t is a vector of additional variables influencing consumption. Also, let D_t be the level of domestic disappearance from the primary sector while I_t is the level of secondary–tertiary stocks. If exports are negligible, D_t must then conform to the equation

$$D_t = X_t\ (P_t, \mathrm{ZX}_t) + I_t - I_{t-1} \tag{14.2}$$

Plainly, a build-up in hidden stocks increases the disappearance volume, while a drawdown reduces it.

To complete the demand specification, we require a behavioural relation for ending stocks. Here it is quite reasonable to assume that distributors and end-users are price takers and expected profit maximizers. Unfortunately, there is no consensus on the best way to model the convenience yield motives for storage. Here we posit the existence of a vector ZI_t of variables measuring exogenous economic conditions with the property that an increase in the value of any element of ZI_t increases storage convenience. Regardless of the convenience yield specification, stocks are an increasing function of the discounted spread between the current price and that expected in the near future. We denote this by $\mathrm{EPS}_t = \beta E_t(P_{t-1}) + P_t$ where E_t is an expectations operator and β is a discount factor, assumed constant.

Assuming a linear functional form, we may posit the following supply of storage function for the secondary–tertiary sector:

$$I_t = a_0 + a_1^*[\beta E_t(P_{t+1}) - P_t] + a_2^*\mathrm{ZI}_t \tag{14.3}$$

Here a_1 and a_2 are predicted to be positive.

Substituting for the stock variables in 14.2 and 14.3, it is now possible to obtain a behavioural equation that we shall call the total (domestic) demand for primary sector supplies:

$$\begin{aligned} D_t &= X_t(P_t, ZX_t) + \{a_0 + a_1^* [\beta E_t(P_{t+1}) - P_t] + a_2^* ZI_t \\ &\quad - \{a_0 + a_1^*[\beta E_{t-1}(P_t) - P_{t-1}] + a_2^* ZI_{t-1}\} \\ &= X_t(P_t, ZX_t) + a_1^*\{[\beta E_t(P_{t+1}) - P_t] - [\beta E_{t-1}(P_t) - P_{t-1}]\} \\ &\quad + a_2^* (ZI_t - ZI_{t-1}) \qquad (14.4) \end{aligned}$$

This tells us that the disappearance volume depends on the economic conditions determining storage as well as consumption. Moreover, it is the change in conditions promoting storage since last period that is pertinent. For example, last period's expectation of a rapid price increase will reduce the current disappearance if it does not continue.

14.2 THE DISTILLATE FUEL OIL MARKET OF THE EASTERN UNITED STATES

To test the usefulness of this simple enhancement the model was used to study the quarterly domestic disappearance of distillate fuel oil in the eastern market of the USA. The eastern market is defined as the United States east of Petroleum Administration for Defense District (PADD) V, which includes the Pacific states. We shall refer to the eastern market as USA East. The time period examined was the consecutive quarters from 1980.II to 1986.III, a total of 26 observations.

This is a suitable application for several reasons. One is that this is one of the US refinery product markets in which secondary–tertiary storage is most likely to have impact. Primary sector distillate storage is an important market factor and secondary–tertiary stocks appear to be comparable in size.

The composition of distillate demand is also noteworthy. About 30% of all distillate consumed in USA East is used for space heating. Demand in this application has extreme seasonality, falling close to zero in the summer. The logistic system used to distribute heating oil from large bulk terminals to end-users is somewhat separate from that used to distribute diesel fuel. The storage facilities of heating oil users are also specialized since most have no other alternative use for the product. The combination of specialized storage and variable demand promotes variability in convenience yield.

Another reason to study US distillate storage is the quality of the available data. Notable in this regard is the ongoing trade in the NYMEX futures contract for no. 2 heating oil delivered to New York harbour. Trade in this contract has been sizeable since early 1980. The published futures prices constitute a high quality proxy for expectations of near-term future spot market prices in the harbour. This is an important price since New York lies near the centre of the main region of heating oil consumption. Lowry (1988a, b) uses NYMEX heating oil prices to estimate behavioural relations for primary sector distillate storage, with good results.

PADD V was excluded from the data set because of the low volume and infrequency of trade in distillate between this district and those to the east. This is a consequence of the long distance between eastern refineries and west coast markets and the rough balance between west coast consumption and refining capacity. The lack of trade means that west coast distillate prices are not closely linked with those in the rest of the USA in the short run. In contrast, distillate shipments are large and regular between PADD III, with its large surplus of refining capacity, and PADDs I and II, which have large deficits.

14.3 EMPIRICAL WORK: DESCRIPTION

The domestic disappearance function used in econometric work can be summarized as follows:

$$\begin{aligned}\mathrm{DDFO}_t = {} & a_0 + a_1^* \,\mathrm{PDFO}_t + a_2^*\mathrm{PMG}_t + a_3^* \,\mathrm{PNG}_t \\ & + a_4^*\mathrm{GNP}_t + a_5^*\mathrm{DDKDSH}_t \\ & + a_6^*\mathrm{KBTT}_t + a_7^*\mathrm{AP}_t + a_8^*(\mathrm{EPS}_t - \mathrm{EPS}_{t-1}) \\ & + a_9^*(\mathrm{DDFO}_t - \mathrm{DDFO}_{t-1})\end{aligned} \qquad (14.5)$$

Here DDFO_t is the quarterly disappearance volume. Terms 1–7 are those appropriate to a conventional demand for current distillate consumption function. They include real US price indices for distillate fuel oil (PDFO_t) and motor gasoline (PMG_t), the real price of natural gas to residences in the mid-Atlantic states (PNG_t), real US gross national product (GNP_t), US bus and trailer registrations (KBTT_t) and US acreage planted (AP_t). The variable DDKDSH_t is the product of degree days for the mid-Atlantic states and the number of heating oil hookups in the USA.

The signs of the coefficients for these variables cannot be predicted theoretically. However, tradition and common sense suggest that the own-price coefficient should be negative. Those for the other seven variables should be positive.

Terms 8 and 9 are those intended to explain the adjustment in the level of hidden distillate stocks. Term 8 is the difference in the expected discounted price spread for heating oil in New York harbour this period and last. The other is the difference in the volume of USA East distillate disappearance this period and last. Our justification for using the disppearance volume is that it is a proxy for economic conditions that give distillate storage a convenience yield. Economic theory predicts that the coefficients for terms 8 and 9 are positive.

The parameter values were estimated by an instrumental variables method intended to correct for simultaneous equation bias. This involved regressing the apparent endogenous variables—$PDFO_t$, PMG_t, $EPS_t - EPS_{t-1}$ and $DDFO_t - DDFO_{t-1}$—against a set of exogenous variables that would be important in a complete model of the markets for distillate and related products. The fitted values of the variables were then used as regressors. The natural gas price was deemed exogenous in the short run due to governmental controls.

In addition to the other explanatory variables noted above, the set of regressors included a trade-weighted exchange rate index for the US dollar, an index of West European industrial production, crude oil production by OPEC countries, miles per gallon of the US motor vehicle fleet, refining capacity of the USA and selected Caribbean basin countries, Mexican refining product exports, dummy variables for quarter I to quarter III and a trend variable. Some of the variables were available only on an annual basis. In such instances, quarterly data were generated using a conventional smoothing technique.

Regarding the sources of variables used in the econometric work, four kinds of data were used to fashion a proxy for the expected discounted price spread. The price of no. 2 heating oil for New York harbour spot delivery was obtained from various issues of Platt's *Oil Price Handbook*. The price provisions of NYMEX no. 2 fuel oil contracts for delivery next quarter were obtained from various issues of the NYMEX *Daily Futures Report*. The interest rate on bankers' acceptances in New York reported in various issues of the *Survey of Current Business* was used to create a smoothed monthly average discount factor. The resulting price spread proxy was converted to real terms using the US producer price index for all commodities, also report in the *Survey*.

As for the other variables, distillate disappearance and refining capacity in the eastern US market are reported in the EIA's *Petroleum Supply Annual*. OPEC oil production is reported in the EIA's *Monthly Energy Review*. The price indices for distillate and motor gasoline were obtained from the *Survey of Current Business*. Refining capacity in selected Caribbean basin countries is reported in the *BP Statistical Review of World Energy*. Total locomotive horsepower is reported in *Railroad Facts*, a publication of the Association of American Railroads. Mexican refinery product exports are reported in *Petroleum Economist*.

The exchange rate and European production variables are available on CITIBASE. Miles per gallon is reported in EIA's *Annual Energy Review*, and bus the tractor trailer registrations in the Federal Highway Administration's *Highway Statistics*. The National Oceanic and Atmospheric Administration reports degree days in the mid-Atlantic states. The American Gas Association estimates heating oil hookups.

14.4 ECONOMETRIC WORK: RESULTS

The results of the econometric work are reported in Table 14.1. Here specification (a) is the base case. Specifications (b)–(d) are the same save that one or both of the hidden stocks variables are deleted. Specification (e) uses ordinary least squares (OLS) instead of the instrumental variables method.

In evaluating the base case, it is first notable that the data support the competitive theory of secondary–tertiary storage that is embodied in the hidden stock adjustment variables. The coefficients of both variables are correctly signed. To the extent that significance tests are meaningful, the *t* statistics for the price-spread coefficient indicate significance at the 95% confidence level using a one-tailed test. The coefficient for the convenience yield variable is significant at the 90% confidence level using the same test.

The other base case coefficient estimates are also correctly signed. Most have high *t* statistics if a one-tailed test is appropriate. The adjusted R^2 is a substantial 0.943. Hence the base case specification explains the variation in distillate disappearance during the sample period well. Two surprises in the base case results are the estimated price elasticities for distillate (−1.116) and motor gasoline (1.112). There are higher than those obtained in most econometric studies. However, the 90% confidence intervals for each elasticity encompass values in the conventional range.

Let us now compare the base case results with those for specifications (b)–(d), in which the hidden stocks specification is altered. It is noteworthy first that, while the addition of hidden stocks terms increases adjusted R^2,

Table 14.1 Estimated ending stocks of selected refined oil products in the United States, 31 March 1983 (million barrels)

	Motor gasoline	*Kero-Jet fuel*	*Kerosene and distillate*	*Residual fuel oil*
Primary sector	223	35	127	46
Secondary (distribution) sector	35	nr	11	2
bulk plants	9	nr	9	2
retail motor fuel outlets	26	nr	2	nr
Tertiary (end use) sector	42	11	131	86
Secondary tertiary total	77	11	142	88
Total	300	46	269	134
Secondary–tertiary share of total	26	24	53	66
Secondary–tertiary stocks/march 1983 disappearance	36	45	149	180
nr = not reported				

Sources: primary sector: U.S. Energy Information Administration (1984)
Secondary and Tertiary Sectors: National Petroleum Council (1984)

the change is slight. The results from Equations (a) and (d) make possible a test of the hypothesis that the coefficients for both hidden stocks terms equal zero. Using a conventional *F* test, the hypothesis could be rejected only at the 75% confidence level. Apparently then, inclusion of the hidden stocks terms does not significantly improve the explanatory power of the model.

Another interesting result is the sensitivity of demand function coefficient estimates to the inclusion of the hidden stocks terms. The coefficient estimates for distillate, motor gasoline and methane prices and GNP are all substantially larger when the hidden stocks terms are included. However, the estimate of the degree days coefficient is quite robust. The coefficient estimate of each hidden stocks variable is robust with respect to the exclusion of the other.

Regarding the estimation procedure, note first that the Durbin–Watson statistics are close to 2.000 in all cases. This suggests that first-order autocorrelation is not a problem. Comparing specifications (a) and (e) we find that the use of an instrumental variables method has a modest but significant effect on most coefficient estimates. The expected price spread term is one of the most sensitive.

The base case results can be used to calculate the trajectory of hidden stocks over the sample period. The basic approach may be simply described. First, an estimate of the 31 March 1983 level of secondary–tertiary stocks of distillate fuel oil in USA East was made from data presented in National Petroleum Council (1984). The base case results then permit us to calculate how the stock level varied around this value during the sample period.

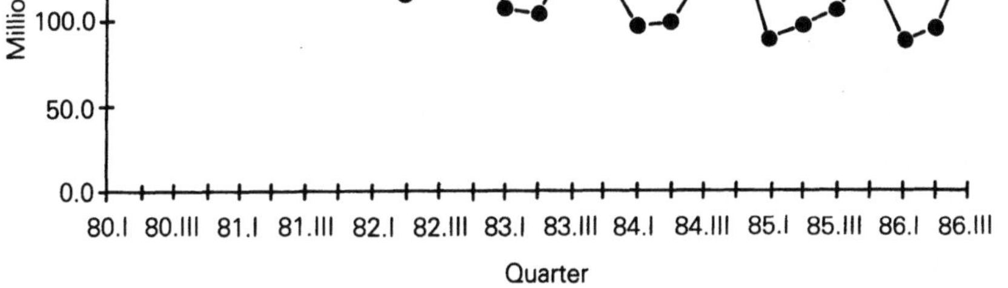

Fig. 14.1 Primary and estimated secondary–tertiary stocks of distillate fuel oil in USA East, 1980–6.

The estimated trajectory is shown in Fig. 14.1, together with the trajectory of primary sector stocks during the sample period. Table 14.2 presents the average quarterly level of primary and estimated secondary–tertiary stocks during the 1983–5 period. These years were chosen as ones of relative oil price stability during the sample period.

It can be seen that the estimated trajectory for hidden stocks displays a pattern of seasonality that is similar to that for primary sector stocks. In both cases, stocks are typically lowest at the end of the winter quarter (March) and build during the rest of the year to a peak at the close of the autumn quarter (December). While the seasonal patterns are similar, the degree of seasonal variation is far smaller in the secondary–tertiary stock estimates. During the 1983–5 period, the average peak-to-trough variation was 38.5 million barrels for primary stocks and 7.3 million barrels for estimated secondary–tertiary stocks.

A similar result holds for the secular trend in stock levels. In each case, the trend is downward, a reflection of the downward movement of distillate prices and stagnant consumption. However, the decline in primary sector

Table 14.2 Uses of distillate fuel oil in USA East, 1984–5

Sectors	*Average Sales (mm gallons)*	*% of Total Sales*
Distillate sales: all categories	38 611	100
Diesel applications predominate	25 720	67
Transportation	18 450	48
On-highway diesel	14 690	38
Railroad (all distillate)	2 431	6
Vessel bunkering (all distillate)	1 330	3
Farm (diesel)	2 495	6
Commercial (no.2 diesel)	2 046	5
Industrial (no.2 diesel)	1 192	3
Construction (diesel)	1 025	3
Military (diesel)	214	1
Other Off-Highway Diesel	298	1
Heating Applications Predominate	11 204	29
Residential (all distillate)	7 671	20
Commercial (all distillate)	2 423	6
Industrial (other than no. 2 diesel)	953	2
Farm (non-diesel)	157	0
Miscellaneous Applications	1 688	4
Oil companies (all distillate)	701	2
Electric utilities (all distillate)	522	1
Military (other than no. 2 diesel)	332	1
All other	134	0

Source: Form E1A 821, 'Annual Fuel Oil and Kerosene Sales Report'

stocks was much more dramatic than that for estimated secondary–tertiary stocks.

The econometric results can also be used to obtain an estimate of the volume of distillate consumed in USA East during the sample period. The formula used for this purpose is

$$\text{XDFO}_t = \text{DDFO}_t - [\hat{a}_8^*(\text{EPS}_t - \text{EPS}_{t-1}) - \hat{a}_9^*(\text{DDFO}_t - \text{DDFO}_{t-1})] \tag{14.6}$$

where $\hat{a}_8$ and $\hat{a}_9$ are the base case coefficient estimates. In effect, we adjust the disappearance trajectory for that portion of variation that appears to result from hidden stock adjustments.

Average quarterly disappearance and estimated consumption for the 1983–5 period are compared in Table 14.3. It is evident that the seasonal patterns are similar. Both disappearance and estimated consumption peak during the winter quarter and are lowest during the summer quarter. Estimated consumption displays greater seasonal variability than disappearance. Storage by distributors and end-users has the net effect of stabilizing the disappearance volume.

Table 14.3 Average quarterly primary sector disappearance and estimated consumption of distillate fuel oil (million barrels) in USA East, 1983–5

Quarter	*Primary sector disappearance*	*Estimated consumption*
I Jan–Mar	254.97	261.40
II Apr–Jun	211.43	208.80
III Jul–Sep	201.63	197.43
IV Oct–Dec	238.93	238.57

Sources: Primary sector disappearance, *Petroleum Supply Annual*; estimated consumption, calculated from base case results and National Petroleum Council stocks data.

14.5 SUMMARY AND CONCLUSIONS

In this chapter we investigated the potential problems posed for petroleum market analysis by the poor quality of data on secondary–tertiary stocks of refinery products. Economic theory was used to fashion a more complete model of refinery product disappearance. Model parameters were estimated econometrically using NYMEX futures prices and other data from the distillate market of the eastern United States.

The results support a competitive theory of secondary–tertiary storage. Hidden stocks display a seasonal trajectory during periods of stable crude oil prices. Like primary sector stocks they are lowest at the end of the winter quarter and rise during the balance of the year to an autumn quarter peak. As a consequence, consumption appears to have greater seasonal variation than disappearance. The data also support the notion that hidden stocks are less variable than primary sector stocks. By implication, it is not evident that they are a market factor of comparable importance.

Since the distillate market was chosen for the potential importance of secondary–tertiary storage, this suggests that the hidden stocks problem may not pose a major obstacle to short-run market analysis. One qualification to this conclusion is that the demand for distillate is substantially more seasonal in other OECD countries where natural gas supplies are less abundant. Hence the results do not preclude the possibility that the hidden stocks problem is significant in some countries but not in the USA.

Another qualification is that the convenience yield specification is quite crude. In effect, we are allowing one variable to serve as an index of conditions promoting convenience yield. Yet we have to acknowledge the existence of several convenience benefits. While the disappearance volume is a reasonable proxy variable, it is quite possible that other stand-alone proxies are substantially better. Moreover, several variables may be needed to capture the full array of convenience benefits.

Note finally that several non-storage variables in the base case model (e.g. $PDFO_t$ and $DDKDSH_t$) display seasonality and trend. Thus, hidden stock adjustments may be partly explained by variables that are already typically included in studies of short-run refinery product demand. In that event, further stock adjustment enhancements will not significantly sharpen the explanatory power of disappearance models. However, they may lead to sharper estimates of the magnitude of hidden stocks and refinery product consumption.

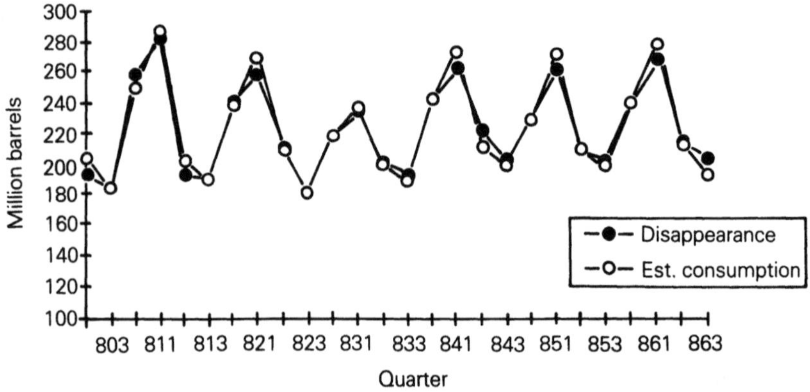

Fig. 14.2 Domestic disappearance and estimated consumption of distillate fuel oil in USA East 1980–86.

REFERENCES AND FURTHER READING

Association of American Railroads (1987) *Railroad Facts, 1987.*

British Petroleum (1987) *BP Statistical Review of World Energy.*

Energy Information Administration. *Monthly Energy Review*, US Government Printing Office, Washington DC, various issues.

Energy Information Administration. *Petroleum Supply Annual*, US Government Printing Office, Washington, DC, various issues.

Federal Highway Administration. *Highway Statistics*, US Government Printing Office, Washington, DC, various issues.

Keynes, J. M. (1964) *The General Theory of Employment, Interest, and Money*, Harcourt, Brace, Jovanovich, New York.

Lowry, M. N. (1988a) Petroleum product storage by competitive crude oil processors. *Resources and Energy*, 10 (2), 95–110.

Lowry, M. N. (1988b) Precautionary storage of refinery products: the case of distillate fuel oil. *Energy Economics*, 10 (4), 254–60.

National Petroleum Council (1984) *Petroleum Inventories and Storage Capacity*, National Petroleum Council, Washington, DC.

NYMEX. *Daily Futures Report*, New York Mercantile Exchange, New York, various issues.

Petroleum Economist, various issues.

Platt's Oil Price Handbook, McGraw-Hill, New York, various issues.

US Department of Commerce, *Survey of Current Business*, various issues.

Verleger, P. K., Jr (1982) *Oil Markets in Turmoil: An Economic Analysis*, Ballinger, Cambridge, MA.

PART FOUR

Application of New Methodologies to Other Commodity Market Issues

15

Post-recession commodity price formation

MONTAGUE J. LORD

The commodity market price fall in the first part of the 1980s was the largest since the Great Depression. In 1986, the commodity price index of the International Monetary Fund (1988) was more than 25% below its 1980 level. Although the drop followed the 1979–80 commodity price boom, the average annual commodity price index relative to the US wholesale price index in 1985–6 was 47% less than that in 1977–8 (Fig. 15.1). The situation was worse than in the 1930s when viewed in the context of movements in commodity prices relative to prices of manufactured goods. According to Grilli and Yang's (1988) recently constructed long-term price indices, the commodity terms of trade, measured as the ratio of the commodity price index to that of manufactured goods, dropped nearly 40% between 1980 and 1986, compared with a 30% drop between 1929 and 1932.

The sharp fall in commodity prices immediately following the 1981 worldwide recession was expected by many economists since low price elasticities of these types of goods cause them to respond strongly to supply or demand shifts in their markets. What was unexpected was the sluggish price movements that followed the 1984 recovery in world economic activity, as well as the subsequent downturn in prices of many commodities in 1985–6. In retrospect it is clear that abnormally large stocks resulting from lower demand and lagged production responses to high 1979–80 prices played a role in holding down prices of many commodities. Several commodity-specific occurrences also had significant effects on the outcomes for prices in the 1980s.

In this chapter we examine some of the major causes of price movements that have taken place in the 1980s based on a set of market models for seven major commodities. The approach used to analyse these movements differs in two ways from other recent studies, such as those by Chu and

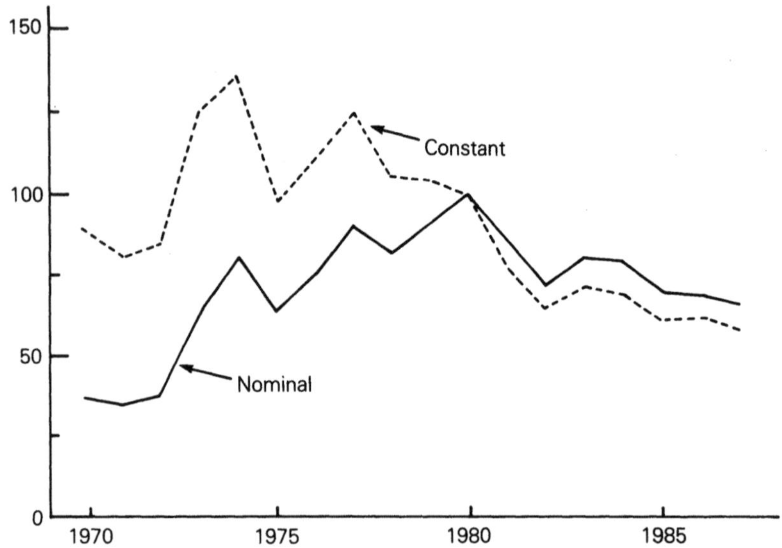

Fig. 15.1 Market price index of non-fuel commodities, 1970–87 (indices of US dollar price quotations, 1980 = 100).

Morrison (1984), Englander (1985) and Morrison and Wattleworth (1987). First, commodity markets are examined separately rather than being treated in the aggregate, and the findings about individual commodity markets are what give rise to generalizations about commodity market behaviour during the 1980s. Second, the analysis has been conducted with a set of structural econometric models. In contrast, other studies have used the reduced form of a system of equations. The problem with reduced-form equations is that, as Klein (1974, p. 16) has noted, the parameter estimates of these types of equations are unrelated to the separate estimates of parameters in the original structural form of the system of equations in the model. Consequently, separate influences on the original supply and demand relationships in the model cannot be identified. Moreover, as Chu and Morrison (1984, p. 131) have recognized, the use of a reduced-form equation prevents the analysis of the dynamics underlying the behaviour of commodity markets, and it is these dynamics that often give rise to observed cyclical movements in commodity markets.

With these considerations in mind, we proceed to analyse the recent commodity price movements in the following way. Taking actual changes in the major conditions influencing the markets as given, we simulate the impact of these changes on prices of each product. Next, we assume stable conditions in each of the major determinants of market prices and estimate their impact on market prices. A comparison of the different simulations shows the impact of each of the major determinants on market prices. The econometric model used for the simulations is described below. The following section presents the results of the analysis.

15.1 THE MODEL

The dynamic specification of the relationships characterizing the adjustment processes in the present set of commodity market models adopts recent work on time-series models that explains observed disequilibria in the context of steady state solutions that are consonant with the theory underlying price formation in commodity markets. In particular, the error correction mechanism (Hendry *et al.*, 1984) has been used to specify the relationship between those variables that are cointegrated (Engle and Granger, 1987), and the formation of market prices is based on a stock adjustment mechanism in which expectations are assumed to be formed through a rational process.

The error correction mechanism representation of the relationship relating consumption, denoted C, of a commodity to economic activity Y and that commodity's price P relative to the general price index D is

$$\Delta c_t = \alpha_0 + \alpha_1 \Delta y_t + \alpha_2 (c - y)_{t-1} + \alpha_3 y_{t-1} + \alpha_4 \Delta (p - d)_t + \alpha_5 (p - d)_{t-1} + v_{1t} \tag{15.1}$$

where lower-case letters denote logarithms of corresponding capital letters, e.g. $p - d = \ln (P/D)$, and the expected signs are $\alpha_1 > 0$, $-1 < \alpha_2 < 0$, $\alpha_3 > \alpha_2$ and α_4, $\alpha_5 < 0$. In Equation 15.1 Δc and Δy are cointegrated if $\alpha_2 \neq 0$, and the term $\alpha_3 y_{t-1}$ accounts for any non-proportional response of demand for a commodity as a result of a change in the level of income.

On a steady-state growth path, $\Delta c = g_1$ and $\Delta y = g_2$ are the growth rates of consumption and income respectively. Since market prices would not be expected to have any long-run dynamic influence, their effect is constrained to zero so that $\Delta p_t = 0$ in the long run. Hence, following Currie (1981), the long-run dynamic relationship implicit in Equation 15.1 is

$$C = k_1 Y^{1-\alpha_3/\alpha_2} (P/D)^{-\alpha_5/\alpha_2} \tag{15.2}$$

where $k_1 = \exp (-\alpha_0/\alpha_2 + \kappa_1 g_2)$, such that $\kappa_1 = (1 - \alpha_1)/\alpha_2$ is the income growth elasticity. Thus, on a steady state growth path, the level of consumption depends on the rate of growth of income, g_2, as well as on price and income levels.

The production relationship is often characterized by long lags since the effects of price changes usually take a long time to work themselves through to supply. Moreover, transmission of the price effects can be complex. The lag structure of production can be represented by a stochastic difference equation whereby 'differences' formulations of the variables are nested in their levels form. The advantage to this transformation is that it avoids multicollinearity between lagged values of the price variable when the dynamics are of a relatively high order, whereas reformulations with only differences variables meant to avoid serial correlation when the true relationship is in terms of levels will introduce an additional moving-average term into the disturbance (Plosser and Schwert, 1977). The expression for production Q in terms of own market price P relative to the general price deflator D, major disturbances W and a secular trend T measuring technological changes in the production process is

$$\Delta q_t = \beta_0 + \beta_1 q_{t-1} + \sum_{k=0}^{n-1} \beta_{2+k}\Delta(p - d)_{t-k} + \gamma_3(p - d)_{t-n} + \gamma_4 T + \gamma_5 W_t + v_{2t} \qquad (15.3)$$

where, as before, lower-case letters denote logarithms of corresponding capital letters, and where the expected signs are $-1 < \beta_1 < 0$; β_{2+k}, $\gamma_3 > 0$; γ_4, $\gamma_5 \gtreqless 0$. Since production of a commodity has a transient response to the rate of change of its constant dollar price, the long-run dynamic solution of the dynamic specification in Equation 15.3 is

$$Q = k_2(P/D)^{-\gamma_3/\beta_1} \qquad (15.4)$$

where $k_2 = \exp[-(\beta_0 + \gamma_4 T + \gamma_5 W)/\beta_1]$.

Equilibrium in the world market of a commodity is attained when the demand for stocks, K^d, equals the supply of stocks, K^s:

$$K^d = K^s \qquad (15.5)$$

The supply of stocks is simply defined to be actual stocks on hand. The change in the supply of stocks, or inventories, is equal to the difference between production and consumption:

$$\Delta K_t^s = Q_t - C_t \qquad (15.6)$$

Demand for stocks arising from transactions and precautionary motives is expected to have a proportional response to changes in consumption or production of the commodity in the long run, while the speculative demand for stocks, which is motivated by the desire of economic agents to profit from future price changes, is related to expected prices. Accordingly, the demand for stocks, K^d, depends on production Q (or consumption C) and on expected prices P^e:

$$k_t^d = \psi_0 + \psi_1 q_t + \psi_2 p_t^e + v_{3t} \qquad (15.7)$$

where the expected signs are ψ_1, $\psi_2 > 0$.

The expected price is assumed to be formed through a rational process (Muth, 1961). At period t the expected price P^e of a commodity is the expected value of the price, conditional on all information I, available at period $t - m$ to agents concerned with stocks in the commodity market:

$$P_t^e = E_{t-m}(P_t|I) \tag{15.8}$$

where E_{t-m} is the expectation operator for expectations formed at $t - m$. The characterization of the data-generating process for expected prices in the stock demand relationship in Equation 15.7 is based on the work of Wallis (1980). The price of the commodity is assumed to follow a general autoregressive moving-average ARMA(p, q) process:

$$\phi(L)p_t = \theta(L)u_t \tag{15.9}$$

where $\phi(L)$ and $\theta(L)$ are polynomials in the lag operator L. Then P_t can be represented by an autoregressive process:

$$P_t = \omega(L)p_{t-1} + u_t \tag{15.10}$$

where $\omega(L) = [\theta(L) - \phi(L)]/L\theta(L)$. The error-correction mechanism form of the stock demand relationship with a first-order polynomial in the lag operator is

$$\Delta k_t^d = \psi_0' + \psi_1 \Delta q_t + \xi_2(k^d - q)_{t-1} + \psi_2 \omega(L)p_{t-1} + v_{3t}' \tag{15.11}$$

where the expected signs are $\psi_1 > 0$, $-1 < \xi_2 < 0$, and $\psi_2 > 0$. Hence, the dynamic specification of the relationship for the demand for stocks yields a unitary elasticity of demand for stocks with respect to production (or consumption) of the commodity in its long-run dynamic equilibrium solution:

$$K^d = k_3 Q(P^e)^{-\psi_2/\xi_2} \tag{15.12}$$

where $k_3 = \exp(-\psi_0'/\xi_2 + \kappa_2 g_3)$, such that $\kappa_2 = (1 - \psi_1)/\xi_2$ is the production or consumption growth elasticity. Thus the demand for stocks of the commodity is related to both the level and the rate of growth of production or consumption, as well as its expected price level.

The aforementioned system of equations has been applied to the markets for seven commodities: coffee, cocoa, copper, sugar, cotton, maize and soybeans. The equations were initially estimated by the method of two-stage least squares; when consumption or production was found to have a non-contemporaneous response to price changes, the equations were estimated by the method of ordinary least squares since there exists a trade-off between consistency and efficiency in the method of instrumental variables (for details, see Judge *et al.* (1985, pp. 167–79) and Harvey (1981, pp. 77–80)). A test of parameter constancy following the 1981 world-wide recession was performed with the Chow test (Maddala, 1977, pp. 198–201) to examine whether the estimated coefficients were stable. The empirical results indicate that the general dynamic specification provides a good representation of the data-generating process in the commodity markets.

The test of parameter constancy showed the coefficients to be stable at the 5% level of significance in all the estimated relationships.

Details of the estimated parameter values of the relationships in the model are presented in Lord (1988). The general findings provide an indication of the magnitude of the response of the markets to key explanatory variables and consequently have a bearing on the analysis of price movements in the 1980s. In the consumption function, the average income elasticity of demand for the seven commodities was unity, where income has been measured on the basis of total, rather than per capita, gross domestic product (GDP). In addition, there are generally strong dynamic effects on the demand for the commodities arising from changes in the rate of growth of economic activity. As expected, the price elasticities of demand for all the commodities are less than unity, the average elasticity for the commodities being −0.2. Production is also inelastic with respect to prices in the long run in all the commodities, the average elasticity of the estimated production functions being 0.2. Finally, in the stock demand function, the ratio of stocks to production or consumption was found to be significantly affected by expected prices, the mean average elasticity being 1.3. Nevertheless, the range of elasticity estimates is fairly wide and these differences, for the most part, accounted for the different responses of market prices to the changes in economic conditions that took place in 1980–6.

15.2 MEASURING INFLUENCES ON COMMODITY PRICES

As Table 15.1 shows, considerable changes occurred in the macroeconomic conditions of the world economy during the first part of the 1980s. The cyclical swings in economic activity undoubtedly had the most pervasive influence on commodity markets. At the same time, large interest rate movements influenced the cost of inventory holding. It is less clear whether

Table 15.1 Movements of key economic indicators, 1981–6 (percentages)

	1981	*1982*	*1983*	*1984*	*1985*	*1986*
Change in real GDP						
World of which	1.6	−0.5	2.7	4.9	3.2	2.8
USA	1.9	−2.5	3.6	6.5	3.4	2.9
EEC	0.1	0.7	1.6	2.6	2.4	2.6
Japan	3.9	2.8	3.2	5.0	4.5	2.5
Inflation	9.1	2.0	1.3	2.4	−0.4	−2.9
Interest rates	14.1	10.7	8.6	9.6	7.5	6.0

See Appendix for data sources.

the deceleration of inflation in the industrialized countries affected commodity prices since Beckerman and Jenkinson (1986) have found that the deceleration of inflation was to a large extent attributable to the fall in prices of primary commodities imported into the industrialized countries. Thus, inflation cannot be considered to be weakly exogenous to the set of market models used in this chapter, and commodity market prices cannot be made conditional on inflation (for a discussion of exogeneity in econometric models, see Engle *et al*., 1983). We have accordingly limited the scope of the analysis to the effects of economic activity changes, interest rate movements and commodity-specific influences on market prices.

15.2.1 Economic activity

Changes in the demand for primary commodities resulting from changes in economic activity tend to produce sharp price movements since production often responds with a lag to price changes and the short-term price elasticity of supply tends to be small. The questions to be addressed here are how much changes in economic activity in the post-recession period contributed to price movements and how much the early price declines contributed to later price movements.

We have measured the effect of the 1982 deceleration in economic growth on commodity prices by comparing actual price movements with those that would have occurred had economic growth been stable in that year. In particular, the levels of economic growth of the USA, the EEC, Japan and the rest of the world were set at the averages of the years preceding and following 1982. This assumption implied an overall rate of economic growth equal to 2.1%, rather than −0.5%, in 1982. The economic growth rates in all other years were maintained at their historical rates.

As expected, the results show that commodity prices would have advanced in 1982, rather than dropping by an average of 15% (Table 15.2). However, the lagged response of production to those higher prices would also have led to lower prices in the following 4 years. Prices would have been somewhat lower in 1983 and they would have dropped sharply in 1984, rather than remaining nearly unchanged in that year.

The lagged production response to higher prices suggests that the downturn that actually occurred in commodity prices in 1985 could have been initiated by the economic recovery in 1983–4. Higher levels of economic activity in 1983–4 increased demand for most of the commodities. However, higher prices led to expectations of continued improvements in prices and greater production. Much of this additional output may not have entered the market until 1985–6 when demand decreased as a result of another slowdown in economic growth.

To test this theory, another set of simulations was performed in which 1983 economic activity growth rates in all the markets were set at

Table 15.2 Commodity price movements resulting from different economic growth rate assumptions (annual percentage change of commodity price index[a])

	1981	*1982*	*1983*	*1984*	*1985*	*1986*
With actual economic growth	−15.2	−15.3	11.8	−1.0	−12.1	−1.7
With stable 1982 economic growth	−15.2	3.3	9.7	−19.5	−13.3	−4.6
With lower 1983 economic growth	−15.2	−15.3	2.9	1.0	−9.1	−2.4
With stable 1985–6 economic growth	−15.2	−15.3	11.8	−1.0	2.5	10.2

[a]Unweighted price index of cocoa, coffee, copper, cotton, maize, soybeans and sugar.
See text for details about alternative growth rate assumptions.

two-thirds of their actual rates. In this case, world economic growth would have equalled 1.8% in 1983, rather than 2.7% The results demonstrate that the lower expansion in the demand for the commodities would have resulted in a smaller price rise in that year, but it would have kept prices from falling as much as they actually did in 1985. Nevertheless, the results also show that lower prices in 1983 would not have prevented the downturn in prices of most commodities that occurred in 1985–6.

These findings suggested that the downturn in commodity prices in 1985–6 could have been caused by the deceleration of economic activity in that period. A third set of simulations was therefore performed in which economic growth during 1985–6 in the USA, the EEC, Japan and the rest of the world was assumed to have been 50% higher than the actual growth. Overall, this would have meant that economic growth registered by the world in 1985 would have been 4.8% rather than 3.2%, and in 1986 it would have been 4.2% rather than 2.8%. These higher economic growth rates would have led to the early recovery of market prices of the commodities. The price index for these commodities would have risen by 2.5% in 1985 and by 10.2% in 1986.

The results for the aggregate of the seven commodities generally reflect the responses of individual commodity markets. Differences between commodities occur in the way shifts in demand resulting from income changes affect the magnitude of the price change and in the extent of the lagged response of output to price changes. For example, short-term price changes of copper and sugar are in general much larger than those of cotton and soybeans because of their lower short-term price elasticities of demand and supply, while the lagged responses of coffee, cocoa and copper production to price changes result in slow adjustments of market prices from one equilibrium state to another.

15.2.2 Inventory build-ups

Another factor which may have contributed to the intensity of the price decline in the post-1981 recession period and which may have inhibited an earlier recovery in prices was the high levels of stocks that existed in many commodity markets. Table 15.3 presents the 1981–2 actual stock to production or consumption ratios compared with estimates of desired stock ratios calculated from Equation 15.11 using 1961–87 data. Four of the seven commodities—cocoa, copper, maize and sugar—had stock ratios in 1981–2 that were significantly higher than the estimated desired stock ratios.

We estimated the effect of the high stock ratios on prices by comparing actual outcomes with those that would have occurred had actual stocks been equal to desired stocks. The calculations showed that all four commodity markets would have had high prices in 1982–3. However, the price advances would have led to lower prices in 1985–6 since the earlier high prices would have stimulated investment and stimulated production in later years. In the case of sugar, the large stock accumulation had a particularly strong damping effect on price. More normal stock levels at the turn of the decade would have led to an earlier recovery in the price of this commodity. Moreover, the price rise following a reduction to more normal levels of stocks would have been greater than in other commodities since the low price elasticity of demand for the product tends to cause large swings in its price when changes occur in the key determinants of the market.

15.2.3 Interest rates

The fall in interest rates in 1982–3 and again in 1985–6 coincided with the drop in commodity prices. Interest rates were found to be statistically

Table 15.3 Actual 1981–2 versus desired stock to production or consumption ratio

	Actual stock ratio 1981–2	*Desired stock ratio at g*[a]
Cocoa	0.41	0.35 (g = 1.6%)
Coffee	0.27	0.40 (g = 3.3%)
Copper	0.21	0.16 (g = 2.9%)
Cotton	0.28	0.38 (g = 2.8%)
Maize	0.18	0.16 (g = 5.0%)
Soybeans	0.09	0.16 (g = 5.9%)
Sugar	0.66	0.59 (g = 3.4%)

[a]The ratio of stocks to production or consumption is given at the level when the production or consumption growth rate g equals the average rate during the period in which the equation was estimated.
The ratio of desired stocks was estimated on the basis of Equation 15.11 with 1961–87 data.

significant in explaining the demand for copper stocks, but not that of other products covered in this study. The fall in interest rates would have been expected to have increased the demand for stocks of copper as the cost of holding stocks declined. Consequently, commodity prices should have been higher had interest rates not fallen as much as they did during the period.

To test this theory, interest rates were increased by one percentage point in each year during 1982–6, and the results were then compared with actual outcomes. As expected, copper prices would have been higher in 1983–4. However, higher prices would have reduced demand and lowered prices in 1985–6. In effect, higher interest rates would have reversed the cyclical swing in prices that took place during 1982–6.

15.2.4 Price stabilization schemes

Two of the seven products covered in this study were subject to price stabilization schemes during the first part of the 1980s. The International Cocoa Agreement failed to support prices within the established range since the financial resources of the buffer stock manager were quickly exhausted. In contrast, the International Coffee Agreement, which sought to stabilize the Composite Indicator Price at between \$1.20 and \$1.40 a pound by means of export quotas, succeeded in maintaining prices.

As Fig. 15.2 demonstrates, coffee prices were kept above the levels that would otherwise have been expected in the absence of a quota system through 1986. The effect of higher prices prior to 1986 was to stimulate production above what it would have been without export quotas. Several years are required for price changes to work themselves through to output changes—about 4 years for Robusta-type coffee and around 7 years for Arabica-type coffee. Thus, higher prices in the first half of the 1980s will affect the market in later years if the International Coffee Agreement fails to be renegotiated in 1990. In contrast, the smaller amount of coffee entering the market without the quota system would have helped to raise prices in the second half of the 1980s.

15.2.5 Supply disturbances

At the end of 1985, Brazil experienced a severe drought which led to a fall of nearly 60% in its 1986 coffee output. The indicator price of coffee rose from under \$1.40 a pound before October 1985 to \$2.38 a pound in March 1986. The sharp rise was caused by a temporary imbalance in the market as the quantity of coffee purchased and sold did not have time to adjust when the fall in coffee production of Brazil was announced. In response to the price rise, buyers began to reduce the quantity demanded while sellers released stocks and thereby increased the quantity of coffee available in the market. Moreover, Brazil announced that the bumper crop it anticipated in

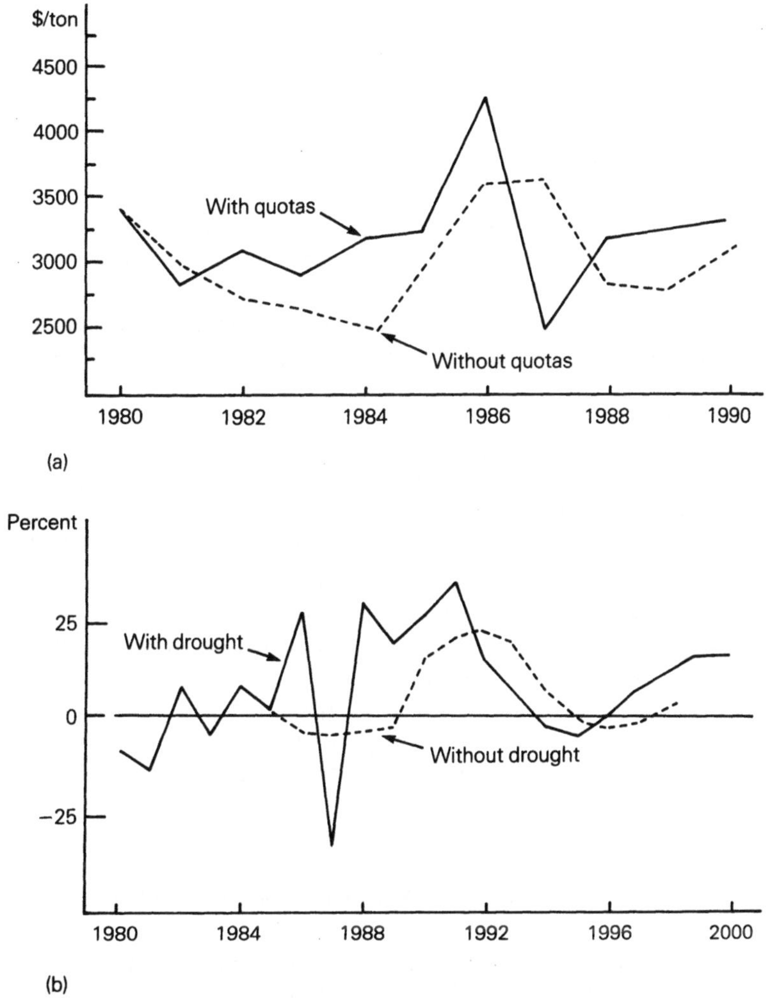

Fig. 15.2 Estimated response of coffee prices (a) to reintroduction of quotas under ICA and (b) to 1986 shortfall in Brazilian production due to drought.

1987–8 would raise total world output by 27% this season. Quotas on coffee exports under the International Coffee Aggrement were suspended in February 1986 when prices rose above $1.60 a pound. Without the International Coffee Agreement, higher output and lower consumption levels in response to the high prices put downward pressure on prices.

Figure 15.2 shows the estimated differences in price movements with the supply disruption and those that would have occurred without it. Had the Brazilian drought not taken place, prices would have continued to fall through 1989. Low prices would have caused producers to lose their incentives to expand plantings and would have given rise to an upswing in prices during the first half of the next decade as growth of production fell below that of consumption. Instead, the drop in prices that followed the drought will probably lead to a downturn in prices at the turn of the decade. In both cases, however, the price movements reflect the familiar cyclical behaviour of the coffee market. The price of coffee would eventually return to its long-term level without the Brazilian drought, after the quantities of coffee produced and consumed adjusted to the price hike of 1986. The gradual return of the price of coffee to its long-term level began in mid-1986 when the quantity of coffee demanded declined, large stockholdings were placed in the market and producers decided to expand output.

Weather-related disturbances in supplies also occurred in other commodities during the early 1980s, and these effects can also be directly assessed with structural market models. In cocoa, poor crops resulting from unusual weather conditions and political problems in Ghana reduced expected world output levels by 11% in 1983 and kept output unchanged in 1984. The effect of this disturbance has been measured against price movements which would have occurred had 1983 world production increased by 3%, as was anticipated prior to the time the disruptions occurred. The results indicate that there would have been a reversal in the cycle which actually took place after 1983. Instead of rising by nearly 40% in 1983–4 and then falling through 1987, prices would have dropped in 1984 and then would have risen in 1985 as a result of increased demand due to lower prices in the previous year.

15.3 SUMMARY AND CONCLUSIONS

In this chapter we have examined demand and supply influences on prices of seven major commodities in 1981–6 with a set of structural econometric models. A common modelling framework has been applied to the characterization of the underlying data-generating processes in commodity markets, while market-specific features have been retained through the parameters included in the model and the values assigned to those parameters. Since the nature of the response to price changes is central to the dyamic specification of the supply relationship, the stochastic difference equation framework provides a convenient means with which to move from a general to a specific lag structure. For the consumption and stock demand relationships, the error-correction mechanism offers an appropriate means by which to characterize data-generating processes within this class of equations. A rational expectations process has also been built into the dynamic specification of the stock demand function in order to account for speculative inventory behaviour.

This approach differs from other recent approaches which have attempted to explain commodity price movements on the basis of reduced-form equations to analyse aggregate price movements. By examining commodity markets separately, we have been able to identify separate influences on the original supply and demand relationships of commodity markets using parameter estimates from the original structural form of the system of equations in the models. The results of the analysis indicate that the initial worldwide deceleration of economic activity was the main cause of the downturn in commodity prices, but that the strength and duration of the downturn was caused by other factors as well. These other factors included the deceleration of economic growth in the industrialized countries in 1985–6 and the build-up of inventories in several commodity markets that took place at the beginning of the decade. The results also demonstrate that there were offsetting influences from price stabilization schemes and temporary supply shortages in some commodities. These circumstances have altered the timing, but not the occurrence, of the cyclical behaviour underlying the data-generating processes that characterize commodity markets.

ACKNOWLEDGEMENTS

I am grateful to Greta Boye for undertaking the estimation and simulations of the models described herein. The views expressed in this paper do not necessarily reflect those of the Inter-American Development Bank.

REFERENCES

Beckerman, W. and Jenkinson, T. (1986) What stopped the inflation? Unemployment or commodity prices? *Economic Journal*, **96**, 39–54.

Chu, K. and Morrison, T. K. (1984) The 1981–82 recession and non-oil primary commodity prices. *IMF Staff Papers*, **31** (1), 93–140.

Currie, D. (1981) Some long-run features of dynamic time-series models. *Economic Journal*, **91**, 704–15.

Englander, A. S. (1985) Commodity prices in the current recovery. *Federal Reserve Bank of New York Quarterly Review*, **10** (1), 11–19.

Engle, R. F. and Granger, C. W. J. (1987) Co-integration and error correction: representation, estimation, and testing. *Econometrica*, **55** (2), 251–76.

Engle, R. F., Hendry, D. F. and Richard, J. F. (1983) Exogeneity. *Econometrica*, **51** (2), 277–304.

Grilli, E. and Yang, C. (1988) Primary commodity prices, manufactured goods prices, and the terms of trade of developing countries: what the long run shows. *World Bank Economic Review*, **2** (1).

Harvey, A. C. (1981) *The Econometric Analysis of Time Series*, Phillip Allen, London.

Hendry, D. F., Pagan, A. R. and Sargan, J. D. (1984) Dynamic specification, in

Handbook of Econometrics (eds Z. Griliches and M. Intriligator), Vol. II, North-Holland, Amsterdam.
International Monetary Fund (1988) *International Financial Statistics*, October.
Judge, G. G., Griffiths, W. E., Carter-Hill, R. and Lee, T. (1985) *The Theory and Practice of Econometrics*, Wiley, New York.
Klein, L. R. (1974) *A Textbook of Econometrics*, Prentice-Hall, Englewood Cliffs, NJ.
Lord, M. J. (1988) *Price Formation in Commodity Markets with Inventories* Inter-American Development Bank, Washington DC, (*Mimeo*).
Maddala, G. S. (1977) *Econometrics*, McGraw-Hill, New York.
Morrison, T. and Wattleworth, M. (1987) The 1984–86 commodity recession: an analysis of the underlying causes. *IMF Staff Papers*, **35**.
Muth, J. F. (1961) Rational expectations and the theory of price movements. *Econometrica*, **29**, 315–35.
Plosser, C. I. and Schwert, G. W. (1977) Estimation of a non-invertible moving average process: the case of overdifferencing. *Journal of Econometrics*, **6**, 199–224.
Wallis, K. F. (1980) Econometric implications of the rational expectations hypothesis. *Econometrica*, **48** (1), 49–73.

APPENDIX

Data sources and definitions

Sources for production, consumption and stock data, and Standard International Trade Classification (SITC) (Rev. 2) description of commodities

Cocoa (SITC 072): Gill and Duffus Group Ltd *Cocoa Market Report.*
Coffee (SITC 071.1): United States Department of Agriculture (USDA), *World Coffee Situation.*
Copper (SITC 287.1; 682): World Bureau of Metal Statistics, *Metal Statistics.*
Cotton (SITC 263.1): International Cotton Advisory Committee, *Cotton: World Statistics.*
Maize (SITC 044): USDA, *Grains.*
Soybeans (081.31; 222.2): USDA, *Oilseeds and Products.*
Sugar (SITC 061.1; 062.2): International Sugar Organization, *Sugar Yearbook.*

Description and sources of market prices

Cocoa: Ghanaian, ICCO daily price, average, New York and London, nearest three future trading months.
Coffee: ICO indicator price, other mild arabicas, average NY Bremen/Hamburg markets, ex-dock; prior to January 1984, Guatemalan prime washed, ex-dock NY for prompt shipment.

Copper: London Metal Exchange, grade A cathodes, settlement price; prior to July 1986, high grade.

Cotton: 'A' Index, Middling (1–3/32″), c.i.f. Europe; prior to January 1984, Mexican, middling (1–3/32″), c.i.f. N. Europe.

Maize: USA, No. 2 yellow, f.o.b. Gulf ports.

Soybeans: USA, c.i.f. Rotterdam.

Sugar: ISA daily price, f.o.b. and stowed at greater Caribbean ports; prior to 1961, New York World Contract No. 4, f.a.s. Cuba.

All price data are from the World Bank, *Commodity Trade and Price Trends*, International Bank for Reconstruction and Development, Washington, DC, 1987.

Exogenous variables

Real gross domestic product data from Organization for Economic Cooperation and Development, diskettes.

Interest Rates are measured by the Treasury bill rates. The data source is International Monetary Fund, *International Financial Statistics*.

Inflation is measured by the US wholesale price index. The data source is International Monetary Fund, *International Financial Statistics*.

16

Trade-offs between short-run stability and long-run risk when stabilizing a commodity market

BRUCE L. DIXON AND ANDREW J. HUGHES HALLETT

16.1 SEPARABLE AND NON-SEPARABLE OBJECTIVES

The focus of attention in agricultural and commodity markets has always been the price at which trade takes place. But the stability of prices is seldom a goal in itself. In fact, price support schemes have been widely used as a means raising prices, stabilizing producers' revenues, stabilizing consumers' expenditures or improving the functioning of the market. Thus, although producers and governments often claim that the volatility of prices constitutes a prima facie case for stabilization, price stabilization in practice usually serves as an intermediate target for a wider range of objectives.

Assuming conditions exist in the economy to justify market intervention or that a decision has been made to intervene, the orthodox specification of the objective of market management is the maximization of the discounted sum of producers' and consumers' surplus (see, for example, Just *et al.* (1978) or Newbery and Stiglitz (1981)). Cochrane (1980) among others, has attacked this formulation, arguing that consumers and producers are more concerned with price levels because they identify prices, and hence revenues, with welfare. Certainly policy goals can only be established by consensus (Chambers, 1985), and for agricultural policy this is generally interpreted as meaning the survival of some set of farms as a goal. A minimal goal is therefore the stabilization of producer income levels and/or consumer expenditures in the long term.

Economic surplus also has its limitations as a measure of social welfare under uncertainty (Turnovsky, 1978). Indeed, producers' surplus only gives a satisfactory measure of producers' benefits from stabilization if producers' marginal utilities of income are constant, and that holds only under risk neutrality. Similarly, consumers' surplus only provides an adequate measure of their benefits if consumers' marginal utilities of income are independent of the commodity price. The latter condition might be satisfied approximately in developed countries, but the former is implausible for many and, perhaps, most agricultural or commodity producers. Ideally welfare comparisons should analyse the variations of producers' marginal utilities directly, but this is difficult to do because it involves imposing explicit utility functions (or risk aversion coefficient values) in order to be operational. Instead, we use a very simple objective function here (involving just aggregate producer income (API) and aggregate consumer expenditures (ACE)) as a means of identifying optimal market interventions. The resulting stabilization programme can then be judged by whatever measures of financial cost, stability and risk are considered appropriate to the case in hand. This has the important advantage of focusing the debate on how income support schemes should be evaluated rather than on how they should be created.

16.2 THE TARGETS OF STABILIZATION POLICY

Stabilizing the targets of aggregate long-term producers' revenues or consumers' expenditures, however, is quite different from trying to minimize the fluctuation of an endogenous variable from its target value in each period. The fundamental difference is that the latter objective is (additively) separable with respect to time whereas minimizing the deviations of the sum of each period's revenues (or expenditures) from a pre-assigned level is not separable with respect to time.

In market management problems, this lack of separability is a virtue. Suppose the objective of buffer stock actions is to keep prices near the target trajectory in each period. If there is a stochastic event that drives prices down in one period, then the objective for the remainder of the horizon will be to keep prices near the originally specified target trajectory. However, with the given price decline, producers' revenues have been reduced and the only way the aggregate revenue level can be maintained is if an offsetting random event occurs to restore revenues to their originally expected path. While this may be satisfactory for infinite horizon problems, it will be disastrous in practical situations if producers are unable to survive until that offsetting event occurs at some point in the future. A non-separable objective which minimizes the variations in aggregate revenues, however, would require counterbalancing control actions in the periods following the first random shock, so that producers or consumers hurt by an unexpected fluctuation in the current period receive assistance immediately in succeeding periods.

In what follows two approaches are examined with respect to income and expenditure stabilization. The first method takes the conventional (i.e. separable) approach of minimizing the sum of squared deviations of endogenous variables from their individual period targets. The other approach minimizes the variation of ACE and API for the planning horizon. In this chapter we show how optimal control techniques can be adapted to handle that kind of non-separable objective in a simple way. The two approaches are then applied to an empirical model of the Taiwan rice market and compared in terms of their implications for revenue and expenditure stability. We find the difference in performance to be quite marked in terms of stability.

16.3 MARKET STABILIZATION AS AN OPTIMIZATION PROBLEM

16.3.1 The basic model

We assume a standard dynamic linear econometric model of the commodity market of the form

$$y_{t+1} = Ay_t + Bx_t + Cz_t + e_t \tag{16.1}$$

where y_t is an $n \times 1$ vector of endogenous variables, x_t is an $m \times 1$ vector of control variables, z_t is a $b \times 1$ vector of uncontrollable exogenous variables and e_t is a vector of stochastic error terms with convariance matrix W_t. The matrices A, B and C are of conformable dimensions and are assumed to be composed of known parameters. The subscript t denotes time period.

For deriving the solution to a non-separable programme problem, it is preferable to use the final form of an econometric model rather than the reduced form. The final form model of 16.1 can be readily obtained by repeated back substitutions. Successive target vectors y_t for the entire time horizon are determined from

$$y = Rx + s \tag{16.2}$$

where, assuming a planning horizon of T periods, $y' = (y'_1, \ldots, y'_T)$ is a $Tn \times 1$ vector of stacked endogenous variables y_t, and $x' = (x'_1, \ldots, x'_T)$ is a $Tm \times 1$ vector of similarly stacked control variables. The matrix

$$R = \begin{bmatrix} R_1 & \cdot & 0 \\ \vdots & \ddots & \\ R_T & \cdots & R_1 \end{bmatrix}$$

contains the model's dynamic multipliers, $R_j = \partial y_{t+j-1}/\partial x_t$ for $j \geqslant 1$ and all t, and $s' = (s'_1, \ldots, s'_T)$ defines the impact of the initial conditions, non-controllable events and error terms on the endogenous variables:

$$s_t = A^t y_o + \sum_{j=1}^{t} A^{j-1} (Cz_{t-j+1} + e_{t-j+1}) \tag{16.3}$$

Because of the inclusion of the e_t, s is stochastic with mean $E_o(s)$ and convariance P_o where the subscripts indicate conditioning on the information available at the end period 0.[1]

16.3.2 Certainty equivalent strategies

In the standard separable linear quadratic Gaussian (LQG) control problem the objective function L is specified as

$$L = \sum_{t=1}^{T} (y_t - d_t)'Q_t(y_t - d_t) = (y - d)'Q(y - d) \tag{16.4}$$

where d is a $Tn \times 1$ stacked vector of target values for the endogenous variables and Q is a block diagonal positive-semidefinite matrix with the diagonal blocks having the same dimensions as y_t in (16.1).

Obtaining the optimal controls for minimizing the expected value of 16.4 subject to 16.2 can be accomplished by either Theil's certainty equivalence rule (Theil, 1964) or by backward recursion using dynamic programming as given in Duchan (1975) or Chow (1975). Theil's certainty equivalence rule is written as

$$x^* = (R'QR)^{-1}R'Q[d - E_o(s)] \tag{16.5}$$

The first m components of x^* are the first-period optimal controls for immediate implementation. The remaining $m(T-1)$ components are current forecasts of future interventions. These are updated between periods as realizations on s_t become available, as discussed below.

For the analysis which follows it is important to note that if the target values d are set equal to $E_o(s)$ then the optimal paths of the target variables are the *free-market solution* since $x^* = 0$.

16.3.3 Optimal policy revisions

A disadvantage of the final form approach is that the decision rule is not derived in feedback form. In fact 16.5 is optimal only for the first period if $E_o(s)$ is used. For periods $t \geqslant 2$, the control vector must be updated to reflect past information errors ($E_t(s_i)$ instead of $E_{t-1}(s_i)$) for $i \leqslant t-1$ and to reflect the revised expectations of future s_i. The exact form of this rule is

[1]The covariance of s can be derived from 16.3 as given in Hughes Hallett (1984).

given in Dixon and Hughes Hallett (1988) or Hughes Hallett and Rees (1983). The rule is linear and similar to 16.5 except that the conditional expectation of s_t is updated to reflect past observations on s_t and a term that includes the off-diagonal blocks of Q. This off-diagonal block is multiplied by the difference of the targets for past variables and the realized levels of past y_t. If the objective function is separable with respect to time, the off-diagonal blocks of Q are null. Thus in the separable case x^* is only updated because new information changes the expected value of s_i. But in the non-separable case x^* also has to be adjusted to compensate for past policy 'failures', i.e. discrepancies between past targets and the corresponding realizations of the y_t. Thus past differences are immediately counteracted in non-separable problems, whereas with a separable objective the only concern is how to deal with the current and future expected uncontrollable events and not with compensating for any disasters (or bonanzas) which may have occurred in the past. Since a responsible agricultural policy is directed at long-run objectives and stability, it will be essential to cater for this feature of a minimum non-separable objective function which we noted at the start was the minimum policy goal.

16.3.4 Risk sensitivity

One important criticism of the certainty equivalence approach is that the controls are risk neutral. Under additive uncertainties, L is replaced by $E(L)$ to yield

$$E(L) = [E(y) - d]'Q[E(y) - d] + \operatorname{tr} QP_0 \qquad (16.6)$$

where it is assumed that Q, R and $E(s)$ are known. Therefore, under certainty equivalence, the same decision values will be recommended regardless of the magnitude of P_0. Yet the larger P_0 is for a given mean, the greater the risk which the decision maker is being asked to accept. This has proven highly unsatisfactory to most policy makers. However, 16.6 does not imply that all choices of the control vector x involve the same risks. In fact they do not, as we show below. In particular, pursuing short-run as opposed to long-run income or expenditure targets can involve quite different risks even when the average income/expenditure levels yielded by the two strategies would be expected to be virtually the same. If the average income/expenditure position were the same, the strategy involving lower risks would certainly be preferable.

16.4 AGGREGATED TARGET VARIABLES

Suppose the decision maker wishes to target aggregate revenues and expenditures (API and ACI) over a certain planning horizon as to minimize the fluctuations around those target values. Such objectives are fundamental, and they presuppose some knowledge about the feasible supply and demand relationships in the model. For example, consumer expenditure targets must be consistent with necessary aggregated food levels, while producers' revenues must ensure a minimum net return to keep a reasonable number of farmers in production. These requirements are in addition to any of the reasons why one would be interested in aggregate targets, which were given in the introduction.

16.4.1 Optimal intervention

Assume, for the moment, that d consists solely of the income targets for each period. The aggregate policy goal considered above is to minimize the (squared) deviation of the sum of realizing income from the sum of the targeted incomes over the entire planning horizon. This is to be distinguished from the coventional goal of minimizing the sum of squared deviations of income levels from their targeted values in each period over the same horizon. In the latter (conventional) case the objective is

$$\min_{X_t} \sum_{t=1}^{T} (y_y - d_t)^2 \tag{16.7}$$

whereas the former (aggregated targets) case has

$$\min_{x_t} \left[\sum_{t=1}^{T} (y_t - d_t) \right]^2 \tag{16.8}$$

as its objective. In contrast with 16.7, the objective in 16.8 is clearly non-separable with respect to time. In fact, both can be written as special cases of 16.4. If y is a stacked $T \times 1$ vector of incomes in each period, then 16.7 can be written as $(y - d)'Q(y - d)$ with $Q = I$. But 16.8 can be written the same way with $Q = ll'$, where l is a column vector of units. Hence, 16.8 implies that Q has every element equal to unity. The distinction is therefore between a diagonal Q matrix in the conventional formulation and a non-diagonal Q in the aggregate targets case.[2]

Optimal decision rules for 16.8 follow as before. Substituting 16.2 into 16.8 and taking expectations implies

$$L = \min \{x'R'll'Rx + 2x'R'll'[E_0(s) - d] + \\ \operatorname{tr}(ll'P_0 + \operatorname{tr} ll'[E_0(s) - d][E_0(s) - d]'\} \tag{16.9}$$

[2]The extension to several aggregated targets is straightforward. If, as in the empirical section, there are two targets and the objective is to minimize the fluctuations of each of their total values from certain desired levels, then Q would be a matrix where the odd numbered rows correspond to the first aggregated target and the even numbered rows correspond to the second one. Then the i, j element of Q is unity if $i + j$ is even and zero otherwise. Appropriate designs for Q can be obtained as needed. The distinction remains between a separable objective where Q is block diagonal (with respect to time) and a non-separable objective where Q is not block diagonal.

It is clear from 16.9 that there are an infinite number of optimal control vectors x because $R'll'R$ is of rank 1. One set of optimal decisions can be derived using the Moore–Penrose pseudo-inverse:

$$x^* = (R'll'R)^+R'll'[d - E_0(s)] \tag{16.10}$$

where + denotes the pseudo-inverse. This is a particular case of 16.5. Moreover, inserting 16.10 into 16.9 yields

$$L^* = -[E^o(s) - d]'ll'R(R'll'R)^+R'll'[E_o(s) - d] + \\ \operatorname{tr} ll'P_0 + \operatorname{tr} ll'[E_o(s) - d]\,[E_o(s) - d]' \tag{16.11}$$

It is not difficult to show that 16.11 is equal to $\operatorname{tr} ll'P_o$ as long as $R'll'R$ is of rank 1. Hence (a) there are an infinite number of equally good solutions, one of which uses the Moore–Penrose pseudo-inverse while others are associated with different generalized inverses and, hence, norms on x^*; (b) all these solutions meet the expected aggregated target exactly, so we need some other criterion in order to discriminate between them; and (c) the free-market solution is just as good as, but not better than, any of the optimal solutions.

The criterion chosen here to discriminate between possible control vectors is minimum length. That is desirable for buffer stock interventions because minimizing buffer stock activity will reduce the warehousing or storage costs and transactions costs, as well as general interference in the market.

16.4.2 The free-market solution

Suppose that the free-market solution was initially the preferred target trajectory. In the first period the optimal intervention would be $x_o = 0$. In the second period the expectations of the remaining components of s are updated given the observed value of s_o. In general, unless $s_o = E_o(s_o)$, $E_1(s_1) \neq E_o(s_1)$. In addition, if $s_o \neq E_o(s_o)$, then $y_o - d_o \neq 0$. Thus the optimal interventions for the second period will not, in general, be zero activity. As the results show, strict adherence to a free-market rule, all $x = 0$, results in much larger variance and greater instability than intervening as necessary around the free-market trajectory. The point here is not that the free-market solution is worse in terms of expected outcomes, but it does involve substantially more uncertainty about these future outcomes than do other strategies.

16.5 EMPIRICAL APPLICATION

In this section the strategies in section 16.4 are applied to a model of the Taiwan rice economy to demonstrate, in an empirical setting, the impact of the non-separable approach to minimizing the variation of both aggregates. First the rice model is briefly described. The layout for the various stochastic experiments is discussed. Finally the results of these experiments are analysed.

16.5.1 Taiwan rice market model

The model used for the experiment is a monthly supply and demand model although there is no harvesting in the first four months of the year. The demand model consists of five endogenous variables: retail price, farm price, quantities demanded for final consumption, private stocks and unreported stocks. The government also possesses stocks although these stocks are not an explicit part of the model. The supply component consists of acreage and yield equations, a pair for summer harvests and a pair for autumn harvests. The harvest proportion of acres per month is considered as fixed since time of harvest is largely controlled by weather patterns. The two years simulated are 1979–80. Since our interest lies in controlling the product of prices and quantities, and the model is linear in price and quantities, it is clear that variables we wish to control are non-linear in the controls and predetermined variables so an approximate linear model must be obtained to use the problem format of 16.2 and 16.4.[3]

16.5.2 Experimental design

A total of six experiments are conducted. Each experiment consists of 30 stochastic simulations where each simulation computes the optimal solution sequentially for each of the 24 periods (months). The s_t are obtained using a random number generator in each simulation. Since a free-market solution had the minimum variation in the solutions computed in Dixon and Chen (1982), it is one of the three target trajectories used for the purposes of empirical comparison. The other two target trajectories each have a set of targets for consumer expenditures in each period and a set of targets for gross revenues of producers. These trajectories are labelled B for the trajectory favouring producers and C for the trajectory favouring consumers.

The experiments are labelled by a number followed by a letter. The numeral 1 indicates that Q is an identity matrix and a 2 indicates that Q is structured to minimize the sum of the variability of ACE and API. The letters A, B and C identify the trajectory of the targets. For example, experiment 2A minimizes the sum of the variability of ACE and API about the free-market incomes and expenditures. Trajectory A is the free-market

[3]The methodology and goodness of fit for the linear approximation scheme is reported in detail in Dixon and Hughes Hallett (1988). The convariance matrix P_0 was quite sensitive to the target path and whether the objective function was separable or not.

trajectory where the buying and selling activities are zero in every period for the deterministic solution. Total production varies negligibly over the three trajectories.

16.6 RESULTS

Table 16.1 displays several summary statistics describing the behaviour of API and ACE in the stochastic simulations. As we would expect, there is a reduction in the variability of API and ACE by using the non-separable preferences. In fact, the degree of reduction is in excess of 90% in all cases. Note that in experiments 1A and 2A the mean API and ACE are almost identical but their standard deviations differ considerably. Thus there is no trade-off of less variablity for higher expected returns, the criterion of choice is just the variability of API and ACE.

On the surface it appears that a different phenomenon holds for trajectories B and C. For both trajectories consumers pay less in the non-separable schemes as well as experiencing considerably less variability. A similar phenomenon occurs for producers. The explanation is that the target trajectories d between 1B and 2B (or 1C and 2C) are different for the stochastic models.[4] None the less, there is no monotonic relationship

Table 16.1 Summary statistics for stochastic experiments[a]

Experiment	*API*[b]	*ACE*[b]	*OBJ-SEP*	*OBJ-NONSEP*
1A	69.2 (0.686)	84.2 (0.971)	0.4829×10^{12} (0.3562×10^{12})	0.1469×10^{13} (0.1766×10^{13})
1B	83.6 (0.744)	98.2 (0.958)	0.6059×10^{12} (0.3499×10^{12})	0.1424×10^{13} (0.1553×10^{13})
1C	56.9 (0.535)	72.9 (0.713)	0.3504×10^{12} (0.4224×10^{12})	0.88088×10^{12} (0.1013×10^{12})
2A	69.1 (0.016)	84.5 (0.034)	0.5325×10^{13} (0.5161×10^{13})	0.1356×10^{10} (0.1606×10^{10})
2B	80.1 (0.036)	94.2 (0.021)	0.9902×10^{13} (0.6313×10^{13})	0.1679×10^{10} (0.1500×10^{10})
2C	56.3 (0.018)	56.3 (0.024)	0.7047×10^{13} (0.2983×10^{13})	0.8545×10^{9} (0.8650×10^{9})

[a]All figures are means over the 30 simulations. Figures in parentheses are standard deviations.
[b]Millions of the new Taiwanese dollars.

[4]The product of the expectations is not equal to the expectations of the products. The deterministic target trajectories have to be adjusted after the stochastic simulation of the error terms for the linear approximations. Hence, because the target trajectories of the controls for

between the level of API and its variance (or ACE and its variance) for experiments 2A–2C. However, some trajectories are clearly superior to others. For example, risk-averse producers would prefer 2A to 2C since 2A has both higher income and lower variance than 2C. 2B has even higher income but significantly higher variance. Thus there is a trade-off between API and its variance. A similar situation arises for consumers. Hence decision makers must be prepared to examine a number of trajectories and select that trajectory which adequately balances expected API and ACE and their associated risks.

The results reported in Table 16.1 also point to important trade-offs between separable and non-separable perferences. By pursuing targets period by period, stability of the aggregates is sacrificed. This can be seen in the observed average variablility of the sum of the API and ACE variables shown in the column headed 'OBJ-NONSEP'. By switching from a separable to a non-separable objective function the variability of the aggregates falls by a factor of 1000 while the increase in the separable objective function in going to non-separable preferences is a factor of between 10 and 20. Thus, trying to hit targets in each period induces substantial variability of the aggregates. But minimizing the variation of the aggregates requires a comparatively modest addition of variation of the variables from their target values in individual time periods. The latter approach is more attractive.

An interesting aspect of Table 16.1 is the relationship of the variances of API and ACE to the target trajectory chosen. The last column (OBJ-NONSEP) gives the mean of the sum of the variability of API and ACE. For the non-separable approach the $E(L)$ is significantly less for 2C than for 2B or 2A and any of the separable strategies. This indicates that there is a trade-off between path chosen for the targets and their stability. Furthermore, the free-market trajectory does not give the path with least variability for API and ACE taken together.

A final and perhaps most important comparison is that of the variances under the free-market solution 2A with the variance under 2B and 2C. The earlier results in Dixon and Chen (1982, 1984) show that stabilizing a trajectory of a large variety of price targets induced excess variability in ACE and API over the free-market solution. The results of their experiments showed that tracking any of a number of price trajectories resulted in higher variance of ACE and API than a simple free-market solution. The numerical results in Table 16.3 indicate a different result. It can be computed by analytical methods that the standard deviation of API is 1.85 and the standard deviation of ACI is 2.52 million new Taiwanese dollars under a free-market solution, i.e. where $x^* = 0$ for all periods. Both these figures are

simulating the non-linear model to get P_0 and the errors for s are different, the resulting targets are different between the separable and non-separable experiments so that the mean API and ACE are different.

considerably in excess of the standard deviations shown in Table 16.1 for experiment 1A and, of course, 2A. Thus, some stability of income aggregates is induced by tracking the particular trajectory in each period. However, pursuing non-separable strategies around a free-market trajectory reduces income variability even more. Thus our conclusion is that one of the social costs of a free market is unnecessary variability in incomes and expenditures. This is true even if the free-market trajectory becomes the target for each period and it is tracked with time-separable strategies.

As discusses in Hughes Hallett (1984), the variance of the objective function as well as its mean is likely to be of interest to a risk-averse decision maker. By minimizing the mean of L the decision maker is essentially risk neutral and therefore willing to tolerate wide variations of L from its mean for any particular realization of the planning horizon. Thus it is likely that many decision makers would find solutions attractive that had low variance of L. Observe that for 2A, 2B and 2C the standard deviations of L are much lower than any of the other objective functions, indicating that the likelihood of large variations in income aggregates over the planning horizon is much less using the non-separable specification than the separable specification.

16.7 CONCLUSIONS

A method has been demonstrated whereby the variance of income and expenditure aggregates can be minimized using a non-separable optimization approach. While some stability in terms of period-to-period targets is sacrificed, the numerical results indicate that there is an important trade-off in terms of increasing the overall stability of aggregate variables which decision makers should consider. Higher variability in the aggregate targets of course means more uncertainty about the future welfare for both producers and consumers. A reduction in variation of API and ACE is obtained by the buffer stock actions compensating for past shocks instead of letting such shocks be evened out solely by the long-run dynamic forces in the market.

A free-market solution therefore implies much greater variability of income and expenditures than would occur by a non-separable stabilization scheme about the free-market trajectory. Thus proponents of strict free-market policies must be prepared to accept greater risk to their future survival than would otherwise be necessary. There is also a trade-off in target levels and variability. Decision makers must determine those levels of variability and mean income/expenditures which are in some sense optimal. This may require extensive experiments with numerous target levels until the 'ideal' set is identified.

REFERENCES

Chambers, R. A. (1985) Least cost subsidization alternatives. *American Journal of Agricultural Economics*, **67**, 251–6.

Chow, G. C. (1975) *Analysis and Control of Dynamic Economic Systems*, Wiley, New York.

Cochrane, W. W. (1980) Some nonconformist thoughts on welfare economics and commodity stabilization policy. *American Journal of Agricultural Economics*, **62**, 508–11.

Dixon, B. L. and Chen, W. H. (1982) A stochastic control approach to buffer stock management in the Taiwan rice market. *Journal of Development Economics*, **10**, 187–207.

Dixon, B. L. and Chen, W. H. (1984) Endogenous versus exogenous targets for commodity price stabilization, in *Applied Decision Analysis and Economic Behaviour* (ed. A. J. Hughes Hallett), Nijhoff, The Hague.

Dixon, B. L. and Hughes Hallett, A. J. (1988) Tradeoffs between short run stability and long run risk when stabilizing a commodity market, Staff Paper SP 89, Department of Agricultural Economics and Rural Sociology, University of Arkansas; also presented at the XXV International Conference of the Applied Econometrics Association, Washington, DC, 24–26 October.

Duchan, A. I. (1975) A clarification and a new proof of the certainty equivalence theorem. *International Economic Review*, **15**, 216–24.

Hughes Hallett, A. J. (1984) Optimal stockpiling in a high risk commodity market: the case of copper. *Journal of Economic Dynamics and Control*, 211–38.

Hughes Hallett, A. J. and Rees, H. J. B. (1983) *Quantitative Economic Policies and Interactive Planning*, Cambridge University Press, Cambridge and New York.

Just, R., Lutz, E., Schmitz, A. and Turnovsky, S. J. (1978) The distribution of welfare gains from pure stabilization: an international perspective. *Journal of International Economics*, **8**, 551–63.

Newbery, D. M. G. and Stiglitz, J. E. (1981) *The Theory of Commodity Price Stabilization*, Oxford University Press, Oxford.

Theil, H. (1964) *Optimal Decision Rules for Government and Industry*, North-Holland, Amsterdam.

Turnovsky, S. J. (1978) The distribution of welfare gains from price stabilization: a survey of some theoretical issues, in *Stabilizing World Commodity Markets*, (eds F. G. Adams and S. A. Klein), Lexington Books, Lexington, MA.

17

Are commodity prices leading indicators of OECD prices?

MARTINE DURAND

AND SVENBJÖRN BLÖNDAL

17.1 INTRODUCTION AND SUMMARY

Commodity price series have long been scrutinized for indications of the future course of inflation or deflation. Recently, there has been particular interest in their usefulness as an advance indicator of inflation in the process of multilateral surveillance of economic policies. US Treasury Secretary James Baker and UK Chancellor of the Exchequer Nigel Lawson both made proposals along these lines at the September 1987 annual meetings of the International Monetary Fund and the World Bank (Baker, 1987; Lawson, 1987). Secretary Baker suggested the use of commodity prices, including gold, in assessing the inflation outlook for each country. Chancellor Lawson proposed the use of a commodity price index as a global indicator or early warning signal of the risks of inflation and disinflation for major industrial countries as a group. Both ministers were responding to the need for a way to gauge the appropriate overall balance of macroeconomic policies between stimulus and restraint in the context of coordinated efforts to reduce and contain imbalances among the major economies and promote greater stability of exchange rates.

In recent years a number of other mechanisms have been proposed to counter any tendency for coordinated policies to have either an inflationary or a deflationary bias. For example, McKinnon (1984) advocated targeting world money supply to anchor nominal world prices but has since abandoned his earlier proposal (McKinnon, 1988) in favour of targeting the average price of traded goods directly, thus allowing for velocity shifts. To the same end, Williamson and Miller (1987) have suggested a nominal expenditure target in conjunction with stable exchange rates. However, a nominal expenditure target as an operational guide for policy is of limited practical use owing to lags in the availability of national income accounts and frequent revisions, as well as to uncertainty concerning its future evolution for any given policy setting—hence the search for indicators around which policies could be coordinated.

A number of economic arguments can be advanced in support of the hypothesis that commodity price developments pre-date movements in the general price level. Many commodity prices are determined in auction markets which respond quickly to changes in supply and demand, in contrast with prices in customer markets for processed goods. This distinction is emphasized by, among others, Bosworth and Lawrence (1982) and Beckerman and Jenkinson (1986). Provided that conditions in commodity markets reflect aggregate supply and demand in the whole economy, an increase in aggregate demand, which might eventually translate into higher price inflation, might be expected to show up much earlier in commodity prices. Another characteristic of commodity prices which may make them suitable leading indicators is their forward-looking element, arising from their storability—including, in the case of minerals, the existence of unextracted reserves. Hence, commodity stocks, and claims on them which are traded in futures markets, are similar to financial assets in the sensitivity of their prices to expectations of future economic conditions. For example, provided that such markets are efficient, an increase in expectations of inflation should immediately be reflected in higher commodity prices. If these expectations are rational, such price movements could provide information about future prices of use to policy makers. Van Duyne (1979) and Frankel and Hardouvelis (1983) have emphasized this asset character of commodities. Finally, a more traditional argument for relying on commodity prices in predicting price developments is that they enter, generally with a lag, as costs in output price equations.[1] Thus current commodity price movements have a direct effect on future movements of general price indices.

Commodity prices, however, also have potential weaknesses as indicators of general price developments. Supply conditions in commodity markets can deviate significantly from aggregate supply in an individual economy or the world economy because of specific factors such as climate. In addition, movements in non-commodity costs, particularly labour costs, can dominate the influence of past commodity prices on current prices. And either market inefficiency, expectational aberrations or both could reduce the indicator value of commodity price movements. In the end, the quality of commodity prices as a leading indicator of general price developments can only be determined by reference to empirical evidence as to how they would have performed in the past.

[1]Such a specification is used in the Secretariat's Interlink model (Stiehler, 1987).

An important distinction needs to be made between the hypothesis of commodity price levels being leading indicators for general price *levels*, and the hypothesis that commodity price inflation leads general price *inflation* (i.e. rates of change in price levels). The suggestion that the trend in commodity prices could serve as a leading indicator for general prices, referred to above, appears to require a relationship between the *levels* of the two series. This in turn implies that the long-run terms of trade are fixed, and completely independent of economic factors such as differential productivity growth. This is a very restrictive assumption, but one which would strengthen the possible role of commodity prices as indicators of general price developments since observed deviations from a long-run level relationship could be expected to be closed in the future.

Even if a level relationship cannot be established between commodity prices and aggregate prices, a relationship between their changes could still make commodity price changes informative as to future inflation movements. In either case, however, at least some of the disturbances that affect both time series must appear earlier in the commodity series than in the aggregate price series. Tests for such leads in innovations, generally referred to as 'Granger-causality' tests, are not tests for causality in a structural sense. Moreoever, in the present context evidence that disturbances in commodity price do or do not lead aggregate price developments does not settle the question of whether commodity prices contain useful information about future general price developments. It may be that monetary authorities are already making full use of this information to attempt to stabilize the price level. In this case, disturbances in commodity prices would call forth a monetary response that would neutralize the disturbance before it could be reflected in the general inflation rate. Nevertheless, the tests are useful in establishing the extent to which commodity prices have been given too little weight by policy makers in the past. Thus they are appropriate for examining the question of whether policy making might be improved by giving them a more important or more formal role as indicators.

In this chapter we examine this question by exploring bivariate relationships between a number of commodity price indices, both for individual commodities and aggregates of commodities, and consumer price indices in the OECD area.[2] The objective is to test whether commodity prices are useful for predicting consumer prices, and whether this refers to level or rates of change relationships. Although some existing aggregate commodity price indices have been considered, no attempt has been made to define a new 'basket' of commodity prices that could predict OECD consumer prices. As such, it is a reconnaissance rather than an in-depth investigation. Many interesting questions are not addressed. Among them are what

[2]In contrast with previous work by the OECD Secretariat (Holtham and Durand, 1987; Holtham *et al.*, 1985 where multivariate relationships have been tested.

information commodity prices might add in the context of a larger set of indicators—a question for multivariate analysis. It seemed that the place to start was to examine bivariate relationships.

The results of these tests can be summarized as follows.

1. The levels of all the commodity and consumer price indices are non-stationary, while their inflation rates are without apparent trends.
2. There is little evidence for a level relationship between commodity and consumer price indices, the only exceptions being consumer and gold price indices in the UK and, in the 1960–73 period, France.
3. Innovations in the inflation of some important metal, food and agricultural raw material prices lead aggregate OECD inflation rates, but the autoregressive relationships are in general unstable over time and subject to two-way intertemporal causation.

Thus there does appear to be useful information about the future course of OECD inflation to be gleaned from the growth of some commodity prices. Developments in some metals, food and agricultural raw materials prices may therefore be a useful supplement to the existing indicators used in the multilateral surveillance process.

Some methodological issues related to tests for discrimination between level and rate of change bivariate relationships are discussed in section 17.2. The results from this analysis are presented in section 17.3 together with Granger-causality tests and stability tests of the specification most favoured by the data.

17.2 METHODOLOGICAL ISSUES

Secular movements in finite time series with apparent trends either can be represented as deterministic time trends or can be characterized as being driven by stochastic processes with no tendency to return to a predetermined path, i.e. as integrated processes in which errors are cumulative. Despite the common use of deterministic time trends to describe the tendency for macroeconomic time series to deviate cumulatively from their starting level, there is growing evidence that their secular components are better represented as integrated stochastic processes, such as a random walk (Nelson and Plosser, 1982). In this study, we assume that the secular behaviour of both commodity and consumer prices is stochastic and note deterministic.

In this context, a necessary condition for commodity price levels to forecast future levels of consumer prices is that they are driven by a common integrated process, and therefore the time series, if appropriately scaled, do not drift too far apart, at least in the long run. In other words, non-contemporaneous levels of consumer and commodity prices must form an equilibrium relationship. Under these circumstances, non-contemporaneous values of the same variable should form an equilibrium relationship between themselves, which in turn implies that a contemporaneous relationship must exist between the levels of commodity prices and consumer prices.[3] It has been shown that standard test statistics, such as t and F ratios, do not possess limiting distributions in the case of mutually trending variables except under special circumstances, thus invalidating standard statistical inference (Phillips, 1986). Recently developed tests for cointegration, however, offer a framework to test for the existence of a level relationship among jointly trending variables (Granger, 1986; Hendry, 1986; Engle and Granger, 1987). This strategy is employed below.[4]

The cointegration testing strategy starts from the premise that a precondition for the levels of two variables to form an equilibrium relationship is that they both share the same intertemporal characteristics. A dynamic property of a single series can be described by how often it needs to be differenced in order to provide a stationary process in which any tendency for errors to cumulate has been eliminated, i.e. by its order of integration (denoted I(d) where d is the order of integration). Two variables can form an equilibrium relationship only if both are I(1). The order of integration can be inferred either from the sample autocorrelation function (Box and Jenkins, 1976) or by testing for unit roots. The latter method tests the null hypothesis of a process driven by a random walk against the alternative of stationarity. A number of unit root tests have been suggested in the literature (Dickey and Fuller, 1979; Evans and Savin, 1981, 1984; Sargan and Bhargava, 1983; Phillips, 1987). In this study we have employed the Durbin–Watson test of Sargan and Bhargava (CRDW) and the adjusted Dickey–Fuller test (ADF). The relevant test statistics can be obtained from standard ordinary least squares (OLS) regression ouput by running the following regressions:

CRDW $$X_t = C + u_t$$

ADF $$\Delta x_T = \alpha X_{t-1} + \sum_{j=1}^{p} \gamma_i \Delta X_{t-j} + e_t \qquad (17.1)$$

where C is a constant, u and e are error terms and Δ is the first difference operator. The test statistic for CRDW is the Durbin–Watson coefficient for the regression. Its distribution has been derived by Sargan and Bhargava (1983) who also computed critical values. The distribution of the ADF test

[3]Econometric models which incorporate long-run relationships between jointly drifting variables typically assume their existence without testing for them. Instead, they impose prior constraints in the long-run in order to avoid relying on often spurious results from regressions with non-stationary variables.

[4]Cointegration tests have been applied, for example, by Hall (1986), Jenkinson (1986) and Campbell and Schiller (1987). In Holtham and Durand (1987) cointegration tests have been conducted to specify the relationship between commodity prices and their main determinants.

statistic, which is the t ratio associated with α, was derived by Dickey and Fuller (1979).[5] The two different test statistics can lead to conflicting inferences since the CRDW test presupposes that both the null and the alternative hypotheses are of first order, while the ADF test is adjusted for higher order (the ΔX terms). Which of the two test statistics is more appropriate will thus differ from case to case. If the null hypothesis of a random walk cannot be rejected, the I(1) hypothesis can be tested against a non-stationary first difference (i.e. I(2)) by differencing the X and ΔX terms in 17.1

After establishing that two series X and Y are both integrated of the first order, they are candidates for an equilibrium relationship. To gauge whether the variables form a level relationship, the following cointegration regression is run:

$$X_t = a_0 + a_1 Y_t + u_t \tag{17.2}$$

An equilibrium relationship is said to exist if the residual from this regression is stationary. As discussed above, standard statistical inference is not valid when the variables are non-stationary and regressions are prone to generate spurious relationships among variables. However, contrary to the spurious regression case, the estimated parameters do converge in probability to constants provided that the variables cointegrate (Phillips, 1986). Moreover, it has been shown that the convergence is faster than in ordinary regression (Stock, 1987). These properties suggest that the cointegration regression provides a good estimate of the long-run equilibrium relationship.

The intertemporal properties of the residual in 17.2, in particular whether it is stationary, can again be determined by any of the unit root test procedures. However, the distribution of the test statistics differ from the univariate case. The critical significance levels now become functions of the data generation process, which is generally unknown. Sargan and Bhargava (1983) computed upper and lower critical bounds for the null hypothesis that the residual from the cointegration regression is driven by a random walk process. Unfortunately, this leaves a large region where the test would be indeterminate. Exact critical values can be derived by computing probability distributions for each data-generating process. Alternatively, it is possible to compute exact critical values using Monte Carlo methods for a given data-generating process. Engle and Granger (1987) provide exact critical values for the CRDW and ADF test statistics for two different generating processes. Since exact CRDW critical levels are shown to be very sensitive to the data-generating process while the critical levels for the ADF are relatively robust, the ADF test seems more appropriate when the CRDW falls within the indeterminate range.

[5]Distribution tables for the CRDW and ADF statistics are provided in Sargan and Bhagava (1983, p. 157) and Fuller (1976, p. 373).

17.3 EMPIRICAL RESULTS

17.3.1 Description of the data

The tests reported below were conducted on consumer price indices (CPIs) for the major seven OECD countries and their gross domestic product weighted average (hereafter somewhat loosely referred to as OECD inflation), and individual and aggregate commodity price indices. Aggregate commodity price indices from three different sources have been used: the HWWA (Hamburg Institüt für Weltwirtschaftforschung), the UNCTAD and the IMF price indices. Although they all include total commodity, foodstuff, tropical beverages, agricultural raw materials and metals and minerals price indices,[6] they differ in construction, in terms of both weights and commodity coverage, and may thus exhibit different movements over time. In particular, commodities in the UNCTAD and IMF indices are weighted according to trade patterns based on exports whereas the HWWA indices use weights based on imports of primary commodities.[7] Since aggregate indices include primary commodity prices which have very different characteristics, a number of individual commodity prices used in the construction of the aggregate indices were also considered.[8] These are listed in Table 17.1. For consistency, all CPIs and commodity price indices were expressed in US dollars and were not seasonally adjusted. Given the extreme volatility of the monthly data, quarterly data were used.

When considered at the aggregate level, the first oil shock represents a break in the time profile of the series between 1960 and 1987 (Fig. 17.1). For this reason, the study of the intertemporal properties of commodity prices was conducted over two subperiods: before and after 1973. The same was done for the CPIs, in order to establish possible relationships between CPIs and commodity prices limited to subperiods of the sample.

Finally, the analysis of the time series properties conducted here was on the logarithms of the levels of the series.[9] The reason for choosing these rather than the levels themselves is because differenced log levels of prices are an approximation of inflation rates, so that while testing for level equilibrium relationships it is also possible to analyse relationships between inflation rates in CPIs and commodity prices.

[6]IMF total commodity index is not included in this study because of availability problems.

[7]For a detailed analysis of the construction of aggregate commodity price indices see Siddique (1984).

[8]For instance, the aggregate metals and minerals price index includes aluminium and tin prices. The tin market is very different from the aluminium market. Large surpluses in tin led to the creation of a buffer stock to regulate prices. It is therefore possible that aluminium and tin price indices move quite differently over time, although they are included in the same aggregate index.

[9]Integration and cointegration tests were also carried out on untransformed values of the variables. This led to the same results as those described below.

Table 17.1 First-order integration tests

	Test statistics	
	CRDW	ADF
A Aggregate price indices		
HWWA		
Total commodities	0.076	−1.67
Food	0.133	−1.40
Tropical beverages	0.132	−1.60
Metals and minerals	0.147	−1.87
Agricultural raw materials	0.073	−1.49
UNCTAD		
Total commodities	0.094	−1.39
Food	0.142	−1.42
Tropical beverages	0.121	−1.53
Metals and minerals	0.166	−1.76
Agricultural raw materials	0.093	−1.41
IMF		
Food	0.029	0.53
Tropical beverages	0.025	0.36
Metals and minerals	0.036	0.77
Agricultural raw materials	0.014	1.22
Consumer price indices		
Unites States	0.005	1.53
Japan	0.011	3.31
Germany	0.020	2.35
France	0.030	2.45
United Kingdom	0.027	2.25
Italy	0.008	2.47
Canada	0.009	2.32
Average for above countries	0.006	2.87
B Individual price indices		
Precious metals		
Gold	0.068	0.85
Silver[a]	0.064	0.35
Metals		
Aluminium	0.027	0.99
Bauxite[a]	0.015	1.54
Copper	0.055	0.86
Iron	0.054	0.24
Lead	0.049	0.79
Nickel	0.021	1.18
Tin	0.021	0.51
Zinc	0.049	0.69
Agricultural raw materials		
Cotton[a]	0.096	0.43
Jute[a]	0.236	−0.12
Rubber[a]	0.048	0.29
Wool	0.042	0.43

Table 17.1 (*Cont.*)

	Test statistics	
	CRDW	*ADF*
Food and tropical beverages		
Cocoa	0.029	0.99
Coffee	0.038	0.33
Maize[a]	0.086	0.26
Rice[a]	0.061	0.22
Soybeans	0.075	0.56
Sugar	0.038	1.05
Tea	0.139	−0.05
Wheat	0.046	0.32

Regressions were done from 1962.II to 1987.II unless otherwise indicated, except for gold prices (from 1965.I and 1987.II).

The critical value for the CRDW test below which first-order integration is rejected is 0.259 for 100 degrees of freedom (0.493 for 50 degrees of freedom). The critical value for the AFD test below which at least first-order integration is rejected is −3.45 for 100 degrees of freedom (−3.50 for 50 degrees of freedom).

[a]Regressions were done from 1970.I to 1987.II.

Sources: HWWA indices, *Intereconomics*, Hamburg Weltwirtschaft Forschung Institüt; UNCTAD indices, *U.N. Monthly Commodity Price Bulletin*; IMF indices, IMF, *IFS Statistics*.

17.3.2 Integration and cointegration tests

Table 17.1 reports the CRDW and ADF test statistics; the latter were done with four lags on the ΔX term ($j = 4$ in 17.1). The test indicate that, without exception, the individual and aggregate commodity price indices are estimated to be at least integrated of first order. They all comfortably pass the tests performed at the 5% confidence level on regressions done over the entire period. When the period is split into two subperiods, the same results hold with the one exception that the price index of tea is found to be I(0) in the first subperiod but at least I(1) in the second. The CRDW and ADF tests for a higher degree of integration, with no allowance for lags, clearly reject the hypothesis that commodity prices are I(2).[10] Thus, in general, the indices have to be differenced once to become stationary processes. The results of the tests for the CPIs also lead to the conclusion that country and aggregate CPIs are integrated of first order in the whole period, as well as in the two subperiods.

The CRDW and ADF tests for cointegration show that, taken two by two, individual and aggregate CPIs and commodity price indices are generally *not* cointegrated, in either the whole period or the two subperiods.

[10]For a small number of commodity prices the ADF tests are close to the critical value, but in testing for integration of second order the CRDW test and Dickey–Fuller test with no lags may be more appropriate than the ADF test. Results for the subperiods and for the second-order integration tests are not reported here but are available on request.

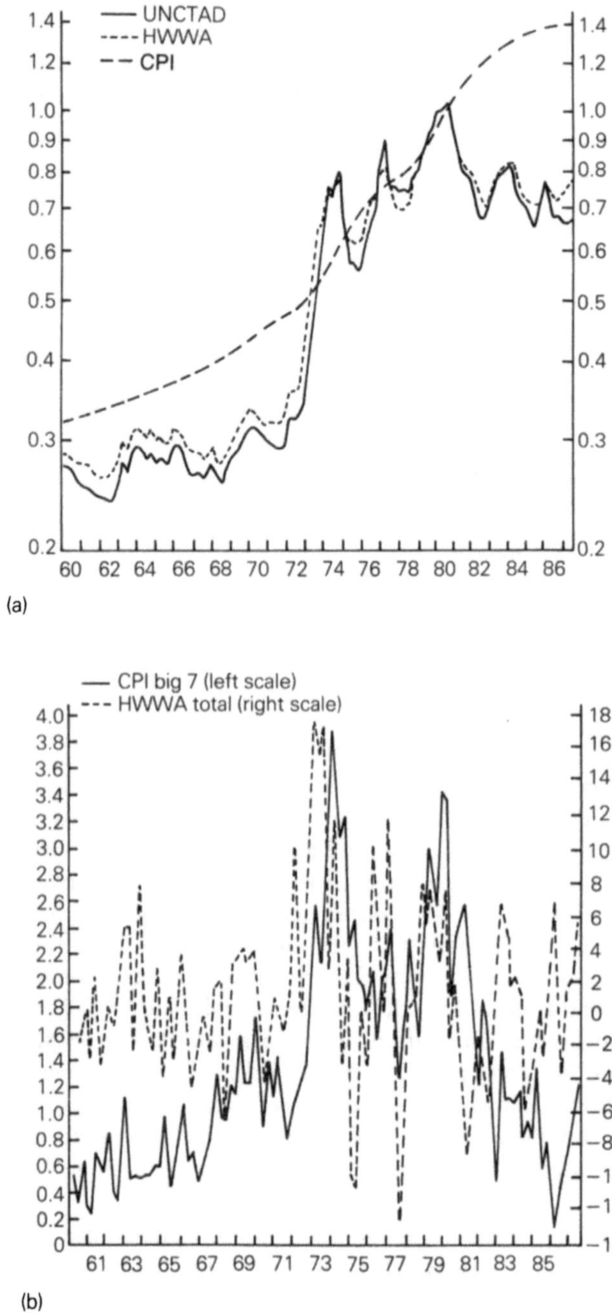

Fig. 17.1 Developments in commodity and consumer price indicies: (a) commodity prices and major seven CPI; (b) Change in commodity prices and OECD inflation.

Table 17.2 Cointegration tests between aggregate commodity and OECD price levels

Commodity prices	OECD CPI	
	Values of the CRDW test	*Value of the ADF test*
UNCTAD		
Total	0.060	−1.32
Food	0.055	−1.52
Tropical beverages	0.091	−1.29
Agricultural raw materials	0.099	−2.15
Metals and minerals	0.092	−2.10
HWWA		
Total	0.049	−1.65
Food	0.082	−1.38
Tropical beverages	0.082	−1.27
Agricultural raw materials	0.055	−1.86
Metals and minerals	0.120	−2.30
IMF		
Food	0.065	−1.73
Tropical beverages	0.089	−1.35
Agricultural raw materials	0.113	−3.04
Metals and minerals	0.096	−1.69

The critical value for the CRDW test below which cointegration is rejected is 0.257.
The critical value for the ADF test above which cointegration is rejected is −3.17.

There are only two exceptions to this general pattern: the UK CPI and the gold price index are cointegrated over the whole period and the French CPI and the gold price index are cointegrated in the first subperiod. Table 17.2 presents the results of cointegration tests between OECD and aggregate commodity prices; results for the individual commodity price indices and individual country CPIs are available on request.

The conclusion is that there is no evidence of equilibrium level relationships between most of the commodity prices considered here and the CPIs of the major seven countries. The hypothesis that in the long run the price levels of consumer goods and primary commodities do not significantly depart from each other is therefore rejected.

17.3.3 Granger-causality and stability tests

Although it was not possible to find any level relationship between CPIs and commodity price indices, the fact that all indices are I(1) is nevertheless an interesting result. Granger-causality (Granger, 1968) tests on the logarithm of the difference of the CPIs and commodity price indices may provide information on possible temporal causality and feedback effects between rates of inflation in CPIs and commodity prices. If changes in commodity prices are useful indicators of future developments in CPI inflation, it is important to establish that commodity prices Granger-cause CPIs. However, if the tests reveal that commodity prices and CPIs Granger-cause each other, then it could be argued that they are both responding to a third variable such as a monetary aggregate.[11]

The Granger-causality regressions took the standard vector autoregressive form:

$$
\begin{aligned}
\Delta \ln \mathrm{CPI}_t = {} & a_0 + a_1 \Delta \ln \mathrm{PC}_{i,t-1} + a_2 \Delta \ln \mathrm{PC}_{i,t-2} \\
& + a_3 \Delta \ln \mathrm{PC}_{i,t-3} + a_4 \Delta \ln \mathrm{PC}_{i,t-4} \\
& + b_1 \Delta \ln \mathrm{CPI}_{t-1} + b_4 \Delta \ln \mathrm{CPI}_{t-2} \\
& + b_3 \Delta \ln \mathrm{CPI}_{t-3} + b_4 \Delta \ln \mathrm{CPI}_{t-4} \\
& + d_l\, \mathrm{DQ}_l + d_2 \mathrm{DQ}_2 + d_3 \mathrm{DQ}_3 + \varepsilon_t \\
\Delta \ln \mathrm{PC}_{i,t} = {} & c_0 + c_1 \Delta \ln \mathrm{CPI}_{t-1} + c_2 \Delta \ln \mathrm{CPI}_{t-2} \\
& + c_3 \Delta \ln \mathrm{CPI}_{t-3} + c_4 \Delta \ln \mathrm{CPI}_{t-4} \\
& + c_1 \Delta \ln \mathrm{PC}_{i,t-1} + c_2 \Delta \ln \mathrm{PC}_{i,t-2} \\
& + e_3 \Delta \ln \mathrm{PC}_{i,t-1} + e_4 \Delta \ln \mathrm{PC}_{i,t-4} \\
& + d_4\, \mathrm{DQ}_1 + d_5\, \mathrm{DQ}_2 + d_6\, \mathrm{DQ}_3 + \mu_t
\end{aligned}
$$

where PC_i are commodity price indices, ε and μ are error terms and DQ_1, DQ_2 and DQ_3 are quarterly intercept dummies included to take care of seasonal variations which may be present in the series. Necessary conditions for PC_i to Granger-cause CPI without feedback from CPI to PC_i are the following: either a_1, a_2, a_3 or a_4 is non-zero and $c_1 = c_2 = c_3 = c_4 = 0$. The first condition was tested using standard t statistics and the second restriction was tested using the standard F ratio test. If the former condition is accepted but the latter is rejected, then the variables are Granger-causing each other.

Table 17.3 summarizes the results of Granger-causality tests over the entire period.[12] Average CPI inflation for the major seven countries can be predicted using available information in the past evolution of inflation in the aggregate prices of the UNCTAD agricultural raw materials and the HWWA metals and minerals. At the disaggregated levels, only changes in

[11]In Holtham *et al.* (1985) causality tests had been done to identify commodity prices determinants such as OECD prices, OECD activity and interest rates and a number of other variables.

[12]Granger-causality has been tested over the two subperiods with similar results to those reported in the text.

Table 17.3 Granger-causality tests on commodity price inflation and major seven OECD countries CPI inflation

	Consumer price index							
	USA	*Japan*	*Germany*	*France*	*UK*	*Italy*	*Canada*	*Major seven*
A Aggregate commodity prices								
UNCTAD								
Total	N	N	Y	Y	N	Y	N	N
Food	N	Y	Y*	Y	Y	Y	N	N
Tropical beverages	N	N	N	N	N	N	N	N
Metals and minerals	Y	N	N	N	N	N	N	N
Agricultural raw materials	Y	N	N	N	Y	N	N	Y
HWWA								
Total	N	N	N	N	N	N	N	N
Food	Y	Y	N	N	N	N	N	N
Tropical beverages	N	N	Y	Y	N	N	N	N
Metals and minerals	N	N	N	N	N	N	N	Y
Agricultural raw materials Y	Y*	N	N	N	Y	N	N	
IMF								
Food	Y*	Y*	Y	N	N	N	N	Y*
Tropical beverages	N	N	N	N	N	N	N	N
Metals and minerals	N	N	N	N	N	N	Y	N
Agricultural raw materials	Y*	Y	N	N	N	Y	N	Y*

Table 17.3 (*Cont.*)

	Consumer price index							
	USA	*Japan*	*Germany*	*France*	*UK*	*Italy*	*Canada*	*Major seven*
B Individual commodity prices								
Gold	Y	N	N	N	N	Y	N	N
Aluminium	N	N	Y	N	N	Y*	N	Y
Coffee	N	N	Y	N	N	N	N	N
Copper	Y*	N	N	Y	Y	N	Y	Y*
Iron	N	N	Y	Y	N	N	Y*	N
Lead	N	Y	N	N	N	Y	Y	N
Nickel	Y*	N	N	N	Y*	N	N	N
Sugar	N	N	N	Y	N	Y*	Y*	Y
Tea	N	N	N	N	N	N	N	N
Tin	Y	N	Y	Y	Y*	N	N	N
Wheat	N	N	Y	N	N	N	N	Y*
Wool	N	Y	N	N	N	Y	N	Y*
Zinc	N	N	Y	N	N	N	N	Y*
Bauxite	N	N	N	N	N	Y*	Y	N
Cotton	Y*	N	N	N	N	N	N	Y*
Maize	Y	Y	Y	N	N	Y	N	Y
Rice	N	N	N	N	N	N	Y	Y
Rubber	N	N	N	N	Y*	N	N	N
Silver	Y*	N	N	N	N	N	Y	N
Soybeans	Y*	N	Y*	Y*	N	N	N	N
Cocoa	N	N	N	N	N	Y	N	N
Jute	Y	N	N	N	N	Y	N	N

Y, a commodity price Granger-causes the country CPI; Y*, presence of feedback effects; N, commodity prices do not Granger-cause CPIs.

aluminium, sugar, maize and rice prices appear to be useful in predicting OECD inflation. Gold, which represents a particular case since its market is probably more speculative than those for other primary commodities, does not appear to be a good indicator of aggregate OECD inflation. It can provide information in the prediction of the US and the Italian inflation rates, however.

Finally, Hendry (1980) predictive failure tests have been carried out to determine whether the CPI forecasting equations were stable over time. This test involves comparing the in-sample and out-of-sample variances of the first Granger-causality regressions; this was done for all the bivariate relationships summarized in Table 17.3. Based on regressions from 1962 to 1973 and from 1974 to 1984, predictions were calculated for the periods after 1973 and 1984 respectively. For most commodity price indices which Granger-cause a country CPI without feedback in each of the two subsamples, the stability test rejects the hypothesis of no predictive failure. This means that in most cases when a relationship between inflation rates in commodity price indices and CPIs exists, this relationship is generally unstable over time. For example, over the full sample period, the UNCTAD agricultural raw materials and the HWWA metals and minerals indices Granger-caused OECD inflation (as indicated by the Y in Table 17.3); however, in both cases the Hendry tests indicate that predictive failure cannot be rejected (Table 17.4). Over the 1974–84 subsample, both the UNCTAD and HWWA indices of agricultural raw materials and metals and minerals Granger-cause OECD inflation; but predictive failure is rejected only for HWWA metals and minerals (Table 17.4).

Table 17.4 Stability tests for OECD inflation

	Hendry χ^2 test statistic	
	1962–84[a]	*1974 –84*
OECD inflation determined by inflation in		
UNCTAD		
Agricultural raw materials	26.41	20.76
Metals and minerals	–	20.49
HWWA		
Agricultural raw materials	–	22.70
Metals and minerals	22.37	14.66

The critical value of the Hendry χ^2 test below which predictive failure is rejected is 18.3 for ten degrees of freedom.

[a]Regressions were done from 1962.I to 1984.IV and from 1974.I to 1984.IV. Predictions were calculated from 1985.I to 1987.II.

ACKNOWLEDGEMENTS

We are very grateful to Jeffrey Shafer who initiated this project and gave many helpful comments. We also wish to acknowledge David Coe for discussions and comments and Anick Lotrous for statistical assistance.

REFERENCES AND FURTHER READING

Baker, J. A. (1987) Statement, IMF-IBRD Board of Governors, 1987 Annual Meetings, Press Release 50.

Beckerman, W. and Jenkinson, T. (1986) What stopped the inflation? unemployment or commodity prices? *Economic Journal*, **96**, 39–54.

Bosworth, B. P. and Lawrence, R. Z. (1982) *Commodity Prices and the New Inflation*, Brookings Institution, Washingon, DC.

Box, G. E. P. and Jenkins, G. M. (1976) *Time Series Analysis. Forecasting and Control*, revised edition, Holden-Day, San Francisco, CA.

Campbell, J. Y. and Shiller, R. J. (1987) Cointegration and tests of present value models. *Journal of Political Economy*, **95**, 1062–88.

Chow, G. C. (1960) Test of equality between sets of coefficients in two linear regressions. *Econometrica*, **28**, 591–605.

Dickey, D. A. and Fuller, W. A. (1979) Distribution of the estimators for autoregressive time series with a unit root. *Journal of the American Statistical Association*, **74**, 427–31.

Engle, R. F. and Granger, C. W. J. (1987) Cointegration and error correction: representation, estimation, and testing, *Econometrica*, **55**, 251–76.

Evans, G. B. A. and Savin, N. E. (1981) Testing for unit roots: 1. *Econometrica*, **49**, 753–79.

Evans, G. B. A. and Savin, N. E. (1984) Testing for unit roots: 2. *Econometrica*, **52**, 1241–69.

Frankel, J. A. and Hardouvelis, G. A. (1983) Commodity prices, overshooting, money surprises and Fed credibility, Working Paper 1121, National Bureau of Economic Research.

Fuller, W. A. (1976) *Introduction to Statistical Time Series*, Wiley, New York.

Granger, C. W. J. (1969) Investigating causal relations by econometric models and cross-spectral methods. *Econometrica*, **37**, 424–38.

Granger, C. W. J. (1986) Developments in the study of cointegrated economic variables. *Oxford Bulletin of Economics and Statistics*, **48**, 213–28.

Hall, S. G. (1986) An application of the Granger and Engle two-step estimation procedure to United Kingdom aggregate wage data. *Oxford Bulletin of Economics and Statistics*, **48**, 229–39.

Hendry, D. F. (1980) Predictive failure and econometric modelling in macroeconomies: the transactions demands for money, in *Modelling the Economy* (ed. P. Ormerod), Heineman Educational, London.

Hendry, D. F. (1986) Econometric modelling with cointegrated variables: an overview. *Oxford Bulletin of Economics and Statistics*, **48**, 201–12.

Holtham, G. and Durand, M. (1987) OECD economic activity and non-oil commodity prices: reduced-form equations for INTERLINK, OECD Department of Economics and Statistics Working Paper 42, June.

Holtham, G., Saavalaien, T., Saunders, P. and Sutch, H. (1985) Commodity prices in

INTERLINK, OECD Department of Economics and Statistics Working Paper 27, November.

Jenkinson, T. J. (1986) Testing neo-classical theories of labour demand: an application of cointegration techniques. *Oxford Bulletin of Economics and Statistics*, **48**, 241–51.

Lawson, N. (1987) Statement, IMF-IBRD Board of Governors, 1987 Annual Meetings, Press Release 44.

McKinnon, R. I. (1984) An international standard for monetary stabilization. *Policy Analyses in International Economics* No. 8, Institute for International Economics, Washington, DC.

McKinnon, R. I. (1988) Monetary and exchange rate policies for international financial stability: a proposal. *Journal of Economic Perspectives*, **2** (1), 83–103.

Nelson, C. R. and Plosser, C. I. (1982) Trends and random walks in macroeconomic time series. Some evidence and implications. *Journal of Monetary Economics*, **10**, 139–62.

Phillips, P. C. B. (1986) Understanding spurious regressions in econometrics. *Journal of Econometrics*, **33**, 311–40.

Phillips, P. C. B. (1987) Time series regression with a unit root. *Econometrica*, **55**, 277–301.

Sargan, J. D. and Bhargava, A. (1983) Testing residuals from least squared regression for being generated by the gaussian random walk. *Econometrica*, **51**, 153–74.

Siddique, A. K. M. (1984) Commodity price indices: a historical and methodological review. *Statistical Journal of the United Nations*, **2**, 255–83.

Stiehler, U. (1987) Price determination in the major seven country models in INTERLINK, OECD Department of Economics and Statistics Working Paper 44, July.

Stock, J. H. (1987) Asymptotic properties of least squares estimators of cointegrated vectors. *Econometrica*, **55**, 1035–56.

Van Duyne, C. (1979) The macroeconomic effects of commodity market disruptions in open economies. *Journal of International Economics*, **9**, 559–82.

Williamson, J. and Miller, M. (1987) Targets and indicators: A blueprint for the international co-ordination of economic policy. *Policy Analyses in International Economics* No. 22, Institute for International Economics, Washington, DC.

18

Conclusion

ORHAN GUVENEN, WALTER C. LABYS AND

JEAN-BAPTISTE LESOURD

This book provides a perspective on some of the latest developments in international commodity markets modelling. It attempts to show where these developments stand in terms of new methodologies, future forecasting needs and economic analysis.

Commodity market modelling techniques and approaches are manifold, since they include standard econometric methods, time series analysis methods, linear and non-linear programming techniques, process and engineering modelling approaches, among others, together with combinations of several of these approaches. As noted by R. Duncan in the Foreword, substantial advances have occurred over the past ten years that concern many of these aspects of commodity market modelling.

Furthermore, commodity market modelling—as emphasized in the Introduction by Nobel Laureate L. Klein—often relies on interdisciplinary approaches, because it interacts with a number of fields: agronomy for agricultural commodity modelling and various engineering fields for energy and mineral commodity modelling, especially when process models or linear and non-linear programming models are developed. Applied mathematics and statistics, computer software and algorithmics, and mathematical economics or operations research are, of course, nearly always used as basic tools in commodity market modelling. Thus the modelling techniques derive from different economic fields or subdisciplines, such as agricultural economics, energy economics, marine economics and natural resource economics, that are related to various neighbouring fields or disciplines. In addition, institutional and historical knowledge is very often indispensable, especially when market structures, regulated price mechanisms, externalities and other non-market phenomena are being analysed in quantitative market models.

With regards to methodologies in commodity market model building, this book provides us with a wealth of information on the most recent developments and advances. It brings together new fields of application, new methodologies and novel economic analyses based on quantitative or econometric models.

The first part begins with a detailed survey of new international commodity market modelling issues and methodologies (Labys, Lesourd, Uri and Güvenen). It then features special advances in modelling methodologies, such as new methods for the analysis of imperfect competition (Kolstad and Mathiesen) and developments in spatial analysis with linear and non-linear programming tools (Takayama and MacAulay). Furthermore, new methods for evaluating shadow prices, in terms of environmental and externality costs, for natural resource goods are discussed by Pillet.

The second part of the book focuses on new fields of application for the above methodologies. These include agricultural commodity markets with an analysis, in terms of rational expectations, of the international coffee agreement (Palm and Vogelvang) and a World Bank international fibre model (Thigpen and Mitchell) which describes the world textile industry. As far as mineral commodity markets are concerned, a new analysis of intermaterial substitution resulting from technical progress (Considine) is developed; new ideas on regression analysis in the frequency domain are applied to the analysis of cyclical behaviour in the copper, lead and zinc markets (Afriasabi, Moallem and Labys). Finally, examples of novel energy commodity market models are presented in the second part of our book. These include an analysis of linkages between crude oil and refined oil product markets, which is based on a combination of linear programming and econometric approaches (Adams, Kroch and Dzidziulis); a new model of contract behaviour on natural gas markets, as applied to the European natural gas market, is also featured (Lesourd, Percebois and Ruiz).

The third part provides applications of new modelling ideas and methods to commodity futures markets. It begins with a general equilibrium model with imperfect futures markets (Rausser and Walraven). An econometric analysis of possibly destabilizing properties of futures markets (a topical and key issue in futures markets analysis) is also presented (Artus). This is followed by an analysis of how features markets may be used as hedging tools by producers such as developing countries (Daloz). The third part concludes with an application to the futures markets for refined oil products (Lowry).

The fourth and last part is devoted to other macroeconomic issues related to commodity market behaviour, especially in terms of world price variations. It opens with a model of post-recession commodity price formation (Lord) which is followed by a complete discussion of trade-offs between short-run stability and long-run risk (Dixon and Hughes Hallett). Finally, the book ends on a topical note with an analysis of international commodity prices as possible leading indicators of OECD prices (Durand and Blöndal).

It is our hope that this volume provides a detailed insight into new methodologies and recent trends in international commodity market modelling. It opens up new frontiers in the application of these modelling approaches to a variety of commodity markets and commodity-related industries. We also hope that it demonstrates how these models can be employed to analyse important issues related to economic decision making planning and forecasting in these markets.

Index

MIX
Papier aus verantwortungsvollen Quellen
Paper from responsible sources
FSC® C105338

If you have any concerns about our products,
you can contact us on
ProductSafety@springernature.com

In case Publisher is established outside the EU,
the EU authorized representative is:
Springer Nature Customer Service Center GmbH
Europaplatz 3, 69115 Heidelberg, Germany

Printed by Libri Plureos GmbH
in Hamburg, Germany